Springer Texts in Statistics

Advisors:
George Casella   Stephen Fienberg   Ingram Olkin

T0207116

Springer Texts in Statistics

(continued after index)

Larry Wasserman

All of Statistics

A Concise Course in Statistical Inference

With 95 Figures

 Springer

Larry Wasserman
Department of Statistics
Carnegie Mellon University
Baker Hall 228A
Pittsburgh, PA 15213-3890
USA
larry@stat.cmu.edu

Library of Congress Cataloging-in-Publication Data
Wasserman, Larry A. (Larry Alan), 1959–
   All of statistics: a concise course in statistical inference / Larry a. Wasserman.
     p. cm. — (Springer texts in statistics)
   Includes bibliographical references and index.

   1. Mathematical statistics.   I. Title.   II. Series.
QA276.12.W37 2003
519.5—dc21   2003062209

ISBN 978-1-4419-2322-6   e-ISBN 978-0-387-21736-9

Printed in the United States of America.   (MVY)

9 8 7 6 5 4 3   (Corrected second printing, 2005)

springeronline.com

To Isa

Preface

Taken literally, the title "All of Statistics" is an exaggeration. But in spirit, the title is apt, as the book does cover a much broader range of topics than a typical introductory book on mathematical statistics.

This book is for people who want to learn probability and statistics quickly. It is suitable for graduate or advanced undergraduate students in computer science, mathematics, statistics, and related disciplines. The book includes modern topics like nonparametric curve estimation, bootstrapping, and classification, topics that are usually relegated to follow-up courses. The reader is presumed to know calculus and a little linear algebra. No previous knowledge of probability and statistics is required.

Statistics, **data mining**, and **machine learning** are all concerned with collecting and analyzing data. For some time, statistics research was conducted in statistics departments while data mining and machine learning research was conducted in computer science departments. Statisticians thought that computer scientists were reinventing the wheel. Computer scientists thought that statistical theory didn't apply to their problems.

Things are changing. Statisticians now recognize that computer scientists are making novel contributions while computer scientists now recognize the generality of statistical theory and methodology. Clever data mining algorithms are more scalable than statisticians ever thought possible. Formal statistical theory is more pervasive than computer scientists had realized.

Students who analyze data, or who aspire to develop new methods for analyzing data, should be well grounded in basic probability and mathematical statistics. Using fancy tools like neural nets, boosting, and support vector machines without understanding basic statistics is like doing brain surgery before knowing how to use a band-aid.

But where can students learn basic probability and statistics quickly? Nowhere. At least, that was my conclusion when my computer science colleagues kept asking me: "Where can I send my students to get a good understanding of modern statistics quickly?" The typical mathematical statistics course spends too much time on tedious and uninspiring topics (counting methods, two dimensional integrals, etc.) at the expense of covering modern concepts (bootstrapping, curve estimation, graphical models, etc.). So I set out to redesign our undergraduate honors course on probability and mathematical statistics. This book arose from that course. Here is a summary of the main features of this book.

1. The book is suitable for graduate students in computer science and honors undergraduates in math, statistics, and computer science. It is also useful for students beginning graduate work in statistics who need to fill in their background on mathematical statistics.

2. I cover advanced topics that are traditionally not taught in a first course. For example, nonparametric regression, bootstrapping, density estimation, and graphical models.

3. I have omitted topics in probability that do not play a central role in statistical inference. For example, counting methods are virtually absent.

4. Whenever possible, I avoid tedious calculations in favor of emphasizing concepts.

5. I cover nonparametric inference before parametric inference.

6. I abandon the usual "First Term = Probability" and "Second Term = Statistics" approach. Some students only take the first half and it would be a crime if they did not see any statistical theory. Furthermore, probability is more engaging when students can see it put to work in the context of statistics. An exception is the topic of stochastic processes which is included in the later material.

7. The course moves very quickly and covers much material. My colleagues joke that I cover all of statistics in this course and hence the title. The course is demanding but I have worked hard to make the material as intuitive as possible so that the material is very understandable despite the fast pace.

8. Rigor and clarity are not synonymous. I have tried to strike a good balance. To avoid getting bogged down in uninteresting technical details, many results are stated without proof. The bibliographic references at the end of each chapter point the student to appropriate sources.

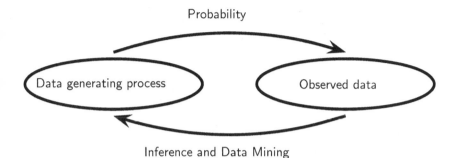

FIGURE 1. Probability and inference.

9. On my website are files with R code which students can use for doing all the computing. The website is:

   http://www.stat.cmu.edu/~larry/all-of-statistics

   However, the book is not tied to R and any computing language can be used.

Part I of the text is concerned with probability theory, the formal language of uncertainty which is the basis of statistical inference. The basic problem that we study in probability is:

Given a data generating process, what are the properties of the outcomes?

Part II is about statistical inference and its close cousins, data mining and machine learning. The basic problem of statistical inference is the inverse of probability:

Given the outcomes, what can we say about the process that generated the data?

These ideas are illustrated in Figure 1. Prediction, classification, clustering, and estimation are all special cases of statistical inference. Data analysis, machine learning and data mining are various names given to the practice of statistical inference, depending on the context.

Part III applies the ideas from Part II to specific problems such as regression, graphical models, causation, density estimation, smoothing, classification, and simulation. Part III contains one more chapter on probability that covers stochastic processes including Markov chains.

I have drawn on other books in many places. Most chapters contain a section called Bibliographic Remarks which serves both to acknowledge my debt to other authors and to point readers to other useful references. I would especially like to mention the books by DeGroot and Schervish (2002) and Grimmett and Stirzaker (1982) from which I adapted many examples and exercises.

As one develops a book over several years it is easy to lose track of where presentation ideas and, especially, homework problems originated. Some I made up. Some I remembered from my education. Some I borrowed from other books. I hope I do not offend anyone if I have used a problem from their book and failed to give proper credit. As my colleague Mark Schervish wrote in his book (Schervish (1995)),

> "...the problems at the ends of each chapter have come from many sources. ...These problems, in turn, came from various sources unknown to me ...If I have used a problem without giving proper credit, please take it as a compliment."

I am indebted to many people without whose help I could not have written this book. First and foremost, the many students who used earlier versions of this text and provided much feedback. In particular, Liz Prather and Jennifer Bakal read the book carefully. Rob Reeder valiantly read through the entire book in excruciating detail and gave me countless suggestions for improvements. Chris Genovese deserves special mention. He not only provided helpful ideas about intellectual content, but also spent many, many hours writing LATEXcode for the book. The best aspects of the book's layout are due to his hard work; any stylistic deficiencies are due to my lack of expertise. David Hand, Sam Roweis, and David Scott read the book very carefully and made numerous suggestions that greatly improved the book. John Lafferty and Peter Spirtes also provided helpful feedback. John Kimmel has been supportive and helpful throughout the writing process. Finally, my wife Isabella Verdinelli has been an invaluable source of love, support, and inspiration.

Larry A. Wasserman
Pittsburgh, Pennsylvania
July 2003

Statistics/Data Mining Dictionary

Statisticians and computer scientists often use different language for the same thing. Here is a dictionary that the reader may want to return to throughout the course.

Statistics	Computer Science	Meaning
estimation	learning	using data to estimate an unknown quantity
classification	supervised learning	predicting a discrete Y from X
clustering	unsupervised learning	putting data into groups
data	training sample	$(X_1, Y_1), \ldots, (X_n, Y_n)$
covariates	features	the X_i's
classifier	hypothesis	a map from covariates to outcomes
hypothesis	—	subset of a parameter space Θ
confidence interval	—	interval that contains an unknown quantity with given frequency
directed acyclic graph	Bayes net	multivariate distribution with given conditional independence relations
Bayesian inference	Bayesian inference	statistical methods for using data to update beliefs
frequentist inference	—	statistical methods with guaranteed frequency behavior
large deviation bounds	PAC learning	uniform bounds on probability of errors

Contents

Part I

Probability

1
Probability

1.1 Introduction

Probability is a mathematical language for quantifying uncertainty. In this Chapter we introduce the basic concepts underlying probability theory. We begin with the sample space, which is the set of possible outcomes.

1.2 Sample Spaces and Events

The **sample space** Ω is the set of possible outcomes of an experiment. Points ω in Ω are called **sample outcomes**, **realizations**, or **elements**. Subsets of Ω are called **Events**.

1.1 Example. If we toss a coin twice then $\Omega = \{HH, HT, TH, TT\}$. The event that the first toss is heads is $A = \{HH, HT\}$. ∎

1.2 Example. Let ω be the outcome of a measurement of some physical quantity, for example, temperature. Then $\Omega = \mathbb{R} = (-\infty, \infty)$. One could argue that taking $\Omega = \mathbb{R}$ is not accurate since temperature has a lower bound. But there is usually no harm in taking the sample space to be larger than needed. The event that the measurement is larger than 10 but less than or equal to 23 is $A = (10, 23]$. ∎

1.3 Example. If we toss a coin forever, then the sample space is the infinite set

$$\Omega = \Big\{ \omega = (\omega_1, \omega_2, \omega_3, \dots,) : \ \omega_i \in \{H, T\} \Big\}.$$

Let E be the event that the first head appears on the third toss. Then

$$E = \Big\{ (\omega_1, \omega_2, \omega_3, \dots,) : \ \omega_1 = T, \omega_2 = T, \omega_3 = H, \ \omega_i \in \{H, T\} \text{ for } i > 3 \Big\}. \ \blacksquare$$

Given an event A, let $A^c = \{ \omega \in \Omega : \ \omega \notin A \}$ denote the complement of A. Informally, A^c can be read as "not A." The complement of Ω is the empty set $\emptyset$. The union of events A and B is defined

$$A \bigcup B = \{ \omega \in \Omega : \ \omega \in A \text{ or } \omega \in B \text{ or } \omega \in \text{both} \}$$

which can be thought of as "A or B." If $A_1, A_2, \dots$ is a sequence of sets then

$$\bigcup_{i=1}^{\infty} A_i = \Big\{ \omega \in \Omega : \ \omega \in A_i \text{ for at least one i} \Big\}.$$

The intersection of A and B is

$$A \bigcap B = \{ \omega \in \Omega : \ \omega \in A \text{ and } \omega \in B \}$$

read "A and B." Sometimes we write $A \bigcap B$ as AB or (A, B). If $A_1, A_2, \dots$ is a sequence of sets then

$$\bigcap_{i=1}^{\infty} A_i = \Big\{ \omega \in \Omega : \ \omega \in A_i \text{ for all i} \Big\}.$$

The set difference is defined by $A - B = \{ \omega : \ \omega \in A, \omega \notin B \}$. If every element of A is also contained in B we write $A \subset B$ or, equivalently, $B \supset A$. If A is a finite set, let $|A|$ denote the number of elements in A. See the following table for a summary.

Summary of Terminology	
Ω	sample space
ω	outcome (point or element)
A	event (subset of Ω)
A^c	complement of A (not A)
$A \bigcup B$	union (A or B)
$A \bigcap B$ or AB	intersection (A and B)
$A - B$	set difference (ω in A but not in B)
$A \subset B$	set inclusion
$\emptyset$	null event (always false)
Ω	true event (always true)

We say that $A_1, A_2, \ldots$ are **disjoint** or are **mutually exclusive** if $A_i \cap A_j = \emptyset$ whenever $i \neq j$. For example, $A_1 = [0, 1), A_2 = [1, 2), A_3 = [2, 3), \ldots$ are disjoint. A **partition** of Ω is a sequence of disjoint sets $A_1, A_2, \ldots$ such that $\bigcup_{i=1}^{\infty} A_i = \Omega$. Given an event A, define the **indicator function of** A by

$$I_A(\omega) = I(\omega \in A) = \begin{cases} 1 & \text{if } \omega \in A \\ 0 & \text{if } \omega \notin A. \end{cases}$$

A sequence of sets $A_1, A_2, \ldots$ is **monotone increasing** if $A_1 \subset A_2 \subset \cdots$ and we define $\lim_{n \to \infty} A_n = \bigcup_{i=1}^{\infty} A_i$. A sequence of sets $A_1, A_2, \ldots$ is **monotone decreasing** if $A_1 \supset A_2 \supset \cdots$ and then we define $\lim_{n \to \infty} A_n = \bigcap_{i=1}^{\infty} A_i$. In either case, we will write $A_n \to A$.

1.4 Example. Let $\Omega = \mathbb{R}$ and let $A_i = [0, 1/i)$ for $i = 1, 2, \ldots$. Then $\bigcup_{i=1}^{\infty} A_i = [0, 1)$ and $\bigcap_{i=1}^{\infty} A_i = \{0\}$. If instead we define $A_i = (0, 1/i)$ then $\bigcup_{i=1}^{\infty} A_i = (0, 1)$ and $\bigcap_{i=1}^{\infty} A_i = \emptyset$. ∎

1.3   Probability

We will assign a real number $\mathbb{P}(A)$ to every event A, called the **probability** of A. [1] We also call $\mathbb{P}$ a **probability distribution** or a **probability measure**. To qualify as a probability, $\mathbb{P}$ must satisfy three axioms:

1.5 Definition. *A function* $\mathbb{P}$ *that assigns a real number* $\mathbb{P}(A)$ *to each event* A *is a* **probability distribution** *or a* **probability measure** *if it satisfies the following three axioms:*
Axiom 1: $\mathbb{P}(A) \geq 0$ *for every* A
Axiom 2: $\mathbb{P}(\Omega) = 1$
Axiom 3: *If* $A_1, A_2, \ldots$ *are disjoint then*

$$\mathbb{P}\left(\bigcup_{i=1}^{\infty} A_i\right) = \sum_{i=1}^{\infty} \mathbb{P}(A_i).$$

---

[1] It is not always possible to assign a probability to every event A if the sample space is large, such as the whole real line. Instead, we assign probabilities to a limited class of set called a σ-field. See the appendix for details.

There are many interpretations of $\mathbb{P}(A)$. The two common interpretations are frequencies and degrees of beliefs. In the frequency interpretation, $\mathbb{P}(A)$ is the long run proportion of times that A is true in repetitions. For example, if we say that the probability of heads is $1/2$, we mean that if we flip the coin many times then the proportion of times we get heads tends to $1/2$ as the number of tosses increases. An infinitely long, unpredictable sequence of tosses whose limiting proportion tends to a constant is an idealization, much like the idea of a straight line in geometry. The degree-of-belief interpretation is that $\mathbb{P}(A)$ measures an observer's strength of belief that A is true. In either interpretation, we require that Axioms 1 to 3 hold. The difference in interpretation will not matter much until we deal with statistical inference. There, the differing interpretations lead to two schools of inference: the frequentist and the Bayesian schools. We defer discussion until Chapter 11.

One can derive many properties of $\mathbb{P}$ from the axioms, such as:

$$
\begin{aligned}
\mathbb{P}(\emptyset) &= 0 \\
A \subset B &\implies \mathbb{P}(A) \leq \mathbb{P}(B) \\
0 \leq \mathbb{P}(A) &\leq 1 \\
\mathbb{P}(A^c) &= 1 - \mathbb{P}(A) \\
A \bigcap B = \emptyset &\implies \mathbb{P}\left(A \bigcup B\right) = \mathbb{P}(A) + \mathbb{P}(B).
\end{aligned}
\tag{1.1}
$$

A less obvious property is given in the following Lemma.

1.6 Lemma. *For any events A and B,*

$$
\mathbb{P}\left(A \bigcup B\right) = \mathbb{P}(A) + \mathbb{P}(B) - \mathbb{P}(AB).
$$

PROOF. Write $A \bigcup B = (AB^c) \bigcup (AB) \bigcup (A^c B)$ and note that these events are disjoint. Hence, making repeated use of the fact that $\mathbb{P}$ is additive for disjoint events, we see that

$$
\begin{aligned}
\mathbb{P}\left(A \bigcup B\right) &= \mathbb{P}\left((AB^c) \bigcup (AB) \bigcup (A^c B)\right) \\
&= \mathbb{P}(AB^c) + \mathbb{P}(AB) + \mathbb{P}(A^c B) \\
&= \mathbb{P}(AB^c) + \mathbb{P}(AB) + \mathbb{P}(A^c B) + \mathbb{P}(AB) - \mathbb{P}(AB) \\
&= \mathbb{P}\left((AB^c) \bigcup (AB)\right) + \mathbb{P}\left((A^c B) \bigcup (AB)\right) - \mathbb{P}(AB) \\
&= \mathbb{P}(A) + \mathbb{P}(B) - \mathbb{P}(AB). \quad \blacksquare
\end{aligned}
$$

1.7 Example. Two coin tosses. Let H_1 be the event that heads occurs on toss 1 and let H_2 be the event that heads occurs on toss 2. If all outcomes are equally likely, then $\mathbb{P}(H_1 \bigcup H_2) = \mathbb{P}(H_1) + \mathbb{P}(H_2) - \mathbb{P}(H_1 H_2) = \frac{1}{2} + \frac{1}{2} - \frac{1}{4} = 3/4.$ ∎

1.8 Theorem (Continuity of Probabilities). *If $A_n \to A$ then*

$$\mathbb{P}(A_n) \to \mathbb{P}(A)$$

as $n \to \infty$.

PROOF. Suppose that A_n is monotone increasing so that $A_1 \subset A_2 \subset \cdots$. Let $A = \lim_{n \to \infty} A_n = \bigcup_{i=1}^{\infty} A_i$. Define $B_1 = A_1$, $B_2 = \{\omega \in \Omega : \omega \in A_2, \omega \notin A_1\}$, $B_3 = \{\omega \in \Omega : \omega \in A_3, \omega \notin A_2, \omega \notin A_1\}, \ldots$ It can be shown that $B_1, B_2, \ldots$ are disjoint, $A_n = \bigcup_{i=1}^{n} A_i = \bigcup_{i=1}^{n} B_i$ for each n and $\bigcup_{i=1}^{\infty} B_i = \bigcup_{i=1}^{\infty} A_i$. (See exercise 1.) From Axiom 3,

$$\mathbb{P}(A_n) = \mathbb{P}\left(\bigcup_{i=1}^{n} B_i\right) = \sum_{i=1}^{n} \mathbb{P}(B_i)$$

and hence, using Axiom 3 again,

$$\lim_{n \to \infty} \mathbb{P}(A_n) = \lim_{n \to \infty} \sum_{i=1}^{n} \mathbb{P}(B_i) = \sum_{i=1}^{\infty} \mathbb{P}(B_i) = \mathbb{P}\left(\bigcup_{i=1}^{\infty} B_i\right) = \mathbb{P}(A). \quad ∎$$

1.4   Probability on Finite Sample Spaces

Suppose that the sample space $\Omega = \{\omega_1, \ldots, \omega_n\}$ is finite. For example, if we toss a die twice, then Ω has 36 elements: $\Omega = \{(i,j); \ i,j \in \{1, \ldots 6\}\}$. If each outcome is equally likely, then $\mathbb{P}(A) = |A|/36$ where $|A|$ denotes the number of elements in A. The probability that the sum of the dice is 11 is 2/36 since there are two outcomes that correspond to this event.

If Ω is finite and if each outcome is equally likely, then

$$\mathbb{P}(A) = \frac{|A|}{|\Omega|},$$

which is called the **uniform probability distribution.** To compute probabilities, we need to count the number of points in an event A. Methods for counting points are called combinatorial methods. We needn't delve into these in any great detail. We will, however, need a few facts from counting theory that will be useful later. Given n objects, the number of ways of ordering these objects is $n! = n(n-1)(n-2)\cdots 3 \cdot 2 \cdot 1$. For convenience, we define $0! = 1$. We also define

$$\binom{n}{k} = \frac{n!}{k!(n-k)!},\tag{1.2}$$

read "n choose k", which is the number of distinct ways of choosing k objects from n. For example, if we have a class of 20 people and we want to select a committee of 3 students, then there are

$$\binom{20}{3} = \frac{20!}{3!17!} = \frac{20 \times 19 \times 18}{3 \times 2 \times 1} = 1140$$

possible committees. We note the following properties:

$$\binom{n}{0} = \binom{n}{n} = 1 \quad \text{and} \quad \binom{n}{k} = \binom{n}{n-k}.$$

1.5   Independent Events

If we flip a fair coin twice, then the probability of two heads is $\frac{1}{2} \times \frac{1}{2}$. We multiply the probabilities because we regard the two tosses as independent. The formal definition of independence is as follows:

1.9 Definition. *Two events A and B are* **independent** *if*

$$\mathbb{P}(AB) = \mathbb{P}(A)\mathbb{P}(B)\tag{1.3}$$

and we write $A \amalg B$. A set of events $\{A_i : i \in I\}$ is independent if

$$\mathbb{P}\left(\bigcap_{i \in J} A_i\right) = \prod_{i \in J} \mathbb{P}(A_i)$$

for every finite subset J of I. If A and B are not independent, we write

$$A \text{ \gg } B$$

Independence can arise in two distinct ways. Sometimes, we explicitly **assume** that two events are independent. For example, in tossing a coin twice, we usually assume the tosses are independent which reflects the fact that the coin has no memory of the first toss. In other instances, we **derive** independence by verifying that $\mathbb{P}(AB) = \mathbb{P}(A)\mathbb{P}(B)$ holds. For example, in tossing a fair die, let $A = \{2, 4, 6\}$ and let $B = \{1, 2, 3, 4\}$. Then, $A \cap B = \{2, 4\}$,

$\mathbb{P}(AB) = 2/6 = \mathbb{P}(A)\mathbb{P}(B) = (1/2) \times (2/3)$ and so A and B are independent. In this case, we didn't assume that A and B are independent — it just turned out that they were.

Suppose that A and B are disjoint events, each with positive probability. Can they be independent? No. This follows since $\mathbb{P}(A)\mathbb{P}(B) > 0$ yet $\mathbb{P}(AB) = \mathbb{P}(\emptyset) = 0$. Except in this special case, there is no way to judge independence by looking at the sets in a Venn diagram.

1.10 Example. Toss a fair coin 10 times. Let A = "at least one head." Let T_j be the event that tails occurs on the jth toss. Then

$$
\begin{aligned}
\mathbb{P}(A) &= 1 - \mathbb{P}(A^c) \\
&= 1 - \mathbb{P}(\text{all tails}) \\
&= 1 - \mathbb{P}(T_1 T_2 \cdots T_{10}) \\
&= 1 - \mathbb{P}(T_1)\mathbb{P}(T_2) \cdots \mathbb{P}(T_{10}) \quad \text{using independence} \\
&= 1 - \left(\frac{1}{2}\right)^{10} \approx .999. \ \blacksquare
\end{aligned}
$$

1.11 Example. Two people take turns trying to sink a basketball into a net. Person 1 succeeds with probability 1/3 while person 2 succeeds with probability 1/4. What is the probability that person 1 succeeds before person 2? Let E denote the event of interest. Let A_j be the event that the first success is by person 1 and that it occurs on trial number j. Note that $A_1, A_2, \ldots$ are disjoint and that $E = \bigcup_{j=1}^{\infty} A_j$. Hence,

$$
\mathbb{P}(E) = \sum_{j=1}^{\infty} \mathbb{P}(A_j).
$$

Now, $\mathbb{P}(A_1) = 1/3$. A_2 occurs if we have the sequence person 1 misses, person 2 misses, person 1 succeeds. This has probability $\mathbb{P}(A_2) = (2/3)(3/4)(1/3) = (1/2)(1/3)$. Following this logic we see that $\mathbb{P}(A_j) = (1/2)^{j-1}(1/3)$. Hence,

$$
\mathbb{P}(E) = \sum_{j=1}^{\infty} \frac{1}{3} \left(\frac{1}{2}\right)^{j-1} = \frac{1}{3} \sum_{j=1}^{\infty} \left(\frac{1}{2}\right)^{j-1} = \frac{2}{3}.
$$

Here we used that fact that, if $0 < r < 1$ then $\sum_{j=k}^{\infty} r^j = r^k/(1-r)$. $\blacksquare$

Summary of Independence

1. A and B are independent if and only if $\mathbb{P}(AB) = \mathbb{P}(A)\mathbb{P}(B)$.

2. Independence is sometimes assumed and sometimes derived.

3. Disjoint events with positive probability are not independent.

1.6   Conditional Probability

Assuming that $\mathbb{P}(B) > 0$, we define the conditional probability of A given that B has occurred as follows:

1.12 Definition. *If* $\mathbb{P}(B) > 0$ *then the* **conditional probability** *of* A *given* B *is*

$$\mathbb{P}(A|B) = \frac{\mathbb{P}(AB)}{\mathbb{P}(B)}. \tag{1.4}$$

Think of $\mathbb{P}(A|B)$ as the fraction of times A occurs among those in which B occurs. For any fixed B such that $\mathbb{P}(B) > 0$, $\mathbb{P}(\cdot|B)$ is a probability (i.e., it satisfies the three axioms of probability). In particular, $\mathbb{P}(A|B) \geq 0$, $\mathbb{P}(\Omega|B) = 1$ and if $A_1, A_2, \ldots$ are disjoint then $\mathbb{P}(\bigcup_{i=1}^{\infty} A_i | B) = \sum_{i=1}^{\infty} \mathbb{P}(A_i|B)$. But it is in general **not** true that $\mathbb{P}(A|B \bigcup C) = \mathbb{P}(A|B) + \mathbb{P}(A|C)$. The rules of probability apply to events on the left of the bar. In general it is **not** the case that $\mathbb{P}(A|B) = \mathbb{P}(B|A)$. People get this confused all the time. For example, the probability of spots given you have measles is 1 but the probability that you have measles given that you have spots is not 1. In this case, the difference between $\mathbb{P}(A|B)$ and $\mathbb{P}(B|A)$ is obvious but there are cases where it is less obvious. This mistake is made often enough in legal cases that it is sometimes called the prosecutor's fallacy.

1.13 Example. A medical test for a disease D has outcomes $+$ and $-$. The probabilities are:

	D	D^c
$+$	.009	.099
$-$	.001	.891

From the definition of conditional probability,

$$\mathbb{P}(+|D) = \frac{\mathbb{P}(+\cap D)}{\mathbb{P}(D)} = \frac{.009}{.009 + .001} = .9$$

and

$$\mathbb{P}(-|D^c) = \frac{\mathbb{P}(-\cap D^c)}{\mathbb{P}(D^c)} = \frac{.891}{.891 + .099} \approx .9.$$

Apparently, the test is fairly accurate. Sick people yield a positive 90 percent of the time and healthy people yield a negative about 90 percent of the time. Suppose you go for a test and get a positive. What is the probability you have the disease? Most people answer .90. The correct answer is

$$\mathbb{P}(D|+) = \frac{\mathbb{P}(+\cap D)}{\mathbb{P}(+)} = \frac{.009}{.009 + .099} \approx .08.$$

The lesson here is that you need to compute the answer numerically. Don't trust your intuition. ∎

The results in the next lemma follow directly from the definition of conditional probability.

1.14 Lemma. *If A and B are independent events then $\mathbb{P}(A|B) = \mathbb{P}(A)$. Also, for any pair of events A and B,*

$$\mathbb{P}(AB) = \mathbb{P}(A|B)\mathbb{P}(B) = \mathbb{P}(B|A)\mathbb{P}(A).$$

From the last lemma, we see that another interpretation of independence is that knowing B doesn't change the probability of A. The formula $\mathbb{P}(AB) = \mathbb{P}(A)\mathbb{P}(B|A)$ is sometimes helpful for calculating probabilities.

1.15 Example. Draw two cards from a deck, without replacement. Let A be the event that the first draw is the Ace of Clubs and let B be the event that the second draw is the Queen of Diamonds. Then $\mathbb{P}(AB) = \mathbb{P}(A)\mathbb{P}(B|A) = (1/52) \times (1/51)$. ∎

Summary of Conditional Probability

1. If $\mathbb{P}(B) > 0$, then

$$\mathbb{P}(A|B) = \frac{\mathbb{P}(AB)}{\mathbb{P}(B)}.$$

2. $\mathbb{P}(\cdot|B)$ satisfies the axioms of probability, for fixed B. In general, $\mathbb{P}(A|\cdot)$ does not satisfy the axioms of probability, for fixed A.

3. In general, $\mathbb{P}(A|B) \neq \mathbb{P}(B|A)$.

4. A and B are independent if and only if $\mathbb{P}(A|B) = \mathbb{P}(A)$.

1.7 Bayes' Theorem

Bayes' theorem is the basis of "expert systems" and "Bayes' nets," which are discussed in Chapter 17. First, we need a preliminary result.

1.16 Theorem (The Law of Total Probability). *Let $A_1, \ldots, A_k$ be a partition of Ω. Then, for any event B,*

$$\mathbb{P}(B) = \sum_{i=1}^{k} \mathbb{P}(B|A_i)\mathbb{P}(A_i).$$

PROOF. Define $C_j = BA_j$ and note that $C_1, \ldots, C_k$ are disjoint and that $B = \bigcup_{j=1}^{k} C_j$. Hence,

$$\mathbb{P}(B) = \sum_j \mathbb{P}(C_j) = \sum_j \mathbb{P}(BA_j) = \sum_j \mathbb{P}(B|A_j)\mathbb{P}(A_j)$$

since $\mathbb{P}(BA_j) = \mathbb{P}(B|A_j)\mathbb{P}(A_j)$ from the definition of conditional probability. ∎

1.17 Theorem (Bayes' Theorem). *Let $A_1, \ldots, A_k$ be a partition of Ω such that $\mathbb{P}(A_i) > 0$ for each i. If $\mathbb{P}(B) > 0$ then, for each $i = 1, \ldots, k$,*

$$\mathbb{P}(A_i|B) = \frac{\mathbb{P}(B|A_i)\mathbb{P}(A_i)}{\sum_j \mathbb{P}(B|A_j)\mathbb{P}(A_j)}. \tag{1.5}$$

1.18 Remark. We call $\mathbb{P}(A_i)$ the **prior probability of** A and $\mathbb{P}(A_i|B)$ the **posterior probability of** A.

PROOF. We apply the definition of conditional probability twice, followed by the law of total probability:

$$\mathbb{P}(A_i|B) = \frac{\mathbb{P}(A_iB)}{\mathbb{P}(B)} = \frac{\mathbb{P}(B|A_i)\mathbb{P}(A_i)}{\mathbb{P}(B)} = \frac{\mathbb{P}(B|A_i)\mathbb{P}(A_i)}{\sum_j \mathbb{P}(B|A_j)\mathbb{P}(A_j)}. \quad \blacksquare$$

1.19 Example. I divide my email into three categories: $A_1 = $ "spam," $A_2 = $ "low priority" and $A_3 = $ "high priority." From previous experience I find that

$\mathbb{P}(A_1) = .7$, $\mathbb{P}(A_2) = .2$ and $\mathbb{P}(A_3) = .1$. Of course, $.7 + .2 + .1 = 1$. Let B be the event that the email contains the word "free." From previous experience, $\mathbb{P}(B|A_1) = .9$, $\mathbb{P}(B|A_2) = .01$, $\mathbb{P}(B|A_1) = .01$. (Note: $.9 + .01 + .01 \neq 1$.) I receive an email with the word "free." What is the probability that it is spam? Bayes' theorem yields,

$$\mathbb{P}(A_1|B) = \frac{.9 \times .7}{(.9 \times .7) + (.01 \times .2) + (.01 \times .1)} = .995. \quad \blacksquare$$

1.8 Bibliographic Remarks

The material in this chapter is standard. Details can be found in any number of books. At the introductory level, there is DeGroot and Schervish (2002); at the intermediate level, Grimmett and Stirzaker (1982) and Karr (1993); at the advanced level there are Billingsley (1979) and Breiman (1992). I adapted many examples and exercises from DeGroot and Schervish (2002) and Grimmett and Stirzaker (1982).

1.9 Appendix

Generally, it is not feasible to assign probabilities to all subsets of a sample space Ω. Instead, one restricts attention to a set of events called a **σ-algebra** or a **σ-field** which is a class $\mathcal{A}$ that satisfies:

(i) $\emptyset \in \mathcal{A}$,

(ii) if $A_1, A_2, \ldots, \in \mathcal{A}$ then $\bigcup_{i=1}^{\infty} A_i \in \mathcal{A}$ and

(iii) $A \in \mathcal{A}$ implies that $A^c \in \mathcal{A}$.

The sets in $\mathcal{A}$ are said to be **measurable**. We call $(\Omega, \mathcal{A})$ a **measurable space.** If $\mathbb{P}$ is a probability measure defined on $\mathcal{A}$, then $(\Omega, \mathcal{A}, \mathbb{P})$ is called a **probability space.** When Ω is the real line, we take $\mathcal{A}$ to be the smallest σ-field that contains all the open subsets, which is called the **Borel σ-field.**

1.10 Exercises

1. Fill in the details of the proof of Theorem 1.8. Also, prove the monotone decreasing case.

2. Prove the statements in equation (1.1).

3. Let Ω be a sample space and let $A_1, A_2, \ldots$, be events. Define $B_n = \bigcup_{i=n}^{\infty} A_i$ and $C_n = \bigcap_{i=n}^{\infty} A_i$.

   (a) Show that $B_1 \supset B_2 \supset \cdots$ and that $C_1 \subset C_2 \subset \cdots$.

   (b) Show that $\omega \in \bigcap_{n=1}^{\infty} B_n$ if and only if ω belongs to an infinite number of the events $A_1, A_2, \ldots$.

   (c) Show that $\omega \in \bigcup_{n=1}^{\infty} C_n$ if and only if ω belongs to all the events $A_1, A_2, \ldots$ except possibly a finite number of those events.

4. Let $\{A_i : i \in I\}$ be a collection of events where I is an arbitrary index set. Show that

$$\left(\bigcup_{i \in I} A_i \right)^c = \bigcap_{i \in I} A_i^c \quad \text{and} \quad \left(\bigcap_{i \in I} A_i \right)^c = \bigcup_{i \in I} A_i^c$$

   Hint: First prove this for $I = \{1, \ldots, n\}$.

5. Suppose we toss a fair coin until we get exactly two heads. Describe the sample space S. What is the probability that exactly k tosses are required?

6. Let $\Omega = \{0, 1, \ldots, \}$. Prove that there does not exist a uniform distribution on Ω (i.e., if $\mathbb{P}(A) = \mathbb{P}(B)$ whenever $|A| = |B|$, then $\mathbb{P}$ cannot satisfy the axioms of probability).

7. Let $A_1, A_2, \ldots$ be events. Show that

$$\mathbb{P} \left(\bigcup_{n=1}^{\infty} A_n \right) \leq \sum_{n=1}^{\infty} \mathbb{P}(A_n).$$

   Hint: Define $B_n = A_n - \bigcup_{i=1}^{n-1} A_i$. Then show that the B_n are disjoint and that $\bigcup_{n=1}^{\infty} A_n = \bigcup_{n=1}^{\infty} B_n$.

8. Suppose that $\mathbb{P}(A_i) = 1$ for each i. Prove that

$$\mathbb{P} \left(\bigcap_{i=1}^{\infty} A_i \right) = 1.$$

9. For fixed B such that $\mathbb{P}(B) > 0$, show that $\mathbb{P}(\cdot | B)$ satisfies the axioms of probability.

10. You have probably heard it before. Now you can solve it rigorously. It is called the "Monty Hall Problem." A prize is placed at random

behind one of three doors. You pick a door. To be concrete, let's suppose you always pick door 1. Now Monty Hall chooses one of the other two doors, opens it and shows you that it is empty. He then gives you the opportunity to keep your door or switch to the other unopened door. Should you stay or switch? Intuition suggests it doesn't matter. The correct answer is that you should switch. Prove it. It will help to specify the sample space and the relevant events carefully. Thus write $\Omega = \{(\omega_1, \omega_2) : \omega_i \in \{1, 2, 3\}\}$ where ω_1 is where the prize is and ω_2 is the door Monty opens.

11. Suppose that A and B are independent events. Show that A^c and B^c are independent events.

12. There are three cards. The first is green on both sides, the second is red on both sides and the third is green on one side and red on the other. We choose a card at random and we see one side (also chosen at random). If the side we see is green, what is the probability that the other side is also green? Many people intuitively answer $1/2$. Show that the correct answer is $2/3$.

13. Suppose that a fair coin is tossed repeatedly until both a head and tail have appeared at least once.

    (a) Describe the sample space Ω.

    (b) What is the probability that three tosses will be required?

14. Show that if $\mathbb{P}(A) = 0$ or $\mathbb{P}(A) = 1$ then A is independent of every other event. Show that if A is independent of itself then $\mathbb{P}(A)$ is either 0 or 1.

15. The probability that a child has blue eyes is $1/4$. Assume independence between children. Consider a family with 3 children.

    (a) If it is known that at least one child has blue eyes, what is the probability that at least two children have blue eyes?

    (b) If it is known that the youngest child has blue eyes, what is the probability that at least two children have blue eyes?

16. Prove Lemma 1.14.

17. Show that
$$\mathbb{P}(ABC) = \mathbb{P}(A|BC)\mathbb{P}(B|C)\mathbb{P}(C).$$

18. Suppose k events form a partition of the sample space Ω, i.e., they are disjoint and $\bigcup_{i=1}^{k} A_i = \Omega$. Assume that $\mathbb{P}(B) > 0$. Prove that if $\mathbb{P}(A_1|B) < \mathbb{P}(A_1)$ then $\mathbb{P}(A_i|B) > \mathbb{P}(A_i)$ for some $i = 2, \ldots, k$.

19. Suppose that 30 percent of computer owners use a Macintosh, 50 percent use Windows, and 20 percent use Linux. Suppose that 65 percent of the Mac users have succumbed to a computer virus, 82 percent of the Windows users get the virus, and 50 percent of the Linux users get the virus. We select a person at random and learn that her system was infected with the virus. What is the probability that she is a Windows user?

20. A box contains 5 coins and each has a different probability of showing heads. Let $p_1, \ldots, p_5$ denote the probability of heads on each coin. Suppose that

$$p_1 = 0, \; p_2 = 1/4, \; p_3 = 1/2, \; p_4 = 3/4 \; \text{ and } \; p_5 = 1.$$

Let H denote "heads is obtained" and let C_i denote the event that coin i is selected.

(a) Select a coin at random and toss it. Suppose a head is obtained. What is the posterior probability that coin i was selected $(i = 1, \ldots, 5)$? In other words, find $\mathbb{P}(C_i|H)$ for $i = 1, \ldots, 5$.

(b) Toss the coin again. What is the probability of another head? In other words find $\mathbb{P}(H_2|H_1)$ where $H_j = $ "heads on toss j."

Now suppose that the experiment was carried out as follows: We select a coin at random and toss it until a head is obtained.

(c) Find $\mathbb{P}(C_i|B_4)$ where $B_4 = $ "first head is obtained on toss 4."

21. (Computer Experiment.) Suppose a coin has probability p of falling heads up. If we flip the coin many times, we would expect the proportion of heads to be near p. We will make this formal later. Take $p = .3$ and $n = 1,000$ and simulate n coin flips. Plot the proportion of heads as a function of n. Repeat for $p = .03$.

22. (Computer Experiment.) Suppose we flip a coin n times and let p denote the probability of heads. Let X be the number of heads. We call X a binomial random variable, which is discussed in the next chapter. Intuition suggests that X will be close to np. To see if this is true, we can repeat this experiment many times and average the X values. Carry

out a simulation and compare the average of the X's to np. Try this for $p = .3$ and $n = 10$, $n = 100$, and $n = 1,000$.

23. (Computer Experiment.) Here we will get some experience simulating conditional probabilities. Consider tossing a fair die. Let $A = \{2, 4, 6\}$ and $B = \{1, 2, 3, 4\}$. Then, $\mathbb{P}(A) = 1/2$, $\mathbb{P}(B) = 2/3$ and $\mathbb{P}(AB) = 1/3$. Since $\mathbb{P}(AB) = \mathbb{P}(A)\mathbb{P}(B)$, the events A and B are independent. Simulate draws from the sample space and verify that $\widehat{\mathbb{P}}(AB) = \widehat{\mathbb{P}}(A)\widehat{\mathbb{P}}(B)$ where $\widehat{\mathbb{P}}(A)$ is the proportion of times A occurred in the simulation and similarly for $\widehat{\mathbb{P}}(AB)$ and $\widehat{\mathbb{P}}(B)$. Now find two events A and B that are not independent. Compute $\widehat{\mathbb{P}}(A), \widehat{\mathbb{P}}(B)$ and $\widehat{\mathbb{P}}(AB)$. Compare the calculated values to their theoretical values. Report your results and interpret.

2
Random Variables

2.1 Introduction

Statistics and data mining are concerned with data. How do we link sample spaces and events to data? The link is provided by the concept of a random variable.

2.1 Definition. *A* **random variable** *is a mapping*[1]

$$X : \Omega \to \mathbb{R}$$

that assigns a real number $X(\omega)$ to each outcome ω.

At a certain point in most probability courses, the sample space is rarely mentioned anymore and we work directly with random variables. But you should keep in mind that the sample space is really there, lurking in the background.

2.2 Example. Flip a coin ten times. Let $X(\omega)$ be the number of heads in the sequence ω. For example, if $\omega = HHTHHTHHTT$, then $X(\omega) = 6$. ∎

---

[1] Technically, a random variable must be measurable. See the appendix for details.

2.3 Example. Let $\Omega = \left\{ (x, y); \; x^2 + y^2 \leq 1 \right\}$ be the unit disk. Consider drawing a point at random from Ω. (We will make this idea more precise later.) A typical outcome is of the form $\omega = (x, y)$. Some examples of random variables are $X(\omega) = x$, $Y(\omega) = y$, $Z(\omega) = x + y$, and $W(\omega) = \sqrt{x^2 + y^2}$. ∎

Given a random variable X and a subset A of the real line, define $X^{-1}(A) = \{\omega \in \Omega : X(\omega) \in A\}$ and let

$$
\begin{aligned}
\mathbb{P}(X \in A) &= \mathbb{P}(X^{-1}(A)) = \mathbb{P}(\{\omega \in \Omega; \; X(\omega) \in A\}) \\
\mathbb{P}(X = x) &= \mathbb{P}(X^{-1}(x)) = \mathbb{P}(\{\omega \in \Omega; \; X(\omega) = x\}).
\end{aligned}
$$

Notice that X denotes the random variable and x denotes a particular value of X.

2.4 Example. Flip a coin twice and let X be the number of heads. Then, $\mathbb{P}(X = 0) = \mathbb{P}(\{TT\}) = 1/4$, $\mathbb{P}(X = 1) = \mathbb{P}(\{HT, TH\}) = 1/2$ and $\mathbb{P}(X = 2) = \mathbb{P}(\{HH\}) = 1/4$. The random variable and its distribution can be summarized as follows:

ω	$\mathbb{P}(\{\omega\})$	$X(\omega)$
TT	1/4	0
TH	1/4	1
HT	1/4	1
HH	1/4	2

x	$\mathbb{P}(X = x)$
0	1/4
1	1/2
2	1/4

Try generalizing this to n flips. ∎

2.2   Distribution Functions and Probability Functions

Given a random variable X, we define the cumulative distribution function (or distribution function) as follows.

2.5 Definition. *The **cumulative distribution function**, or* CDF, *is the function $F_X : \mathbb{R} \to [0, 1]$ defined by*

$$F_X(x) = \mathbb{P}(X \leq x). \tag{2.1}$$

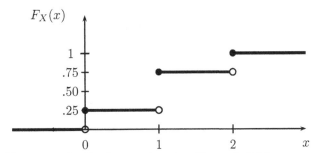

FIGURE 2.1. CDF for flipping a coin twice (Example 2.6.)

We will see later that the CDF effectively contains all the information about the random variable. Sometimes we write the CDF as F instead of F_X.

2.6 Example. Flip a fair coin twice and let X be the number of heads. Then $\mathbb{P}(X = 0) = \mathbb{P}(X = 2) = 1/4$ and $\mathbb{P}(X = 1) = 1/2$. The distribution function is

$$F_X(x) = \begin{cases} 0 & x < 0 \\ 1/4 & 0 \le x < 1 \\ 3/4 & 1 \le x < 2 \\ 1 & x \ge 2. \end{cases}$$

The CDF is shown in Figure 2.1. Although this example is simple, study it carefully. CDF's can be very confusing. Notice that the function is right continuous, non-decreasing, and that it is defined for all x, even though the random variable only takes values $0, 1$, and 2. Do you see why $F_X(1.4) = .75$? ∎

The following result shows that the CDF completely determines the distribution of a random variable.

2.7 Theorem. *Let X have CDF F and let Y have CDF G. If $F(x) = G(x)$ for all x, then $\mathbb{P}(X \in A) = \mathbb{P}(Y \in A)$ for all A.* [2]

2.8 Theorem. *A function F mapping the real line to $[0, 1]$ is a CDF for some probability $\mathbb{P}$ if and only if F satisfies the following three conditions:*
(i) F is non-decreasing: $x_1 < x_2$ implies that $F(x_1) \le F(x_2)$.
(ii) F is normalized:

$$\lim_{x \to -\infty} F(x) = 0$$

---

[2] Technically, we only have that $\mathbb{P}(X \in A) = \mathbb{P}(Y \in A)$ for every measurable event A.

and

$$\lim_{x \to \infty} F(x) = 1.$$

(iii) F is right-continuous: $F(x) = F(x^+)$ *for all x, where*

$$F(x^+) = \lim_{\substack{y \to x \\ y > x}} F(y).$$

PROOF. Suppose that F is a CDF. Let us show that (iii) holds. Let x be a real number and let $y_1, y_2, \ldots$ be a sequence of real numbers such that $y_1 > y_2 > \cdots$ and $\lim_i y_i = x$. Let $A_i = (-\infty, y_i]$ and let $A = (-\infty, x]$. Note that $A = \bigcap_{i=1}^{\infty} A_i$ and also note that $A_1 \supset A_2 \supset \cdots$. Because the events are monotone, $\lim_i \mathbb{P}(A_i) = \mathbb{P}(\bigcap_i A_i)$. Thus,

$$F(x) = \mathbb{P}(A) = \mathbb{P}\left(\bigcap_i A_i\right) = \lim_i \mathbb{P}(A_i) = \lim_i F(y_i) = F(x^+).$$

Showing (i) and (ii) is similar. Proving the other direction — namely, that if F satisfies (i), (ii), and (iii) then it is a CDF for some random variable — uses some deep tools in analysis. ∎

2.9 Definition. X *is* **discrete** *if it takes countably[3] many values* $\{x_1, x_2, \ldots\}$. *We define the* **probability function** *or* **probability mass function** *for X by* $f_X(x) = \mathbb{P}(X = x)$.

Thus, $f_X(x) \geq 0$ for all $x \in \mathbb{R}$ and $\sum_i f_X(x_i) = 1$. Sometimes we write f instead of f_X. The CDF of X is related to f_X by

$$F_X(x) = \mathbb{P}(X \leq x) = \sum_{x_i \leq x} f_X(x_i).$$

2.10 Example. The probability function for Example 2.6 is

$$f_X(x) = \begin{cases} 1/4 & x = 0 \\ 1/2 & x = 1 \\ 1/4 & x = 2 \\ 0 & \text{otherwise.} \end{cases}$$

See Figure 2.2. ∎

---

[3] A set is countable if it is finite or it can be put in a one-to-one correspondence with the integers. The even numbers, the odd numbers, and the rationals are countable; the set of real numbers between 0 and 1 is not countable.

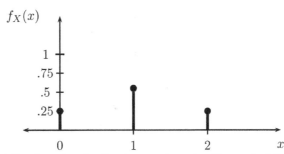

FIGURE 2.2. Probability function for flipping a coin twice (Example 2.6).

2.11 Definition. *A random variable X is* **continuous** *if there exists a function f_X such that $f_X(x) \geq 0$ for all x, $\int_{-\infty}^{\infty} f_X(x)dx = 1$ and for every $a \leq b$,*

$$\mathbb{P}(a < X < b) = \int_a^b f_X(x)dx. \tag{2.2}$$

The function f_X is called the **probability density function** (PDF). *We have that*

$$F_X(x) = \int_{-\infty}^x f_X(t)dt$$

and $f_X(x) = F_X'(x)$ at all points x at which F_X is differentiable.

Sometimes we write $\int f(x)dx$ or $\int f$ to mean $\int_{-\infty}^{\infty} f(x)dx$.

2.12 Example. Suppose that X has PDF

$$f_X(x) = \begin{cases} 1 & \text{for } 0 \leq x \leq 1 \\ 0 & \text{otherwise.} \end{cases}$$

Clearly, $f_X(x) \geq 0$ and $\int f_X(x)dx = 1$. A random variable with this density is said to have a Uniform $(0,1)$ distribution. This is meant to capture the idea of choosing a point at random between 0 and 1. The CDF is given by

$$F_X(x) = \begin{cases} 0 & x < 0 \\ x & 0 \leq x \leq 1 \\ 1 & x > 1. \end{cases}$$

See Figure 2.3. ■

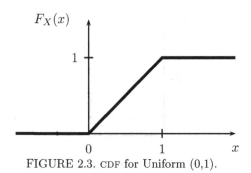

FIGURE 2.3. CDF for Uniform (0,1).

2.13 Example. Suppose that X has PDF

$$f(x) = \begin{cases} 0 & \text{for } x < 0 \\ \frac{1}{(1+x)^2} & \text{otherwise.} \end{cases}$$

Since $\int f(x)dx = 1$, this is a well-defined PDF. ∎

Warning! Continuous random variables can lead to confusion. First, note that if X is continuous then $\mathbb{P}(X = x) = 0$ for every x. Don't try to think of $f(x)$ as $\mathbb{P}(X = x)$. This only holds for discrete random variables. We get probabilities from a PDF by integrating. A PDF can be bigger than 1 (unlike a mass function). For example, if $f(x) = 5$ for $x \in [0, 1/5]$ and 0 otherwise, then $f(x) \geq 0$ and $\int f(x)dx = 1$ so this is a well-defined PDF even though $f(x) = 5$ in some places. In fact, a PDF can be unbounded. For example, if $f(x) = (2/3)x^{-1/3}$ for $0 < x < 1$ and $f(x) = 0$ otherwise, then $\int f(x)dx = 1$ even though f is not bounded.

2.14 Example. Let

$$f(x) = \begin{cases} 0 & \text{for } x < 0 \\ \frac{1}{(1+x)} & \text{otherwise.} \end{cases}$$

This is not a PDF since $\int f(x)dx = \int_0^\infty dx/(1+x) = \int_1^\infty du/u = \log(\infty) = \infty$. ∎

2.15 Lemma. *Let F be the CDF for a random variable X. Then:*

1. $\mathbb{P}(X = x) = F(x) - F(x^-)$ *where* $F(x^-) = \lim_{y \uparrow x} F(y)$;

2. $\mathbb{P}(x < X \leq y) = F(y) - F(x)$;

3. $\mathbb{P}(X > x) = 1 - F(x)$;

4. *If X is continuous then*

$$
\begin{aligned}
F(b) - F(a) &= \mathbb{P}(a < X < b) = \mathbb{P}(a \leq X < b) \\
&= \mathbb{P}(a < X \leq b) = \mathbb{P}(a \leq X \leq b).
\end{aligned}
$$

It is also useful to define the inverse CDF (or quantile function).

2.16 Definition. *Let X be a random variable with* CDF *F. The* **inverse CDF** *or* **quantile function** *is defined by*[4]

$$
F^{-1}(q) = \inf\left\{x : \ F(x) > q\right\}
$$

for $q \in [0, 1]$. If F is strictly increasing and continuous then $F^{-1}(q)$ is the unique real number x such that $F(x) = q$.

We call $F^{-1}(1/4)$ the **first quartile**, $F^{-1}(1/2)$ the **median** (or second quartile), and $F^{-1}(3/4)$ the **third quartile**.

Two random variables X and Y are **equal in distribution** — written $X \overset{d}{=} Y$ — if $F_X(x) = F_Y(x)$ for all x. This does not mean that X and Y are equal. Rather, it means that all probability statements about X and Y will be the same. For example, suppose that $\mathbb{P}(X = 1) = \mathbb{P}(X = -1) = 1/2$. Let $Y = -X$. Then $\mathbb{P}(Y = 1) = \mathbb{P}(Y = -1) = 1/2$ and so $X \overset{d}{=} Y$. But X and Y are not equal. In fact, $\mathbb{P}(X = Y) = 0$.

2.3 Some Important Discrete Random Variables

Warning About Notation! It is traditional to write $X \sim F$ to indicate that X has distribution F. This is unfortunate notation since the symbol $\sim$ is also used to denote an approximation. The notation $X \sim F$ is so pervasive that we are stuck with it. Read $X \sim F$ as "X has distribution F" **not** as "X is approximately F".

---

[4]If you are unfamiliar with "inf", just think of it as the minimum.

THE POINT MASS DISTRIBUTION. X has a point mass distribution at a, written $X \sim \delta_a$, if $\mathbb{P}(X = a) = 1$ in which case

$$F(x) = \begin{cases} 0 & x < a \\ 1 & x \geq a. \end{cases}$$

The probability mass function is $f(x) = 1$ for $x = a$ and 0 otherwise.

THE DISCRETE UNIFORM DISTRIBUTION. Let $k > 1$ be a given integer. Suppose that X has probability mass function given by

$$f(x) = \begin{cases} 1/k & \text{for } x = 1, \ldots, k \\ 0 & \text{otherwise.} \end{cases}$$

We say that X has a uniform distribution on $\{1, \ldots, k\}$.

THE BERNOULLI DISTRIBUTION. Let X represent a binary coin flip. Then $\mathbb{P}(X = 1) = p$ and $\mathbb{P}(X = 0) = 1 - p$ for some $p \in [0, 1]$. We say that X has a Bernoulli distribution written $X \sim \text{Bernoulli}(p)$. The probability function is $f(x) = p^x(1 - p)^{1-x}$ for $x \in \{0, 1\}$.

THE BINOMIAL DISTRIBUTION. Suppose we have a coin which falls heads up with probability p for some $0 \leq p \leq 1$. Flip the coin n times and let X be the number of heads. Assume that the tosses are independent. Let $f(x) = \mathbb{P}(X = x)$ be the mass function. It can be shown that

$$f(x) = \begin{cases} \binom{n}{x} p^x (1 - p)^{n-x} & \text{for } x = 0, \ldots, n \\ 0 & \text{otherwise.} \end{cases}$$

A random variable with this mass function is called a Binomial random variable and we write $X \sim \text{Binomial}(n, p)$. If $X_1 \sim \text{Binomial}(n_1, p)$ and $X_2 \sim \text{Binomial}(n_2, p)$ then $X_1 + X_2 \sim \text{Binomial}(n_1 + n_2, p)$.

Warning! Let us take this opportunity to prevent some confusion. X is a random variable; x denotes a particular value of the random variable; n and p are **parameters**, that is, fixed real numbers. The parameter p is usually unknown and must be estimated from data; that's what statistical inference is all about. In most statistical models, there are random variables and parameters: don't confuse them.

THE GEOMETRIC DISTRIBUTION. X has a geometric distribution with parameter $p \in (0, 1)$, written $X \sim \text{Geom}(p)$, if

$$\mathbb{P}(X = k) = p(1 - p)^{k-1}, \quad k \geq 1.$$

We have that

$$\sum_{k=1}^{\infty} \mathbb{P}(X = k) = p \sum_{k=1}^{\infty} (1-p)^k = \frac{p}{1-(1-p)} = 1.$$

Think of X as the number of flips needed until the first head when flipping a coin.

THE POISSON DISTRIBUTION. X has a Poisson distribution with parameter λ, written $X \sim \text{Poisson}(\lambda)$ if

$$f(x) = e^{-\lambda} \frac{\lambda^x}{x!} \quad x \geq 0.$$

Note that

$$\sum_{x=0}^{\infty} f(x) = e^{-\lambda} \sum_{x=0}^{\infty} \frac{\lambda^x}{x!} = e^{-\lambda} e^{\lambda} = 1.$$

The Poisson is often used as a model for counts of rare events like radioactive decay and traffic accidents. If $X_1 \sim \text{Poisson}(\lambda_1)$ and $X_2 \sim \text{Poisson}(\lambda_2)$ then $X_1 + X_2 \sim \text{Poisson}(\lambda_1 + \lambda_2)$.

Warning! We defined random variables to be mappings from a sample space Ω to $\mathbb{R}$ but we did not mention the sample space in any of the distributions above. As I mentioned earlier, the sample space often "disappears" but it is really there in the background. Let's construct a sample space explicitly for a Bernoulli random variable. Let $\Omega = [0,1]$ and define $\mathbb{P}$ to satisfy $\mathbb{P}([a,b]) = b - a$ for $0 \leq a \leq b \leq 1$. Fix $p \in [0,1]$ and define

$$X(\omega) = \begin{cases} 1 & \omega \leq p \\ 0 & \omega > p. \end{cases}$$

Then $\mathbb{P}(X = 1) = \mathbb{P}(\omega \leq p) = \mathbb{P}([0,p]) = p$ and $\mathbb{P}(X = 0) = 1 - p$. Thus, $X \sim \text{Bernoulli}(p)$. We could do this for all the distributions defined above. In practice, we think of a random variable like a random number but formally it is a mapping defined on some sample space.

2.4 Some Important Continuous Random Variables

THE UNIFORM DISTRIBUTION. X has a Uniform(a,b) distribution, written $X \sim \text{Uniform}(a,b)$, if

$$f(x) = \begin{cases} \frac{1}{b-a} & \text{for } x \in [a,b] \\ 0 & \text{otherwise} \end{cases}$$

where $a < b$. The distribution function is

$$F(x) = \begin{cases} 0 & x < a \\ \frac{x-a}{b-a} & x \in [a, b] \\ 1 & x > b. \end{cases}$$

NORMAL (GAUSSIAN). X has a Normal (or Gaussian) distribution with parameters μ and σ, denoted by $X \sim N(\mu, \sigma^2)$, if

$$f(x) = \frac{1}{\sigma\sqrt{2\pi}} \exp\left\{-\frac{1}{2\sigma^2}(x - \mu)^2\right\}, \quad x \in \mathbb{R} \tag{2.3}$$

where $\mu \in \mathbb{R}$ and $\sigma > 0$. The parameter μ is the "center" (or mean) of the distribution and σ is the "spread" (or standard deviation) of the distribution. (The mean and standard deviation will be formally defined in the next chapter.) The Normal plays an important role in probability and statistics. Many phenomena in nature have approximately Normal distributions. Later, we shall study the Central Limit Theorem which says that the distribution of a sum of random variables can be approximated by a Normal distribution.

We say that X has a **standard Normal distribution** if $\mu = 0$ and $\sigma = 1$. Tradition dictates that a standard Normal random variable is denoted by Z. The PDF and CDF of a standard Normal are denoted by $\phi(z)$ and $\Phi(z)$. The PDF is plotted in Figure 2.4. There is no closed-form expression for Φ. Here are some useful facts:

(i) If $X \sim N(\mu, \sigma^2)$, then $Z = (X - \mu)/\sigma \sim N(0, 1)$.

(ii) If $Z \sim N(0, 1)$, then $X = \mu + \sigma Z \sim N(\mu, \sigma^2)$.

(iii) If $X_i \sim N(\mu_i, \sigma_i^2)$, $i = 1, \ldots, n$ are independent, then

$$\sum_{i=1}^{n} X_i \sim N\left(\sum_{i=1}^{n} \mu_i, \sum_{i=1}^{n} \sigma_i^2\right).$$

It follows from (i) that if $X \sim N(\mu, \sigma^2)$, then

$$\begin{aligned} \mathbb{P}(a < X < b) &= \mathbb{P}\left(\frac{a - \mu}{\sigma} < Z < \frac{b - \mu}{\sigma}\right) \\ &= \Phi\left(\frac{b - \mu}{\sigma}\right) - \Phi\left(\frac{a - \mu}{\sigma}\right). \end{aligned}$$

Thus we can compute any probabilities we want as long as we can compute the CDF $\Phi(z)$ of a standard Normal. All statistical computing packages will

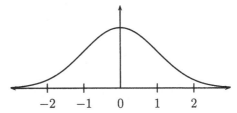

FIGURE 2.4. Density of a standard Normal.

compute $\Phi(z)$ and $\Phi^{-1}(q)$. Most statistics texts, including this one, have a table of values of $\Phi(z)$.

2.17 Example. Suppose that $X \sim N(3,5)$. Find $\mathbb{P}(X > 1)$. The solution is

$$\mathbb{P}(X > 1) = 1 - \mathbb{P}(X < 1) = 1 - \mathbb{P}\left(Z < \frac{1-3}{\sqrt{5}}\right) = 1 - \Phi(-0.8944) = 0.81.$$

Now find $q = \Phi^{-1}(0.2)$. This means we have to find q such that $\mathbb{P}(X < q) = 0.2$. We solve this by writing

$$0.2 = \mathbb{P}(X < q) = \mathbb{P}\left(Z < \frac{q-\mu}{\sigma}\right) = \Phi\left(\frac{q-\mu}{\sigma}\right).$$

From the Normal table, $\Phi(-0.8416) = 0.2$. Therefore,

$$-0.8416 = \frac{q-\mu}{\sigma} = \frac{q-3}{\sqrt{5}}$$

and hence $q = 3 - 0.8416\sqrt{5} = 1.1181$. ∎

EXPONENTIAL DISTRIBUTION. X has an Exponential distribution with parameter β, denoted by $X \sim \text{Exp}(\beta)$, if

$$f(x) = \frac{1}{\beta}e^{-x/\beta}, \quad x > 0$$

where $\beta > 0$. The exponential distribution is used to model the lifetimes of electronic components and the waiting times between rare events.

GAMMA DISTRIBUTION. For $\alpha > 0$, the **Gamma function** is defined by $\Gamma(\alpha) = \int_0^\infty y^{\alpha-1}e^{-y}dy$. X has a Gamma distribution with parameters α and

β, denoted by $X \sim \text{Gamma}(\alpha, \beta)$, if

$$f(x) = \frac{1}{\beta^\alpha \Gamma(\alpha)} x^{\alpha-1} e^{-x/\beta}, \quad x > 0$$

where $\alpha, \beta > 0$. The exponential distribution is just a $\text{Gamma}(1, \beta)$ distribution. If $X_i \sim \text{Gamma}(\alpha_i, \beta)$ are independent, then $\sum_{i=1}^n X_i \sim \text{Gamma}(\sum_{i=1}^n \alpha_i, \beta)$.

THE BETA DISTRIBUTION. X has a Beta distribution with parameters $\alpha > 0$ and $\beta > 0$, denoted by $X \sim \text{Beta}(\alpha, \beta)$, if

$$f(x) = \frac{\Gamma(\alpha + \beta)}{\Gamma(\alpha)\Gamma(\beta)} x^{\alpha-1}(1 - x)^{\beta-1}, \quad 0 < x < 1.$$

t AND CAUCHY DISTRIBUTION. X has a t distribution with ν degrees of freedom — written $X \sim t_\nu$ — if

$$f(x) = \frac{\Gamma\left(\frac{\nu+1}{2}\right)}{\Gamma\left(\frac{\nu}{2}\right)} \frac{1}{\left(1 + \frac{x^2}{\nu}\right)^{(\nu+1)/2}}.$$

The t distribution is similar to a Normal but it has thicker tails. In fact, the Normal corresponds to a t with $\nu = \infty$. The Cauchy distribution is a special case of the t distribution corresponding to $\nu = 1$. The density is

$$f(x) = \frac{1}{\pi(1 + x^2)}.$$

To see that this is indeed a density:

$$\begin{aligned}
\int_{-\infty}^{\infty} f(x)dx &= \frac{1}{\pi} \int_{-\infty}^{\infty} \frac{dx}{1 + x^2} = \frac{1}{\pi} \int_{-\infty}^{\infty} \frac{d\tan^{-1}(x)}{dx} \\
&= \frac{1}{\pi} \left[\tan^{-1}(\infty) - \tan^{-1}(-\infty)\right] = \frac{1}{\pi} \left[\frac{\pi}{2} - \left(-\frac{\pi}{2}\right)\right] = 1.
\end{aligned}$$

THE χ^2 DISTRIBUTION. X has a χ^2 distribution with p degrees of freedom — written $X \sim \chi_p^2$ — if

$$f(x) = \frac{1}{\Gamma(p/2)2^{p/2}} x^{(p/2)-1} e^{-x/2}, \quad x > 0.$$

If $Z_1, \ldots, Z_p$ are independent standard Normal random variables then $\sum_{i=1}^p Z_i^2 \sim \chi_p^2$.

2.5   Bivariate Distributions

Given a pair of discrete random variables X and Y, define the **joint mass function** by $f(x, y) = \mathbb{P}(X = x \text{ and } Y = y)$. From now on, we write $\mathbb{P}(X = x \text{ and } Y = y)$ as $\mathbb{P}(X = x, Y = y)$. We write f as $f_{X,Y}$ when we want to be more explicit.

2.18 Example. Here is a bivariate distribution for two random variables X and Y each taking values 0 or 1:

	$Y = 0$	$Y = 1$	
X=0	1/9	2/9	1/3
X=1	2/9	4/9	2/3
	1/3	2/3	1

Thus, $f(1, 1) = \mathbb{P}(X = 1, Y = 1) = 4/9$. ∎

2.19 Definition. *In the continuous case, we call a function $f(x, y)$ a PDF for the random variables (X, Y) if*

(i) $f(x, y) \geq 0$ for all (x, y),

(ii) $\int_{-\infty}^{\infty} \int_{-\infty}^{\infty} f(x, y) dx dy = 1$ and,

(iii) for any set $A \subset \mathbb{R} \times \mathbb{R}$, $\mathbb{P}((X, Y) \in A) = \int \int_A f(x, y) dx dy$.

In the discrete or continuous case we define the joint CDF as $F_{X,Y}(x, y) = \mathbb{P}(X \leq x, Y \leq y)$.

2.20 Example. Let (X, Y) be uniform on the unit square. Then,

$$f(x, y) = \begin{cases} 1 & \text{if } 0 \leq x \leq 1, \ 0 \leq y \leq 1 \\ 0 & \text{otherwise.} \end{cases}$$

Find $\mathbb{P}(X < 1/2, Y < 1/2)$. The event $A = \{X < 1/2, Y < 1/2\}$ corresponds to a subset of the unit square. Integrating f over this subset corresponds, in this case, to computing the area of the set A which is $1/4$. So, $\mathbb{P}(X < 1/2, Y < 1/2) = 1/4$. ∎

2.21 Example. Let (X, Y) have density

$$f(x, y) = \begin{cases} x + y & \text{if } 0 \le x \le 1,\ 0 \le y \le 1 \\ 0 & \text{otherwise.} \end{cases}$$

Then

$$\int_0^1 \int_0^1 (x + y) dx\, dy = \int_0^1 \left[\int_0^1 x\, dx \right] dy + \int_0^1 \left[\int_0^1 y\, dx \right] dy$$

$$= \int_0^1 \frac{1}{2} dy + \int_0^1 y\, dy = \frac{1}{2} + \frac{1}{2} = 1$$

which verifies that this is a PDF ∎

2.22 Example. If the distribution is defined over a non-rectangular region, then the calculations are a bit more complicated. Here is an example which I borrowed from DeGroot and Schervish (2002). Let (X, Y) have density

$$f(x, y) = \begin{cases} c x^2 y & \text{if } x^2 \le y \le 1 \\ 0 & \text{otherwise.} \end{cases}$$

Note first that $-1 \le x \le 1$. Now let us find the value of c. The trick here is to be careful about the range of integration. We pick one variable, x say, and let it range over its values. Then, for each fixed value of x, we let y vary over its range, which is $x^2 \le y \le 1$. It may help if you look at Figure 2.5. Thus,

$$1 = \int \int f(x, y) dy\, dx = c \int_{-1}^1 \int_{x^2}^1 x^2 y\, dy\, dx$$

$$= c \int_{-1}^1 x^2 \left[\int_{x^2}^1 y\, dy \right] dx = c \int_{-1}^1 x^2 \frac{1 - x^4}{2} dx = \frac{4c}{21}.$$

Hence, $c = 21/4$. Now let us compute $\mathbb{P}(X \ge Y)$. This corresponds to the set $A = \{(x, y); 0 \le x \le 1, x^2 \le y \le x\}$. (You can see this by drawing a diagram.) So,

$$\mathbb{P}(X \ge Y) = \frac{21}{4} \int_0^1 \int_{x^2}^x x^2 y\, dy\, dx = \frac{21}{4} \int_0^1 x^2 \left[\int_{x^2}^x y\, dy \right] dx$$

$$= \frac{21}{4} \int_0^1 x^2 \left(\frac{x^2 - x^4}{2} \right) dx = \frac{3}{20}. \quad ∎$$

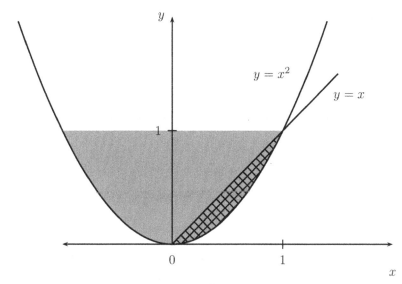

FIGURE 2.5. The light shaded region is $x^2 \leq y \leq 1$. The density is positive over this region. The hatched region is the event $X \geq Y$ intersected with $x^2 \leq y \leq 1$.

2.6   Marginal Distributions

2.23 Definition. *If* (X, Y) *have joint distribution with mass function* $f_{X,Y}$, *then the* **marginal mass function for** X *is defined by*

$$f_X(x) = \mathbb{P}(X = x) = \sum_y \mathbb{P}(X = x, Y = y) = \sum_y f(x, y) \qquad (2.4)$$

and the **marginal mass function for** Y *is defined by*

$$f_Y(y) = \mathbb{P}(Y = y) = \sum_x \mathbb{P}(X = x, Y = y) = \sum_x f(x, y). \qquad (2.5)$$

2.24 Example. Suppose that $f_{X,Y}$ is given in the table that follows. The marginal distribution for X corresponds to the row totals and the marginal distribution for Y corresponds to the columns totals.

	$Y = 0$	$Y = 1$	
X=0	1/10	2/10	3/10
X=1	3/10	4/10	7/10
	4/10	6/10	1

For example, $f_X(0) = 3/10$ and $f_X(1) = 7/10$. ∎

2.25 Definition. *For continuous random variables, the marginal densities are*

$$f_X(x) = \int f(x,y)dy, \quad \text{and} \quad f_Y(y) = \int f(x,y)dx. \qquad (2.6)$$

The corresponding marginal distribution functions are denoted by F_X and F_Y.

2.26 Example. Suppose that

$$f_{X,Y}(x,y) = e^{-(x+y)}$$

for $x, y \geq 0$. Then $f_X(x) = e^{-x} \int_0^\infty e^{-y} dy = e^{-x}$. ∎

2.27 Example. Suppose that

$$f(x,y) = \begin{cases} x+y & \text{if } 0 \leq x \leq 1, \ 0 \leq y \leq 1 \\ 0 & \text{otherwise.} \end{cases}$$

Then

$$f_Y(y) = \int_0^1 (x+y)\,dx = \int_0^1 x\,dx + \int_0^1 y\,dx = \frac{1}{2} + y. \quad ∎$$

2.28 Example. Let (X, Y) have density

$$f(x,y) = \begin{cases} \frac{21}{4}x^2y & \text{if } x^2 \leq y \leq 1 \\ 0 & \text{otherwise.} \end{cases}$$

Thus,

$$f_X(x) = \int f(x,y)dy = \frac{21}{4}x^2 \int_{x^2}^1 y\,dy = \frac{21}{8}x^2(1-x^4)$$

for $-1 \leq x \leq 1$ and $f_X(x) = 0$ otherwise. ∎

2.7   Independent Random Variables

2.29 Definition. *Two random variables X and Y are **independent** if, for every A and B,*

$$\mathbb{P}(X \in A, Y \in B) = \mathbb{P}(X \in A)\mathbb{P}(Y \in B) \qquad (2.7)$$

*and we write $X \amalg Y$. Otherwise we say that X and Y are **dependent** and we write $X \not\amalg Y$.*

In principle, to check whether X and Y are independent we need to check equation (2.7) for all subsets A and B. Fortunately, we have the following result which we state for continuous random variables though it is true for discrete random variables too.

2.30 Theorem. *Let X and Y have joint* PDF $f_{X,Y}$. *Then $X \amalg Y$ if and only if $f_{X,Y}(x,y) = f_X(x)f_Y(y)$ for all values x and y.* [5]

2.31 Example. Let X and Y have the following distribution:

	$Y = 0$	$Y = 1$	
X=0	1/4	1/4	1/2
X=1	1/4	1/4	1/2
	1/2	1/2	1

Then, $f_X(0) = f_X(1) = 1/2$ and $f_Y(0) = f_Y(1) = 1/2$. X and Y are independent because $f_X(0)f_Y(0) = f(0,0)$, $f_X(0)f_Y(1) = f(0,1)$, $f_X(1)f_Y(0) = f(1,0)$, $f_X(1)f_Y(1) = f(1,1)$. Suppose instead that X and Y have the following distribution:

	$Y = 0$	$Y = 1$	
X=0	1/2	0	1/2
X=1	0	1/2	1/2
	1/2	1/2	1

These are not independent because $f_X(0)f_Y(1) = (1/2)(1/2) = 1/4$ yet $f(0,1) = 0$. ∎

2.32 Example. Suppose that X and Y are independent and both have the same density

$$f(x) = \begin{cases} 2x & \text{if } 0 \le x \le 1 \\ 0 & \text{otherwise.} \end{cases}$$

Let us find $\mathbb{P}(X + Y \le 1)$. Using independence, the joint density is

$$f(x,y) = f_X(x)f_Y(y) = \begin{cases} 4xy & \text{if } 0 \le x \le 1, \ 0 \le y \le 1 \\ 0 & \text{otherwise.} \end{cases}$$

---

[5] The statement is not rigorous because the density is defined only up to sets of measure 0.

Now,

$$\begin{aligned}
\mathbb{P}(X + Y \le 1) &= \int \int_{x+y\le 1} f(x,y) dy dx \\
&= 4 \int_0^1 x \left[\int_0^{1-x} y dy \right] dx \\
&= 4 \int_0^1 x \frac{(1-x)^2}{2} dx = \frac{1}{6}. \quad \blacksquare
\end{aligned}$$

The following result is helpful for verifying independence.

2.33 Theorem. *Suppose that the range of* X *and* Y *is a (possibly infinite) rectangle. If* $f(x,y) = g(x)h(y)$ *for some functions* g *and* h *(not necessarily probability density functions) then* X *and* Y *are independent.*

2.34 Example. Let X and Y have density

$$f(x,y) = \begin{cases} 2e^{-(x+2y)} & \text{if } x > 0 \text{ and } y > 0 \\ 0 & \text{otherwise.} \end{cases}$$

The range of X and Y is the rectangle $(0, \infty) \times (0, \infty)$. We can write $f(x,y) = g(x)h(y)$ where $g(x) = 2e^{-x}$ and $h(y) = e^{-2y}$. Thus, $X \amalg Y$. $\blacksquare$

2.8   Conditional Distributions

If X and Y are discrete, then we can compute the conditional distribution of X given that we have observed $Y = y$. Specifically, $\mathbb{P}(X = x|Y = y) = \mathbb{P}(X = x, Y = y)/\mathbb{P}(Y = y)$. This leads us to define the conditional probability mass function as follows.

2.35 Definition. *The* **conditional probability mass function** *is*

$$f_{X|Y}(x|y) = \mathbb{P}(X = x|Y = y) = \frac{\mathbb{P}(X = x, Y = y)}{\mathbb{P}(Y = y)} = \frac{f_{X,Y}(x,y)}{f_Y(y)}$$

if $f_Y(y) > 0$.

For continuous distributions we use the same definitions. [6] The interpretation differs: in the discrete case, $f_{X|Y}(x|y)$ is $\mathbb{P}(X = x|Y = y)$, but in the continuous case, we must integrate to get a probability.

---

[6]We are treading in deep water here. When we compute $\mathbb{P}(X \in A|Y = y)$ in the continuous case we are conditioning on the event $\{Y = y\}$ which has probability 0. We

2.36 Definition. *For continuous random variables, the* **conditional probability density function** *is*

$$f_{X|Y}(x|y) = \frac{f_{X,Y}(x,y)}{f_Y(y)}$$

assuming that $f_Y(y) > 0$. Then,

$$\mathbb{P}(X \in A|Y = y) = \int_A f_{X|Y}(x|y)dx.$$

2.37 Example. Let X and Y have a joint uniform distribution on the unit square. Thus, $f_{X|Y}(x|y) = 1$ for $0 \le x \le 1$ and 0 otherwise. Given $Y = y$, X is Uniform$(0,1)$. We can write this as $X|Y = y \sim$ Uniform$(0,1)$. ∎

From the definition of the conditional density, we see that $f_{X,Y}(x,y) = f_{X|Y}(x|y)f_Y(y) = f_{Y|X}(y|x)f_X(x)$. This can sometimes be useful as in example 2.39.

2.38 Example. Let

$$f(x,y) = \begin{cases} x+y & \text{if } 0 \le x \le 1,\ 0 \le y \le 1 \\ 0 & \text{otherwise.} \end{cases}$$

Let us find $\mathbb{P}(X < 1/4|Y = 1/3)$. In example 2.27 we saw that $f_Y(y) = y + (1/2)$. Hence,

$$f_{X|Y}(x|y) = \frac{f_{X,Y}(x,y)}{f_Y(y)} = \frac{x+y}{y+\frac{1}{2}}.$$

So,

$$P\left(X < \frac{1}{4} \,\Big|\, Y = \frac{1}{3}\right) = \int_0^{1/4} f_{X|Y}\left(x \,\Big|\, \frac{1}{3}\right) dx$$

$$= \int_0^{1/4} \frac{x+\frac{1}{3}}{\frac{1}{3}+\frac{1}{2}} dx = \frac{\frac{1}{32}+\frac{1}{12}}{\frac{1}{3}+\frac{1}{2}} = \frac{11}{80}. \quad ∎$$

2.39 Example. Suppose that $X \sim$ Uniform$(0,1)$. After obtaining a value of X we generate $Y|X = x \sim$ Uniform$(x,1)$. What is the marginal distribution

---

avoid this problem by defining things in terms of the PDF. The fact that this leads to a well-defined theory is proved in more advanced courses. Here, we simply take it as a definition.

of Y? First note that,

$$f_X(x) = \begin{cases} 1 & \text{if } 0 \le x \le 1 \\ 0 & \text{otherwise} \end{cases}$$

and

$$f_{Y|X}(y|x) = \begin{cases} \frac{1}{1-x} & \text{if } 0 < x < y < 1 \\ 0 & \text{otherwise.} \end{cases}$$

So,

$$f_{X,Y}(x,y) = f_{Y|X}(y|x)f_X(x) = \begin{cases} \frac{1}{1-x} & \text{if } 0 < x < y < 1 \\ 0 & \text{otherwise.} \end{cases}$$

The marginal for Y is

$$f_Y(y) = \int_0^y f_{X,Y}(x,y)dx = \int_0^y \frac{dx}{1-x} = -\int_1^{1-y} \frac{du}{u} = -\log(1-y)$$

for $0 < y < 1$. ∎

2.40 Example. Consider the density in Example 2.28. Let's find $f_{Y|X}(y|x)$. When $X = x$, y must satisfy $x^2 \le y \le 1$. Earlier, we saw that $f_X(x) = (21/8)x^2(1-x^4)$. Hence, for $x^2 \le y \le 1$,

$$f_{Y|X}(y|x) = \frac{f(x,y)}{f_X(x)} = \frac{\frac{21}{4}x^2y}{\frac{21}{8}x^2(1-x^4)} = \frac{2y}{1-x^4}.$$

Now let us compute $\mathbb{P}(Y \ge 3/4|X = 1/2)$. This can be done by first noting that $f_{Y|X}(y|1/2) = 32y/15$. Thus,

$$\mathbb{P}(Y \ge 3/4|X = 1/2) = \int_{3/4}^1 f(y|1/2)dy = \int_{3/4}^1 \frac{32y}{15}dy = \frac{7}{15}. ∎$$

2.9   Multivariate Distributions and IID Samples

Let $X = (X_1, \ldots, X_n)$ where $X_1, \ldots, X_n$ are random variables. We call X a **random vector**. Let $f(x_1, \ldots, x_n)$ denote the PDF. It is possible to define their marginals, conditionals etc. much the same way as in the bivariate case. We say that $X_1, \ldots, X_n$ are independent if, for every $A_1, \ldots, A_n$,

$$\mathbb{P}(X_1 \in A_1, \ldots, X_n \in A_n) = \prod_{i=1}^n \mathbb{P}(X_i \in A_i). \qquad (2.8)$$

It suffices to check that $f(x_1, \ldots, x_n) = \prod_{i=1}^n f_{X_i}(x_i)$.

2.41 Definition. *If $X_1, \ldots, X_n$ are independent and each has the same marginal distribution with* CDF *F, we say that $X_1, \ldots, X_n$ are* IID *(independent and identically distributed) and we write*

$$X_1, \ldots X_n \sim F.$$

If F has density f we also write $X_1, \ldots X_n \sim f$. We also call $X_1, \ldots, X_n$ a **random sample of size** n *from F.*

Much of statistical theory and practice begins with IID observations and we shall study this case in detail when we discuss statistics.

2.10 Two Important Multivariate Distributions

MULTINOMIAL. The multivariate version of a Binomial is called a Multinomial. Consider drawing a ball from an urn which has balls with k different colors labeled "color 1, color 2, ..., color k." Let $p = (p_1, \ldots, p_k)$ where $p_j \geq 0$ and $\sum_{j=1}^{k} p_j = 1$ and suppose that p_j is the probability of drawing a ball of color j. Draw n times (independent draws with replacement) and let $X = (X_1, \ldots, X_k)$ where X_j is the number of times that color j appears. Hence, $n = \sum_{j=1}^{k} X_j$. We say that X has a Multinomial (n,p) distribution written $X \sim \text{Multinomial}(n, p)$. The probability function is

$$f(x) = \binom{n}{x_1 \ldots x_k} p_1^{x_1} \cdots p_k^{x_k} \tag{2.9}$$

where

$$\binom{n}{x_1 \ldots x_k} = \frac{n!}{x_1! \cdots x_k!}.$$

2.42 Lemma. *Suppose that $X \sim \text{Multinomial}(n, p)$ where $X = (X_1, \ldots, X_k)$ and $p = (p_1, \ldots, p_k)$. The marginal distribution of X_j is Binomial (n, p_j).*

MULTIVARIATE NORMAL. The univariate Normal has two parameters, μ and σ. In the multivariate version, μ is a vector and σ is replaced by a matrix Σ. To begin, let

$$Z = \begin{pmatrix} Z_1 \\ \vdots \\ Z_k \end{pmatrix}$$

where $Z_1, \ldots, Z_k \sim N(0,1)$ are independent. The density of Z is [7]

$$f(z) \;=\; \prod_{i=1}^{k} f(z_i) = \frac{1}{(2\pi)^{k/2}} \exp\left\{ -\frac{1}{2} \sum_{j=1}^{k} z_j^2 \right\}$$

$$=\; \frac{1}{(2\pi)^{k/2}} \exp\left\{ -\frac{1}{2} z^T z \right\}.$$

We say that Z has a standard multivariate Normal distribution written $Z \sim N(0, I)$ where it is understood that 0 represents a vector of k zeroes and I is the $k \times k$ identity matrix.

More generally, a vector X has a multivariate Normal distribution, denoted by $X \sim N(\mu, \Sigma)$, if it has density [8]

$$f(x; \mu, \Sigma) = \frac{1}{(2\pi)^{k/2} |(\Sigma)|^{1/2}} \exp\left\{ -\frac{1}{2}(x - \mu)^T \Sigma^{-1}(x - \mu) \right\} \qquad (2.10)$$

where $|\Sigma|$ denotes the determinant of Σ, μ is a vector of length k and Σ is a $k \times k$ symmetric, positive definite matrix. [9] Setting $\mu = 0$ and $\Sigma = I$ gives back the standard Normal.

Since Σ is symmetric and positive definite, it can be shown that there exists a matrix $\Sigma^{1/2}$ — called the square root of Σ — with the following properties: (i) $\Sigma^{1/2}$ is symmetric, (ii) $\Sigma = \Sigma^{1/2}\Sigma^{1/2}$ and (iii) $\Sigma^{1/2}\Sigma^{-1/2} = \Sigma^{-1/2}\Sigma^{1/2} = I$ where $\Sigma^{-1/2} = (\Sigma^{1/2})^{-1}$.

2.43 Theorem. *If $Z \sim N(0, I)$ and $X = \mu + \Sigma^{1/2}Z$ then $X \sim N(\mu, \Sigma)$. Conversely, if $X \sim N(\mu, \Sigma)$, then $\Sigma^{-1/2}(X - \mu) \sim N(0, I)$.*

Suppose we partition a random Normal vector X as $X = (X_a, X_b)$ We can similarly partition $\mu = (\mu_a, \mu_b)$ and

$$\Sigma = \begin{pmatrix} \Sigma_{aa} & \Sigma_{ab} \\ \Sigma_{ba} & \Sigma_{bb} \end{pmatrix}.$$

2.44 Theorem. *Let $X \sim N(\mu, \Sigma)$. Then:*

(1) The marginal distribution of X_a is $X_a \sim N(\mu_a, \Sigma_{aa})$.

(2) The conditional distribution of X_b given $X_a = x_a$ is

$$X_b | X_a = x_a \sim N\left(\mu_b + \Sigma_{ba}\Sigma_{aa}^{-1}(x_a - \mu_a), \ \Sigma_{bb} - \Sigma_{ba}\Sigma_{aa}^{-1}\Sigma_{ab} \right).$$

(3) If a is a vector then $a^T X \sim N(a^T \mu, a^T \Sigma a)$.

(4) $V = (X - \mu)^T \Sigma^{-1}(X - \mu) \sim \chi_k^2$.

---

[7] If a and b are vectors then $a^T b = \sum_{i=1}^{k} a_i b_i$.
[8] Σ^{-1} is the inverse of the matrix Σ.
[9] A matrix Σ is positive definite if, for all nonzero vectors x, $x^T \Sigma x > 0$.

2.11   Transformations of Random Variables

Suppose that X is a random variable with PDF f_X and CDF F_X. Let $Y = r(X)$ be a function of X, for example, $Y = X^2$ or $Y = e^X$. We call $Y = r(X)$ a transformation of X. How do we compute the PDF and CDF of Y? In the discrete case, the answer is easy. The mass function of Y is given by

$$
\begin{aligned}
f_Y(y) &= \mathbb{P}(Y = y) = \mathbb{P}(r(X) = y) \\
&= \mathbb{P}(\{x;\ r(x) = y\}) = \mathbb{P}(X \in r^{-1}(y)).
\end{aligned}
$$

2.45 Example. Suppose that $\mathbb{P}(X = -1) = \mathbb{P}(X = 1) = 1/4$ and $\mathbb{P}(X = 0) = 1/2$. Let $Y = X^2$. Then, $\mathbb{P}(Y = 0) = \mathbb{P}(X = 0) = 1/2$ and $\mathbb{P}(Y = 1) = \mathbb{P}(X = 1) + \mathbb{P}(X = -1) = 1/2$. Summarizing:

x	$f_X(x)$
-1	1/4
0	1/2
1	1/4

y	$f_Y(y)$
0	1/2
1	1/2

Y takes fewer values than X because the transformation is not one-to-one. ∎

The continuous case is harder. There are three steps for finding f_Y:

Three Steps for Transformations

1. For each y, find the set $A_y = \{x:\ r(x) \leq y\}$.

2. Find the CDF

$$
\begin{aligned}
F_Y(y) &= \mathbb{P}(Y \leq y) = \mathbb{P}(r(X) \leq y) \\
&= \mathbb{P}(\{x;\ r(x) \leq y\}) \\
&= \int_{A_y} f_X(x)\,dx.
\end{aligned} \tag{2.11}
$$

3. The PDF is $f_Y(y) = F_Y'(y)$.

2.46 Example. Let $f_X(x) = e^{-x}$ for $x > 0$. Hence, $F_X(x) = \int_0^x f_X(s)ds = 1 - e^{-x}$. Let $Y = r(X) = \log X$. Then, $A_y = \{x:\ x \leq e^y\}$ and

$$
\begin{aligned}
F_Y(y) &= \mathbb{P}(Y \leq y) = \mathbb{P}(\log X \leq y) \\
&= \mathbb{P}(X \leq e^y) = F_X(e^y) = 1 - e^{-e^y}.
\end{aligned}
$$

Therefore, $f_Y(y) = e^y e^{-e^y}$ for $y \in \mathbb{R}$. ∎

2.47 Example. Let $X \sim \text{Uniform}(-1, 3)$. Find the PDF of $Y = X^2$. The density of X is

$$f_X(x) = \begin{cases} 1/4 & \text{if } -1 < x < 3 \\ 0 & \text{otherwise.} \end{cases}$$

Y can only take values in $(0, 9)$. Consider two cases: (i) $0 < y < 1$ and (ii) $1 \leq y < 9$. For case (i), $A_y = [-\sqrt{y}, \sqrt{y}]$ and $F_Y(y) = \int_{A_y} f_X(x) dx = (1/2)\sqrt{y}$. For case (ii), $A_y = [-1, \sqrt{y}]$ and $F_Y(y) = \int_{A_y} f_X(x) dx = (1/4)(\sqrt{y} + 1)$. Differentiating F we get

$$f_Y(y) = \begin{cases} \frac{1}{4\sqrt{y}} & \text{if } 0 < y < 1 \\ \frac{1}{8\sqrt{y}} & \text{if } 1 < y < 9 \\ 0 & \text{otherwise.} \end{cases} \blacksquare$$

When r is strictly monotone increasing or strictly monotone decreasing then r has an inverse $s = r^{-1}$ and in this case one can show that

$$f_Y(y) = f_X(s(y)) \left| \frac{ds(y)}{dy} \right|. \tag{2.12}$$

2.12   Transformations of Several Random Variables

In some cases we are interested in transformations of several random variables. For example, if X and Y are given random variables, we might want to know the distribution of X/Y, $X + Y$, $\max\{X, Y\}$ or $\min\{X, Y\}$. Let $Z = r(X, Y)$ be the function of interest. The steps for finding f_Z are the same as before:

Three Steps for Transformations

1. For each z, find the set $A_z = \{(x, y) : r(x, y) \leq z\}$.

2. Find the CDF

$$\begin{aligned} F_Z(z) &= \mathbb{P}(Z \leq z) = \mathbb{P}(r(X, Y) \leq z) \\ &= \mathbb{P}(\{(x, y); \ r(x, y) \leq z\}) = \int \int_{A_z} f_{X,Y}(x, y) \, dx \, dy. \end{aligned}$$

3. Then $f_Z(z) = F_Z'(z)$.

2.48 Example. Let $X_1, X_2 \sim \text{Uniform}(0, 1)$ be independent. Find the density of $Y = X_1 + X_2$. The joint density of (X_1, X_2) is

$$f(x_1, x_2) = \begin{cases} 1 & 0 < x_1 < 1, \ 0 < x_2 < 1 \\ 0 & \text{otherwise.} \end{cases}$$

Let $r(x_1, x_2) = x_1 + x_2$. Now,

$$\begin{aligned} F_Y(y) &= \mathbb{P}(Y \leq y) = \mathbb{P}(r(X_1, X_2) \leq y) \\ &= \mathbb{P}(\{(x_1, x_2) : \ r(x_1, x_2) \leq y\}) = \int\!\!\int_{A_y} f(x_1, x_2) dx_1 dx_2. \end{aligned}$$

Now comes the hard part: finding A_y. First suppose that $0 < y \leq 1$. Then A_y is the triangle with vertices $(0, 0), (y, 0)$ and $(0, y)$. See Figure 2.6. In this case, $\int\int_{A_y} f(x_1, x_2) dx_1 dx_2$ is the area of this triangle which is $y^2/2$. If $1 < y < 2$, then A_y is everything in the unit square except the triangle with vertices $(1, y - 1), (1, 1), (y - 1, 1)$. This set has area $1 - (2 - y)^2/2$. Therefore,

$$F_Y(y) = \begin{cases} 0 & y < 0 \\ \frac{y^2}{2} & 0 \leq y < 1 \\ 1 - \frac{(2-y)^2}{2} & 1 \leq y < 2 \\ 1 & y \geq 2. \end{cases}$$

By differentiation, the PDF is

$$f_Y(y) = \begin{cases} y & 0 \leq y \leq 1 \\ 2 - y & 1 \leq y \leq 2 \\ 0 & \text{otherwise.} \end{cases} \blacksquare$$

2.13 Appendix

Recall that a probability measure $\mathbb{P}$ is defined on a σ-field $\mathcal{A}$ of a sample space Ω. A random variable X is a **measurable** map $X : \Omega \to \mathbb{R}$. Measurable means that, for every x, $\{\omega : \ X(\omega) \leq x\} \in \mathcal{A}$.

2.14 Exercises

1. Show that
$$\mathbb{P}(X = x) = F(x^+) - F(x^-).$$

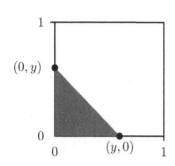
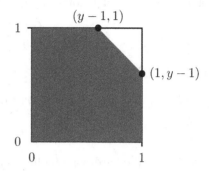

This is the case $0 \leq y < 1$.   This is the case $1 \leq y \leq 2$.

FIGURE 2.6. The set A_y for example 2.48. A_y consists of all points (x_1, x_2) in the square below the line $x_2 = y - x_1$.

2. Let X be such that $\mathbb{P}(X = 2) = \mathbb{P}(X = 3) = 1/10$ and $\mathbb{P}(X = 5) = 8/10$. Plot the CDF F. Use F to find $\mathbb{P}(2 < X \leq 4.8)$ and $\mathbb{P}(2 \leq X \leq 4.8)$.

3. Prove Lemma 2.15.

4. Let X have probability density function

$$f_X(x) = \begin{cases} 1/4 & 0 < x < 1 \\ 3/8 & 3 < x < 5 \\ 0 & \text{otherwise.} \end{cases}$$

(a) Find the cumulative distribution function of X.

(b) Let $Y = 1/X$. Find the probability density function $f_Y(y)$ for Y. Hint: Consider three cases: $\frac{1}{5} \leq y \leq \frac{1}{3}$, $\frac{1}{3} \leq y \leq 1$, and $y \geq 1$.

5. Let X and Y be discrete random variables. Show that X and Y are independent if and only if $f_{X,Y}(x, y) = f_X(x) f_Y(y)$ for all x and y.

6. Let X have distribution F and density function f and let A be a subset of the real line. Let $I_A(x)$ be the indicator function for A:

$$I_A(x) = \begin{cases} 1 & x \in A \\ 0 & x \notin A. \end{cases}$$

Let $Y = I_A(X)$. Find an expression for the cumulative distribution of Y. (Hint: first find the probability mass function for Y.)

7. Let X and Y be independent and suppose that each has a Uniform$(0,1)$ distribution. Let $Z = \min\{X, Y\}$. Find the density $f_Z(z)$ for Z. Hint: It might be easier to first find $\mathbb{P}(Z > z)$.

8. Let X have CDF F. Find the CDF of $X^+ = \max\{0, X\}$.

9. Let $X \sim \text{Exp}(\beta)$. Find $F(x)$ and $F^{-1}(q)$.

10. Let X and Y be independent. Show that $g(X)$ is independent of $h(Y)$ where g and h are functions.

11. Suppose we toss a coin once and let p be the probability of heads. Let X denote the number of heads and let Y denote the number of tails.

    (a) Prove that X and Y are dependent.

    (b) Let $N \sim \text{Poisson}(\lambda)$ and suppose we toss a coin N times. Let X and Y be the number of heads and tails. Show that X and Y are independent.

12. Prove Theorem 2.33.

13. Let $X \sim N(0,1)$ and let $Y = e^X$.

    (a) Find the PDF for Y. Plot it.

    (b) (Computer Experiment.) Generate a vector $x = (x_1, \ldots, x_{10,000})$ consisting of 10,000 random standard Normals. Let $y = (y_1, \ldots, y_{10,000})$ where $y_i = e^{x_i}$. Draw a histogram of y and compare it to the PDF you found in part (a).

14. Let (X, Y) be uniformly distributed on the unit disk $\{(x, y) : x^2 + y^2 \le 1\}$. Let $R = \sqrt{X^2 + Y^2}$. Find the CDF and PDF of R.

15. (A universal random number generator.) Let X have a continuous, strictly increasing CDF F. Let $Y = F(X)$. Find the density of Y. This is called the probability integral transform. Now let $U \sim \text{Uniform}(0,1)$ and let $X = F^{-1}(U)$. Show that $X \sim F$. Now write a program that takes Uniform $(0,1)$ random variables and generates random variables from an Exponential (β) distribution.

16. Let $X \sim \text{Poisson}(\lambda)$ and $Y \sim \text{Poisson}(\mu)$ and assume that X and Y are independent. Show that the distribution of X given that $X + Y = n$ is Binomial(n, π) where $\pi = \lambda/(\lambda + \mu)$.

Hint 1: You may use the following fact: If $X \sim$ Poisson(λ) and $Y \sim$ Poisson(μ), and X and Y are independent, then $X+Y \sim$ Poisson($\mu+\lambda$).

Hint 2: Note that $\{X = x, \ X+Y = n\} = \{X = x, \ Y = n - x\}$.

17. Let
$$f_{X,Y}(x,y) = \begin{cases} c(x + y^2) & 0 \le x \le 1 \text{ and } 0 \le y \le 1 \\ 0 & \text{otherwise.} \end{cases}$$
Find $P\left(X < \frac{1}{2} \mid Y = \frac{1}{2}\right)$.

18. Let $X \sim N(3, 16)$. Solve the following using the Normal table and using a computer package.

    (a) Find $\mathbb{P}(X < 7)$.

    (b) Find $\mathbb{P}(X > -2)$.

    (c) Find x such that $\mathbb{P}(X > x) = .05$.

    (d) Find $\mathbb{P}(0 \le X < 4)$.

    (e) Find x such that $\mathbb{P}(|X| > |x|) = .05$.

19. Prove formula (2.12).

20. Let $X, Y \sim$ Uniform$(0,1)$ be independent. Find the PDF for $X - Y$ and X/Y.

21. Let $X_1, \ldots, X_n \sim$ Exp(β) be IID. Let $Y = \max\{X_1, \ldots, X_n\}$. Find the PDF of Y. Hint: $Y \le y$ if and only if $X_i \le y$ for $i = 1, \ldots, n$.

3
Expectation

3.1 Expectation of a Random Variable

The mean, or expectation, of a random variable X is the average value of X.

3.1 Definition. *The* **expected value**, *or* **mean**, *or* **first moment**, *of* X *is defined to be*

$$\mathbb{E}(X) = \int x \, dF(x) = \begin{cases} \sum_x x f(x) & \text{if } X \text{ is discrete} \\ \int x f(x) dx & \text{if } X \text{ is continuous} \end{cases} \qquad (3.1)$$

assuming that the sum (or integral) is well defined. We use the following notation to denote the expected value of X:

$$\mathbb{E}(X) = \mathbb{E}X = \int x \, dF(x) = \mu = \mu_X. \qquad (3.2)$$

The expectation is a one-number summary of the distribution. Think of $\mathbb{E}(X)$ as the average $\sum_{i=1}^n X_i/n$ of a large number of IID draws $X_1, \ldots, X_n$. The fact that $\mathbb{E}(X) \approx \sum_{i=1}^n X_i/n$ is actually more than a heuristic; it is a theorem called the law of large numbers that we will discuss in Chapter 5.

The notation $\int x \, dF(x)$ deserves some comment. We use it merely as a convenient unifying notation so we don't have to write $\sum_x x f(x)$ for discrete random variables and $\int xf(x)dx$ for continuous random variables, but you should be aware that $\int x\,dF(x)$ has a precise meaning that is discussed in real analysis courses.

To ensure that $\mathbb{E}(X)$ is well defined, we say that $\mathbb{E}(X)$ exists if $\int_x |x|dF_X(x) < \infty$. Otherwise we say that the expectation does not exist.

3.2 Example. Let $X \sim$ Bernoulli(p). Then $\mathbb{E}(X) = \sum_{x=0}^1 xf(x) = (0 \times (1 - p)) + (1 \times p) = p$. ∎

3.3 Example. Flip a fair coin two times. Let X be the number of heads. Then, $\mathbb{E}(X) = \int xdF_X(x) = \sum_x xf_X(x) = (0 \times f(0)) + (1 \times f(1)) + (2 \times f(2)) = (0 \times (1/4)) + (1 \times (1/2)) + (2 \times (1/4)) = 1$. ∎

3.4 Example. Let $X \sim$ Uniform$(-1,3)$. Then, $\mathbb{E}(X) = \int xd\,F_X(x) = \int xf_X(x)dx = \frac{1}{4}\int_{-1}^3 x\,dx = 1$. ∎

3.5 Example. Recall that a random variable has a Cauchy distribution if it has density $f_X(x) = \{\pi(1 + x^2)\}^{-1}$. Using integration by parts, (set $u = x$ and $v = \tan^{-1}x$),

$$\int |x|dF(x) = \frac{2}{\pi} \int_0^\infty \frac{x\,dx}{1 + x^2} = \left[x\,\tan^{-1}(x)\right]_0^\infty - \int_0^\infty \tan^{-1}x\,dx = \infty$$

so the mean does not exist. If you simulate a Cauchy distribution many times and take the average, you will see that the average never settles down. This is because the Cauchy has thick tails and hence extreme observations are common. ∎

From now on, whenever we discuss expectations, we implicitly assume that they exist.

Let $Y = r(X)$. How do we compute $\mathbb{E}(Y)$? One way is to find $f_Y(y)$ and then compute $\mathbb{E}(Y) = \int yf_Y(y)dy$. But there is an easier way.

3.6 Theorem (The Rule of the Lazy Statistician). *Let $Y = r(X)$. Then*

$$\mathbb{E}(Y) = \mathbb{E}(r(X)) = \int r(x)dF_X(x). \tag{3.3}$$

This result makes intuitive sense. Think of playing a game where we draw X at random and then I pay you $Y = r(X)$. Your average income is $r(x)$ times the chance that $X = x$, summed (or integrated) over all values of x. Here is a special case. Let A be an event and let $r(x) = I_A(x)$ where $I_A(x) = 1$ if $x \in A$ and $I_A(x) = 0$ if $x \notin A$. Then

$$\mathbb{E}(I_A(X)) = \int I_A(x) f_X(x) dx = \int_A f_X(x) dx = \mathbb{P}(X \in A).$$

In other words, probability is a special case of expectation.

3.7 Example. Let $X \sim \text{Unif}(0, 1)$. Let $Y = r(X) = e^X$. Then,

$$\mathbb{E}(Y) = \int_0^1 e^x f(x) dx = \int_0^1 e^x dx = e - 1.$$

Alternatively, you could find $f_Y(y)$ which turns out to be $f_Y(y) = 1/y$ for $1 < y < e$. Then, $\mathbb{E}(Y) = \int_1^e y f(y) dy = e - 1$. ∎

3.8 Example. Take a stick of unit length and break it at random. Let Y be the length of the longer piece. What is the mean of Y? If X is the break point then $X \sim \text{Unif}(0, 1)$ and $Y = r(X) = \max\{X, 1 - X\}$. Thus, $r(x) = 1 - x$ when $0 < x < 1/2$ and $r(x) = x$ when $1/2 \le x < 1$. Hence,

$$\mathbb{E}(Y) = \int r(x) dF(x) = \int_0^{1/2} (1 - x) dx + \int_{1/2}^1 x\, dx = \frac{3}{4}. ∎$$

Functions of several variables are handled in a similar way. If $Z = r(X, Y)$ then

$$\mathbb{E}(Z) = \mathbb{E}(r(X, Y)) = \int \int r(x, y) dF(x, y). \tag{3.4}$$

3.9 Example. Let (X, Y) have a jointly uniform distribution on the unit square. Let $Z = r(X, Y) = X^2 + Y^2$. Then,

$$\begin{aligned}
\mathbb{E}(Z) &= \int \int r(x, y) dF(x, y) = \int_0^1 \int_0^1 (x^2 + y^2) \, dx dy \\
&= \int_0^1 x^2 \, dx + \int_0^1 y^2 \, dy = \frac{2}{3}. ∎
\end{aligned}$$

The kth **moment** of X is defined to be $\mathbb{E}(X^k)$ assuming that $\mathbb{E}(|X|^k) < \infty$.

3.10 Theorem. *If the kth moment exists and if $j < k$ then the jth moment exists.*

PROOF. We have

$$\mathbb{E}|X|^j = \int_{-\infty}^{\infty} |x|^j f_X(x) dx$$

$$\begin{aligned}
&= \int_{|x| \leq 1} |x|^j f_X(x) dx + \int_{|x| > 1} |x|^j f_X(x) dx \\
&\leq \int_{|x| \leq 1} f_X(x) dx + \int_{|x| > 1} |x|^k f_X(x) dx \\
&\leq 1 + \mathbb{E}(|X|^k) < \infty. \quad \blacksquare
\end{aligned}$$

The kth central moment is defined to be $\mathbb{E}((X - \mu)^k)$.

3.2   Properties of Expectations

3.11 Theorem. *If $X_1, \ldots, X_n$ are random variables and $a_1, \ldots, a_n$ are constants, then*

$$\mathbb{E}\left(\sum_i a_i X_i\right) = \sum_i a_i \mathbb{E}(X_i). \tag{3.5}$$

3.12 Example. Let $X \sim \text{Binomial}(n, p)$. What is the mean of X? We could try to appeal to the definition:

$$\mathbb{E}(X) = \int x \, dF_X(x) = \sum_x x f_X(x) = \sum_{x=0}^{n} x \binom{n}{x} p^x (1 - p)^{n-x}$$

but this is not an easy sum to evaluate. Instead, note that $X = \sum_{i=1}^{n} X_i$ where $X_i = 1$ if the ith toss is heads and $X_i = 0$ otherwise. Then $\mathbb{E}(X_i) = (p \times 1) + ((1 - p) \times 0) = p$ and $\mathbb{E}(X) = \mathbb{E}(\sum_i X_i) = \sum_i \mathbb{E}(X_i) = np$. $\blacksquare$

3.13 Theorem. *Let $X_1, \ldots, X_n$ be independent random variables. Then,*

$$\mathbb{E}\left(\prod_{i=1}^{n} X_i\right) = \prod_i \mathbb{E}(X_i). \tag{3.6}$$

Notice that the summation rule does not require independence but the multiplication rule does.

3.3   Variance and Covariance

The variance measures the "spread" of a distribution. [1]

---

[1] We can't use $\mathbb{E}(X - \mu)$ as a measure of spread since $\mathbb{E}(X - \mu) = \mathbb{E}(X) - \mu = \mu - \mu = 0$. We can and sometimes do use $\mathbb{E}|X - \mu|$ as a measure of spread but more often we use the variance.

3.14 Definition. *Let X be a random variable with mean μ. The **variance** of X — denoted by σ^2 or σ_X^2 or $\mathbb{V}(X)$ or $\mathbb{V}X$ — is defined by*

$$\sigma^2 = \mathbb{E}(X - \mu)^2 = \int (x - \mu)^2 dF(x) \tag{3.7}$$

*assuming this expectation exists. The **standard deviation** is $\mathrm{sd}(X) = \sqrt{\mathbb{V}(X)}$ and is also denoted by σ and σ_X.*

3.15 Theorem. *Assuming the variance is well defined, it has the following properties:*

1. *$\mathbb{V}(X) = \mathbb{E}(X^2) - \mu^2$.*

2. *If a and b are constants then $\mathbb{V}(aX + b) = a^2\mathbb{V}(X)$.*

3. *If $X_1, \ldots, X_n$ are independent and $a_1, \ldots, a_n$ are constants, then*

$$\mathbb{V}\left(\sum_{i=1}^{n} a_i X_i\right) = \sum_{i=1}^{n} a_i^2 \mathbb{V}(X_i). \tag{3.8}$$

3.16 Example. Let $X \sim \text{Binomial}(n, p)$. We write $X = \sum_i X_i$ where $X_i = 1$ if toss i is heads and $X_i = 0$ otherwise. Then $X = \sum_i X_i$ and the random variables are independent. Also, $\mathbb{P}(X_i = 1) = p$ and $\mathbb{P}(X_i = 0) = 1 - p$. Recall that

$$\mathbb{E}(X_i) = \left(p \times 1\right) + \left((1 - p) \times 0\right) = p.$$

Now,

$$\mathbb{E}(X_i^2) = \left(p \times 1^2\right) + \left((1 - p) \times 0^2\right) = p.$$

Therefore, $\mathbb{V}(X_i) = \mathbb{E}(X_i^2) - p^2 = p - p^2 = p(1 - p)$. Finally, $\mathbb{V}(X) = \mathbb{V}(\sum_i X_i) = \sum_i \mathbb{V}(X_i) = \sum_i p(1 - p) = np(1 - p)$. Notice that $\mathbb{V}(X) = 0$ if $p = 1$ or $p = 0$. Make sure you see why this makes intuitive sense. ∎

If $X_1, \ldots, X_n$ are random variables then we define the **sample mean** to be

$$\overline{X}_n = \frac{1}{n} \sum_{i=1}^{n} X_i \tag{3.9}$$

and the **sample variance** to be

$$S_n^2 = \frac{1}{n-1} \sum_{i=1}^{n} (X_i - \overline{X}_n)^2. \tag{3.10}$$

3.17 Theorem. *Let $X_1, \ldots, X_n$ be* IID *and let $\mu = \mathbb{E}(X_i)$, $\sigma^2 = \mathbb{V}(X_i)$. Then*

$$\mathbb{E}(\overline{X}_n) = \mu, \quad \mathbb{V}(\overline{X}_n) = \frac{\sigma^2}{n} \quad \text{and} \quad \mathbb{E}(S_n^2) = \sigma^2.$$

If X and Y are random variables, then the covariance and correlation between X and Y measure how strong the linear relationship is between X and Y.

3.18 Definition. *Let X and Y be random variables with means μ_X and μ_Y and standard deviations σ_X and σ_Y. Define the* **covariance** *between X and Y by*

$$\mathsf{Cov}(X, Y) = \mathbb{E}\left((X - \mu_X)(Y - \mu_Y) \right) \tag{3.11}$$

and the **correlation** *by*

$$\rho = \rho_{X,Y} = \rho(X, Y) = \frac{\mathsf{Cov}(X, Y)}{\sigma_X \sigma_Y}. \tag{3.12}$$

3.19 Theorem. *The covariance satisfies:*

$$\mathsf{Cov}(X, Y) = \mathbb{E}(XY) - \mathbb{E}(X)\mathbb{E}(Y).$$

The correlation satisfies:

$$-1 \leq \rho(X, Y) \leq 1.$$

If $Y = aX + b$ for some constants a and b then $\rho(X, Y) = 1$ if $a > 0$ and $\rho(X, Y) = -1$ if $a < 0$. If X and Y are independent, then $\mathsf{Cov}(X, Y) = \rho = 0$. The converse is not true in general.

3.20 Theorem. $\mathbb{V}(X + Y) = \mathbb{V}(X) + \mathbb{V}(Y) + 2\mathsf{Cov}(X, Y)$ *and* $\mathbb{V}(X - Y) = \mathbb{V}(X) + \mathbb{V}(Y) - 2\mathsf{Cov}(X, Y)$. *More generally, for random variables $X_1, \ldots, X_n$,*

$$\mathbb{V}\left(\sum_i a_i X_i \right) = \sum_i a_i^2 \mathbb{V}(X_i) + 2 \sum \sum_{i<j} a_i a_j \mathsf{Cov}(X_i, X_j).$$

3.4 Expectation and Variance of Important Random Variables

Here we record the expectation of some important random variables:

Distribution	Mean	Variance
Point mass at a	a	0
Bernoulli(p)	p	$p(1-p)$
Binomial(n,p)	np	$np(1-p)$
Geometric(p)	$1/p$	$(1-p)/p^2$
Poisson(λ)	λ	λ
Uniform(a,b)	$(a+b)/2$	$(b-a)^2/12$
Normal(μ,σ^2)	μ	σ^2
Exponential(β)	β	β^2
Gamma(α,β)	$\alpha\beta$	$\alpha\beta^2$
Beta(α,β)	$\alpha/(\alpha+\beta)$	$\alpha\beta/((\alpha+\beta)^2(\alpha+\beta+1))$
t_ν	0 (if $\nu>1$)	$\nu/(\nu-2)$ (if $\nu>2$)
χ^2_p	p	$2p$
Multinomial(n,p)	np	see below
Multivariate Normal(μ,Σ)	μ	Σ

We derived $\mathbb{E}(X)$ and $\mathbb{V}(X)$ for the Binomial in the previous section. The calculations for some of the others are in the exercises.

The last two entries in the table are multivariate models which involve a random vector X of the form

$$X = \begin{pmatrix} X_1 \\ \vdots \\ X_k \end{pmatrix}.$$

The mean of a random vector X is defined by

$$\mu = \begin{pmatrix} \mu_1 \\ \vdots \\ \mu_k \end{pmatrix} = \begin{pmatrix} \mathbb{E}(X_1) \\ \vdots \\ \mathbb{E}(X_k) \end{pmatrix}.$$

The **variance-covariance matrix** Σ is defined to be

$$\mathbb{V}(X) = \begin{bmatrix} \mathbb{V}(X_1) & \mathrm{Cov}(X_1,X_2) & \cdots & \mathrm{Cov}(X_1,X_k) \\ \mathrm{Cov}(X_2,X_1) & \mathbb{V}(X_2) & \cdots & \mathrm{Cov}(X_2,X_k) \\ \vdots & \vdots & \vdots & \vdots \\ \mathrm{Cov}(X_k,X_1) & \mathrm{Cov}(X_k,X_2) & \cdots & \mathbb{V}(X_k) \end{bmatrix}.$$

If $X \sim$ Multinomial(n,p) then $\mathbb{E}(X) = np = n(p_1,\ldots,p_k)$ and

$$\mathbb{V}(X) = \begin{pmatrix} np_1(1-p_1) & -np_1p_2 & \cdots & -np_1p_k \\ -np_2p_1 & np_2(1-p_2) & \cdots & -np_2p_k \\ \vdots & \vdots & \vdots & \vdots \\ -np_kp_1 & -np_kp_2 & \cdots & np_k(1-p_k) \end{pmatrix}.$$

To see this, note that the marginal distribution of any one component of the vector $X_i \sim$ Binomial(n, p_i). Thus, $\mathbb{E}(X_i) = np_i$ and $\mathbb{V}(X_i) = np_i(1 - p_i)$. Note also that $X_i + X_j \sim$ Binomial$(n, p_i + p_j)$. Thus, $\mathbb{V}(X_i + X_j) = n(p_i + p_j)(1 - [p_i + p_j])$. On the other hand, using the formula for the variance of a sum, we have that $\mathbb{V}(X_i + X_j) = \mathbb{V}(X_i) + \mathbb{V}(X_j) + 2\mathsf{Cov}(X_i, X_j) = np_i(1 - p_i) + np_j(1 - p_j) + 2\mathsf{Cov}(X_i, X_j)$. If we equate this formula with $n(p_i + p_j)(1 - [p_i + p_j])$ and solve, we get $\mathsf{Cov}(X_i, X_j) = -np_i p_j$.

Finally, here is a lemma that can be useful for finding means and variances of linear combinations of multivariate random vectors.

3.21 Lemma. *If a is a vector and X is a random vector with mean μ and variance Σ, then $\mathbb{E}(a^T X) = a^T \mu$ and $\mathbb{V}(a^T X) = a^T \Sigma a$. If A is a matrix then $\mathbb{E}(AX) = A\mu$ and $\mathbb{V}(AX) = A\Sigma A^T$.*

3.5   Conditional Expectation

Suppose that X and Y are random variables. What is the mean of X among those times when $Y = y$? The answer is that we compute the mean of X as before but we substitute $f_{X|Y}(x|y)$ for $f_X(x)$ in the definition of expectation.

3.22 Definition. *The conditional expectation of X given $Y = y$ is*

$$\mathbb{E}(X|Y = y) = \begin{cases} \sum x \, f_{X|Y}(x|y) \, dx & \text{discrete case} \\ \int x \, f_{X|Y}(x|y) \, dx & \text{continuous case.} \end{cases} \qquad (3.13)$$

If $r(x, y)$ is a function of x and y then

$$\mathbb{E}(r(X, Y)|Y = y) = \begin{cases} \sum r(x, y) \, f_{X|Y}(x|y) \, dx & \text{discrete case} \\ \int r(x, y) \, f_{X|Y}(x|y) \, dx & \text{continuous case.} \end{cases}$$
$$(3.14)$$

Warning! Here is a subtle point. Whereas $\mathbb{E}(X)$ is a number, $\mathbb{E}(X|Y = y)$ is a function of y. Before we observe Y, we don't know the value of $\mathbb{E}(X|Y = y)$ so it is a random variable which we denote $\mathbb{E}(X|Y)$. In other words, $\mathbb{E}(X|Y)$ is the random variable whose value is $\mathbb{E}(X|Y = y)$ when $Y = y$. Similarly, $\mathbb{E}(r(X, Y)|Y)$ is the random variable whose value is $\mathbb{E}(r(X, Y)|Y = y)$ when $Y = y$. This is a very confusing point so let us look at an example.

3.23 Example. Suppose we draw $X \sim$ Unif$(0, 1)$. After we observe $X = x$, we draw $Y|X = x \sim$ Unif$(x, 1)$. Intuitively, we expect that $\mathbb{E}(Y|X = x) =$

$(1 + x)/2$. In fact, $f_{Y|X}(y|x) = 1/(1 - x)$ for $x < y < 1$ and

$$\mathbb{E}(Y|X = x) = \int_x^1 y\, f_{Y|X}(y|x)dy = \frac{1}{1-x}\int_x^1 y\, dy = \frac{1+x}{2}$$

as expected. Thus, $\mathbb{E}(Y|X) = (1 + X)/2$. Notice that $\mathbb{E}(Y|X) = (1+X)/2$ is a random variable whose value is the number $\mathbb{E}(Y|X = x) = (1 + x)/2$ once $X = x$ is observed. ∎

3.24 Theorem (The Rule of Iterated Expectations). *For random variables X and Y, assuming the expectations exist, we have that*

$$\mathbb{E}\left[\mathbb{E}(Y|X)\right] = \mathbb{E}(Y) \quad \text{and} \quad \mathbb{E}\left[\mathbb{E}(X|Y)\right] = \mathbb{E}(X). \tag{3.15}$$

More generally, for any function $r(x, y)$ we have

$$\mathbb{E}\left[\mathbb{E}(r(X,Y)|X)\right] = \mathbb{E}(r(X,Y)). \tag{3.16}$$

PROOF. We'll prove the first equation. Using the definition of conditional expectation and the fact that $f(x, y) = f(x)f(y|x)$,

$$\begin{aligned} \mathbb{E}\left[\mathbb{E}(Y|X)\right] &= \int \mathbb{E}(Y|X = x)f_X(x)dx = \int \int y f(y|x)dy f(x)dx \\ &= \int \int y f(y|x)f(x)dxdy = \int \int y f(x,y)dxdy = \mathbb{E}(Y). \quad \blacksquare \end{aligned}$$

3.25 Example. Consider example 3.23. How can we compute $\mathbb{E}(Y)$? One method is to find the joint density $f(x, y)$ and then compute $\mathbb{E}(Y) = \int \int y f(x, y)dxdy$. An easier way is to do this in two steps. First, we already know that $\mathbb{E}(Y|X) = (1 + X)/2$. Thus,

$$\begin{aligned} \mathbb{E}(Y) &= \mathbb{E}\mathbb{E}(Y|X) = \mathbb{E}\left(\frac{(1+X)}{2}\right) \\ &= \frac{(1 + \mathbb{E}(X))}{2} = \frac{(1 + (1/2))}{2} = 3/4. \quad \blacksquare \end{aligned}$$

3.26 Definition. *The **conditional variance** is defined as*

$$\mathbb{V}(Y|X = x) = \int (y - \mu(x))^2 f(y|x)dy \tag{3.17}$$

where $\mu(x) = \mathbb{E}(Y|X = x)$.

3.27 Theorem. *For random variables X and Y,*

$$\mathbb{V}(Y) = \mathbb{E}\mathbb{V}(Y|X) + \mathbb{V}\mathbb{E}(Y|X).$$

3.28 Example. Draw a county at random from the United States. Then draw n people at random from the county. Let X be the number of those people who have a certain disease. If Q denotes the proportion of people in that county with the disease, then Q is also a random variable since it varies from county to county. Given $Q = q$, we have that $X \sim \text{Binomial}(n, q)$. Thus, $\mathbb{E}(X|Q = q) = nq$ and $\mathbb{V}(X|Q = q) = nq(1 - q)$. Suppose that the random variable Q has a Uniform (0,1) distribution. A distribution that is constructed in stages like this is called a **hierarchical model** and can be written as

$$Q \sim \text{Uniform}(0, 1)$$
$$X|Q = q \sim \text{Binomial}(n, q).$$

Now, $\mathbb{E}(X) = \mathbb{E}\mathbb{E}(X|Q) = \mathbb{E}(nQ) = n\mathbb{E}(Q) = n/2$. Let us compute the variance of X. Now, $\mathbb{V}(X) = \mathbb{E}\mathbb{V}(X|Q) + \mathbb{V}\mathbb{E}(X|Q)$. Let's compute these two terms. First, $\mathbb{E}\mathbb{V}(X|Q) = \mathbb{E}[nQ(1 - Q)] = n\mathbb{E}(Q(1 - Q)) = n\int q(1 - q)f(q)dq = n\int_0^1 q(1 - q)dq = n/6$. Next, $\mathbb{V}\mathbb{E}(X|Q) = \mathbb{V}(nQ) = n^2\mathbb{V}(Q) = n^2\int(q - (1/2))^2dq = n^2/12$. Hence, $\mathbb{V}(X) = (n/6) + (n^2/12)$. ∎

3.6   Moment Generating Functions

Now we will define the moment generating function which is used for finding moments, for finding the distribution of sums of random variables and which is also used in the proofs of some theorems.

3.29 Definition. *The* **moment generating function** MGF, *or* **Laplace transform**, *of X is defined by*

$$\psi_X(t) = \mathbb{E}(e^{tX}) = \int e^{tx}dF(x)$$

where t varies over the real numbers.

In what follows, we assume that the MGF is well defined for all t in some open interval around $t = 0$. [2]

When the MGF is well defined, it can be shown that we can interchange the operations of differentiation and "taking expectation." This leads to

$$\psi'(0) = \left[\frac{d}{dt}\mathbb{E}e^{tX}\right]_{t=0} = \mathbb{E}\left[\frac{d}{dt}e^{tX}\right]_{t=0} = \mathbb{E}\left[Xe^{tX}\right]_{t=0} = \mathbb{E}(X).$$

---

[2] A related function is the characteristic function, defined by $\mathbb{E}(e^{itX})$ where $i = \sqrt{-1}$. This function is always well defined for all t.

By taking k derivatives we conclude that $\psi^{(k)}(0) = \mathbb{E}(X^k)$. This gives us a method for computing the moments of a distribution.

3.30 Example. Let $X \sim \text{Exp}(1)$. For any $t < 1$,

$$\psi_X(t) = \mathbb{E}e^{tX} = \int_0^\infty e^{tx}e^{-x}dx = \int_0^\infty e^{(t-1)x}dx = \frac{1}{1-t}.$$

The integral is divergent if $t \geq 1$. So, $\psi_X(t) = 1/(1-t)$ for all $t < 1$. Now, $\psi'(0) = 1$ and $\psi''(0) = 2$. Hence, $\mathbb{E}(X) = 1$ and $\mathbb{V}(X) = \mathbb{E}(X^2) - \mu^2 = 2-1 = 1$. ∎

3.31 Lemma. *Properties of the* MGF.
(1) *If* $Y = aX + b$, *then* $\psi_Y(t) = e^{bt}\psi_X(at)$.
(2) *If* $X_1, \ldots, X_n$ *are independent and* $Y = \sum_i X_i$, *then* $\psi_Y(t) = \prod_i \psi_i(t)$ *where* ψ_i *is the* MGF *of* X_i.

3.32 Example. Let $X \sim \text{Binomial}(n, p)$. We know that $X = \sum_{i=1}^n X_i$ where $\mathbb{P}(X_i = 1) = p$ and $\mathbb{P}(X_i = 0) = 1 - p$. Now $\psi_i(t) = \mathbb{E}e^{X_i t} = (p \times e^t) + ((1 - p)) = pe^t + q$ where $q = 1 - p$. Thus, $\psi_X(t) = \prod_i \psi_i(t) = (pe^t + q)^n$. ∎

Recall that X and Y are equal in distribution if they have the same distribution function and we write $X \stackrel{d}{=} Y$.

3.33 Theorem. *Let X and Y be random variables. If $\psi_X(t) = \psi_Y(t)$ for all t in an open interval around 0, then $X \stackrel{d}{=} Y$.*

3.34 Example. Let $X_1 \sim \text{Binomial}(n_1, p)$ and $X_2 \sim \text{Binomial}(n_2, p)$ be independent. Let $Y = X_1 + X_2$. Then,

$$\psi_Y(t) = \psi_1(t)\psi_2(t) = (pe^t + q)^{n_1}(pe^t + q)^{n_2} = (pe^t + q)^{n_1+n_2}$$

and we recognize the latter as the MGF of a Binomial$(n_1 + n_2, p)$ distribution. Since the MGF characterizes the distribution (i.e., there can't be another random variable which has the same MGF) we conclude that $Y \sim$ Binomial$(n_1 + n_2, p)$. ∎

Moment Generating Functions for Some Common Distributions	
Distribution	MGF $\psi(t)$
Bernoulli(p)	$pe^t + (1-p)$
Binomial(n,p)	$(pe^t + (1-p))^n$
Poisson(λ)	$e^{\lambda(e^t-1)}$
Normal(μ, σ)	$\exp\left\{\mu t + \frac{\sigma^2 t^2}{2}\right\}$
Gamma(α, β)	$\left(\frac{1}{1-\beta t}\right)^\alpha$ for $t < 1/\beta$

3.35 Example. Let $Y_1 \sim \text{Poisson}(\lambda_1)$ and $Y_2 \sim \text{Poisson}(\lambda_2)$ be independent. The moment generating function of $Y = Y_1 + Y + 2$ is $\psi_Y(t) = \psi_{Y_1}(t)\psi_{Y_2}(t) = e^{\lambda_1(e^t-1)}e^{\lambda_2(e^t-1)} = e^{(\lambda_1+\lambda_2)(e^t-1)}$ which is the moment generating function of a Poisson$(\lambda_1 + \lambda_2)$. We have thus proved that the sum of two independent Poisson random variables has a Poisson distribution. ∎

3.7 Appendix

EXPECTATION AS AN INTEGRAL. The integral of a measurable function $r(x)$ is defined as follows. First suppose that r is simple, meaning that it takes finitely many values $a_1, \ldots, a_k$ over a partition $A_1, \ldots, A_k$. Then define

$$\int r(x)dF(x) = \sum_{i=1}^k a_i \, \mathbb{P}(r(X) \in A_i).$$

The integral of a positive measurable function r is defined by $\int r(x)dF(x) = \lim_i \int r_i(x)dF(x)$ where r_i is a sequence of simple functions such that $r_i(x) \le r(x)$ and $r_i(x) \to r(x)$ as $i \to \infty$. This does not depend on the particular sequence. The integral of a measurable function r is defined to be $\int r(x)dF(x) = \int r^+(x)dF(x) - \int r^-(x)dF(x)$ assuming both integrals are finite, where $r^+(x) = \max\{r(x), 0\}$ and $r^-(x) = -\min\{r(x), 0\}$.

3.8 Exercises

1. Suppose we play a game where we start with c dollars. On each play of the game you either double or halve your money, with equal probability. What is your expected fortune after n trials?

2. Show that $V(X) = 0$ if and only if there is a constant c such that $P(X = c) = 1$.

3. Let $X_1, \ldots, X_n \sim \text{Uniform}(0, 1)$ and let $Y_n = \max\{X_1, \ldots, X_n\}$. Find $\mathbb{E}(Y_n)$.

4. A particle starts at the origin of the real line and moves along the line in jumps of one unit. For each jump the probability is p that the particle will jump one unit to the left and the probability is $1 - p$ that the particle will jump one unit to the right. Let X_n be the position of the particle after n units. Find $\mathbb{E}(X_n)$ and $V(X_n)$. (This is known as a **random walk**.)

5. A fair coin is tossed until a head is obtained. What is the expected number of tosses that will be required?

6. Prove Theorem 3.6 for discrete random variables.

7. Let X be a continuous random variable with CDF F. Suppose that $P(X > 0) = 1$ and that $\mathbb{E}(X)$ exists. Show that $\mathbb{E}(X) = \int_0^\infty P(X > x)\,dx$.

   Hint: Consider integrating by parts. The following fact is helpful: if $\mathbb{E}(X)$ exists then $\lim_{x \to \infty} x[1 - F(x)] = 0$.

8. Prove Theorem 3.17.

9. (Computer Experiment.) Let $X_1, X_2, \ldots, X_n$ be $N(0, 1)$ random variables and let $\overline{X}_n = n^{-1} \sum_{i=1}^n X_i$. Plot $\overline{X}_n$ versus n for $n = 1, \ldots, 10,000$. Repeat for $X_1, X_2, \ldots, X_n \sim$ Cauchy. Explain why there is such a difference.

10. Let $X \sim N(0, 1)$ and let $Y = e^X$. Find $\mathbb{E}(Y)$ and $V(Y)$.

11. (Computer Experiment: Simulating the Stock Market.) Let $Y_1, Y_2, \ldots$ be independent random variables such that $P(Y_i = 1) = P(Y_i = -1) = 1/2$. Let $X_n = \sum_{i=1}^n Y_i$. Think of $Y_i = 1$ as "the stock price increased by one dollar", $Y_i = -1$ as "the stock price decreased by one dollar", and X_n as the value of the stock on day n.

    (a) Find $\mathbb{E}(X_n)$ and $V(X_n)$.

    (b) Simulate X_n and plot X_n versus n for $n = 1, 2, \ldots, 10,000$. Repeat the whole simulation several times. Notice two things. First, it's easy to "see" patterns in the sequence even though it is random. Second, you will find that the four runs look very different even though they were generated the same way. How do the calculations in (a) explain the second observation?

12. Prove the formulas given in the table at the beginning of Section 3.4 for the Bernoulli, Poisson, Uniform, Exponential, Gamma, and Beta. Here are some hints. For the mean of the Poisson, use the fact that $e^a = \sum_{x=0}^{\infty} a^x/x!$. To compute the variance, first compute $\mathbb{E}(X(X-1))$. For the mean of the Gamma, it will help to multiply and divide by $\Gamma(\alpha+1)/\beta^{\alpha+1}$ and use the fact that a Gamma density integrates to 1. For the Beta, multiply and divide by $\Gamma(\alpha+1)\Gamma(\beta)/\Gamma(\alpha+\beta+1)$.

13. Suppose we generate a random variable X in the following way. First we flip a fair coin. If the coin is heads, take X to have a Unif(0,1) distribution. If the coin is tails, take X to have a Unif(3,4) distribution.

    (a) Find the mean of X.

    (b) Find the standard deviation of X.

14. Let $X_1, \ldots, X_m$ and $Y_1, \ldots, Y_n$ be random variables and let $a_1, \ldots, a_m$ and $b_1, \ldots, b_n$ be constants. Show that

$$\text{Cov}\left(\sum_{i=1}^{m} a_i X_i, \sum_{j=1}^{n} b_j Y_j\right) = \sum_{i=1}^{m}\sum_{j=1}^{n} a_i b_j \text{Cov}(X_i, Y_j).$$

15. Let

$$f_{X,Y}(x,y) = \begin{cases} \frac{1}{3}(x+y) & 0 \le x \le 1, \ 0 \le y \le 2 \\ 0 & \text{otherwise.} \end{cases}$$

    Find $\mathbb{V}(2X - 3Y + 8)$.

16. Let r(x) be a function of x and let s(y) be a function of y. Show that

$$\mathbb{E}(r(X)s(Y)|X) = r(X)\mathbb{E}(s(Y)|X).$$

    Also, show that $\mathbb{E}(r(X)|X) = r(X)$.

17. Prove that

$$\mathbb{V}(Y) = \mathbb{E}\mathbb{V}(Y \mid X) + \mathbb{V}\mathbb{E}(Y \mid X).$$

    Hint: Let $m = \mathbb{E}(Y)$ and let $b(x) = \mathbb{E}(Y|X = x)$. Note that $\mathbb{E}(b(X)) = \mathbb{E}\mathbb{E}(Y|X) = \mathbb{E}(Y) = m$. Bear in mind that b is a function of x. Now write $\mathbb{V}(Y) = \mathbb{E}(Y - m)^2 = \mathbb{E}((Y - b(X)) + (b(X) - m))^2$. Expand the square and take the expectation. You then have to take the expectation of three terms. In each case, use the rule of the iterated expectation: $\mathbb{E}(\text{stuff}) = \mathbb{E}(\mathbb{E}(\text{stuff}|X))$.

18. Show that if $\mathbb{E}(X|Y = y) = c$ for some constant c, then X and Y are uncorrelated.

19. This question is to help you understand the idea of a **sampling distribution**. Let $X_1, \ldots, X_n$ be IID with mean μ and variance σ^2. Let $\overline{X}_n = n^{-1} \sum_{i=1}^{n} X_i$. Then $\overline{X}_n$ is a **statistic**, that is, a function of the data. Since $\overline{X}_n$ is a random variable, it has a distribution. This distribution is called the *sampling distribution of the statistic*. Recall from Theorem 3.17 that $\mathbb{E}(\overline{X}_n) = \mu$ and $\mathbb{V}(\overline{X}_n) = \sigma^2/n$. Don't confuse the distribution of the data f_X and the distribution of the statistic $f_{\overline{X}_n}$. To make this clear, let $X_1, \ldots, X_n \sim \text{Uniform}(0, 1)$. Let f_X be the density of the Uniform$(0, 1)$. Plot f_X. Now let $\overline{X}_n = n^{-1} \sum_{i=1}^{n} X_i$. Find $\mathbb{E}(\overline{X}_n)$ and $\mathbb{V}(\overline{X}_n)$. Plot them as a function of n. Interpret. Now simulate the distribution of $\overline{X}_n$ for $n = 1, 5, 25, 100$. Check that the simulated values of $\mathbb{E}(\overline{X}_n)$ and $\mathbb{V}(\overline{X}_n)$ agree with your theoretical calculations. What do you notice about the sampling distribution of $\overline{X}_n$ as n increases?

20. Prove Lemma 3.21.

21. Let X and Y be random variables. Suppose that $\mathbb{E}(Y|X) = X$. Show that $\text{Cov}(X, Y) = \mathbb{V}(X)$.

22. Let $X \sim \text{Uniform}(0, 1)$. Let $0 < a < b < 1$. Let

$$Y = \begin{cases} 1 & 0 < x < b \\ 0 & \text{otherwise} \end{cases}$$

and let

$$Z = \begin{cases} 1 & a < x < 1 \\ 0 & \text{otherwise} \end{cases}$$

(a) Are Y and Z independent? Why/Why not?

(b) Find $\mathbb{E}(Y|Z)$. Hint: What values z can Z take? Now find $\mathbb{E}(Y|Z = z)$.

23. Find the moment generating function for the Poisson, Normal, and Gamma distributions.

24. Let $X_1, \ldots, X_n \sim \text{Exp}(\beta)$. Find the moment generating function of X_i. Prove that $\sum_{i=1}^{n} X_i \sim \text{Gamma}(n, \beta)$.

4
Inequalities

4.1 Probability Inequalities

Inequalities are useful for bounding quantities that might otherwise be hard to compute. They will also be used in the theory of convergence which is discussed in the next chapter. Our first inequality is Markov's inequality.

4.1 Theorem (Markov's inequality). *Let X be a non-negative random variable and suppose that $\mathbb{E}(X)$ exists. For any $t > 0$,*

$$\mathbb{P}(X > t) \leq \frac{\mathbb{E}(X)}{t}. \tag{4.1}$$

PROOF. Since $X > 0$,

$$
\begin{aligned}
\mathbb{E}(X) &= \int_0^\infty x f(x)dx = \int_0^t x f(x)dx + \int_t^\infty x f(x)dx \\
&\geq \int_t^\infty x f(x)dx \geq t \int_t^\infty f(x)dx = t\,\mathbb{P}(X > t) \quad \blacksquare
\end{aligned}
$$

4.2 Theorem (Chebyshev's inequality). *Let $\mu = \mathbb{E}(X)$ and $\sigma^2 = \mathbb{V}(X)$.*
Then,

$$\mathbb{P}(|X - \mu| \geq t) \leq \frac{\sigma^2}{t^2} \quad \text{and} \quad \mathbb{P}(|Z| \geq k) \leq \frac{1}{k^2} \tag{4.2}$$

where $Z = (X - \mu)/\sigma$. In particular, $\mathbb{P}(|Z| > 2) \leq 1/4$ and
$\mathbb{P}(|Z| > 3) \leq 1/9$.

PROOF. We use Markov's inequality to conclude that

$$\mathbb{P}(|X - \mu| \geq t) = \mathbb{P}(|X - \mu|^2 \geq t^2) \leq \frac{\mathbb{E}(X - \mu)^2}{t^2} = \frac{\sigma^2}{t^2}.$$

The second part follows by setting $t = k\sigma$. ∎

4.3 Example. Suppose we test a prediction method, a neural net for example, on a set of n new test cases. Let $X_i = 1$ if the predictor is wrong and $X_i = 0$ if the predictor is right. Then $\overline{X}_n = n^{-1} \sum_{i=1}^{n} X_i$ is the observed error rate. Each X_i may be regarded as a Bernoulli with unknown mean p. We would like to know the true — but unknown — error rate p. Intuitively, we expect that $\overline{X}_n$ should be close to p. How likely is $\overline{X}_n$ to not be within ϵ of p? We have that $\mathbb{V}(\overline{X}_n) = \mathbb{V}(X_1)/n = p(1-p)/n$ and

$$\mathbb{P}(|\overline{X}_n - p| > \epsilon) \leq \frac{\mathbb{V}(\overline{X}_n)}{\epsilon^2} = \frac{p(1-p)}{n\epsilon^2} \leq \frac{1}{4n\epsilon^2}$$

since $p(1-p) \leq \frac{1}{4}$ for all p. For $\epsilon = .2$ and $n = 100$ the bound is .0625. ∎

Hoeffding's inequality is similar in spirit to Markov's inequality but it is a sharper inequality. We present the result here in two parts.

4.4 Theorem (Hoeffding's Inequality). *Let $Y_1, \ldots, Y_n$ be independent*
observations such that
$\mathbb{E}(Y_i) = 0$ and $a_i \leq Y_i \leq b_i$. Let $\epsilon > 0$. Then, for any $t > 0$,

$$\mathbb{P}\left(\sum_{i=1}^{n} Y_i \geq \epsilon\right) \leq e^{-t\epsilon} \prod_{i=1}^{n} e^{t^2(b_i - a_i)^2/8}. \tag{4.3}$$

4.5 Theorem. *Let* $X_1, \ldots, X_n \sim$ *Bernoulli*(p). *Then, for any* $\epsilon > 0$,

$$\mathbb{P}\left(|\overline{X}_n - p| > \epsilon\right) \leq 2e^{-2n\epsilon^2} \tag{4.4}$$

where $\overline{X}_n = n^{-1} \sum_{i=1}^{n} X_i$.

4.6 Example. Let $X_1, \ldots, X_n \sim$ Bernoulli(p). Let $n = 100$ and $\epsilon = .2$. We saw that Chebyshev's inequality yielded

$$\mathbb{P}(|\overline{X}_n - p| > \epsilon) \leq .0625.$$

According to Hoeffding's inequality,

$$\mathbb{P}(|\overline{X}_n - p| > .2) \leq 2e^{-2(100)(.2)^2} = .00067$$

which is much smaller than .0625. ∎

Hoeffding's inequality gives us a simple way to create a **confidence interval** for a binomial parameter p. We will discuss confidence intervals in detail later (see Chapter 6) but here is the basic idea. Fix $\alpha > 0$ and let

$$\epsilon_n = \sqrt{\frac{1}{2n} \log\left(\frac{2}{\alpha}\right)}.$$

By Hoeffding's inequality,

$$\mathbb{P}\left(|\overline{X}_n - p| > \epsilon_n\right) \leq 2e^{-2n\epsilon_n^2} = \alpha.$$

Let $C = (\overline{X}_n - \epsilon_n, \overline{X}_n + \epsilon_n)$. Then, $\mathbb{P}(p \notin C) = \mathbb{P}(|\overline{X}_n - p| > \epsilon_n) \leq \alpha$. Hence, $\mathbb{P}(p \in C) \geq 1 - \alpha$, that is, the random interval C traps the true parameter value p with probability $1 - \alpha$; we call C a $1 - \alpha$ confidence interval. More on this later.

The following inequality is useful for bounding probability statements about Normal random variables.

4.7 Theorem (Mill's Inequality). *Let* $Z \sim N(0, 1)$. *Then,*

$$\mathbb{P}(|Z| > t) \leq \sqrt{\frac{2}{\pi}} \frac{e^{-t^2/2}}{t}.$$

4.2 Inequalities For Expectations

This section contains two inequalities on expected values.

4.8 Theorem (Cauchy-Schwartz inequality). *If X and Y have finite variances then*

$$\mathbb{E}|XY| \leq \sqrt{\mathbb{E}(X^2)\mathbb{E}(Y^2)}. \tag{4.5}$$

Recall that a function g is **convex** if for each x, y and each $\alpha \in [0, 1]$,

$$g(\alpha x + (1 - \alpha)y) \leq \alpha g(x) + (1 - \alpha)g(y).$$

If g is twice differentiable and $g''(x) \geq 0$ for all x, then g is convex. It can be shown that if g is convex, then g lies above any line that touches g at some point, called a tangent line. A function g is **concave** if $-g$ is convex. Examples of convex functions are $g(x) = x^2$ and $g(x) = e^x$. Examples of concave functions are $g(x) = -x^2$ and $g(x) = \log x$.

4.9 Theorem (Jensen's inequality). *If g is convex, then*

$$\mathbb{E}g(X) \geq g(\mathbb{E}X). \tag{4.6}$$

If g is concave, then

$$\mathbb{E}g(X) \leq g(\mathbb{E}X). \tag{4.7}$$

PROOF. Let $L(x) = a + bx$ be a line, tangent to $g(x)$ at the point $\mathbb{E}(X)$. Since g is convex, it lies above the line $L(x)$. So,

$$\mathbb{E}g(X) \geq \mathbb{E}L(X) = \mathbb{E}(a + bX) = a + b\mathbb{E}(X) = L(\mathbb{E}(X)) = g(\mathbb{E}X). \quad \blacksquare$$

From Jensen's inequality we see that $\mathbb{E}(X^2) \geq (\mathbb{E}X)^2$ and if X is positive, then $\mathbb{E}(1/X) \geq 1/\mathbb{E}(X)$. Since log is concave, $\mathbb{E}(\log X) \leq \log \mathbb{E}(X)$.

4.3 Bibliographic Remarks

Devroye et al. (1996) is a good reference on probability inequalities and their use in statistics and pattern recognition. The following proof of Hoeffding's inequality is from that text.

4.4 Appendix

PROOF OF HOEFFDING'S INEQUALITY. We will make use of the exact form of Taylor's theorem: if g is a smooth function, then there is a number $\xi \in (0, u)$ such that $g(u) = g(0) + ug'(0) + \frac{u^2}{2}g''(\xi)$.

PROOF of Theorem 4.4. For any $t > 0$, we have, from Markov's inequality, that

$$\mathbb{P}\left(\sum_{i=1}^{n} Y_i \geq \epsilon\right) = \mathbb{P}\left(t\sum_{i=1}^{n} Y_i \geq t\epsilon\right) = \mathbb{P}\left(e^{t\sum_{i=1}^{n} Y_i} \geq e^{t\epsilon}\right)$$

$$\leq e^{-t\epsilon}\mathbb{E}\left(e^{t\sum_{i=1}^{n} Y_i}\right) = e^{-t\epsilon}\prod_{i}\mathbb{E}(e^{tY_i}). \qquad (4.8)$$

Since $a_i \leq Y_i \leq b_i$, we can write Y_i as a convex combination of a_i and b_i, namely, $Y_i = \alpha b_i + (1 - \alpha)a_i$ where $\alpha = (Y_i - a_i)/(b_i - a_i)$. So, by the convexity of e^{ty} we have

$$e^{tY_i} \leq \frac{Y_i - a_i}{b_i - a_i}e^{tb_i} + \frac{b_i - Y_i}{b_i - a_i}e^{ta_i}.$$

Take expectations of both sides and use the fact that $\mathbb{E}(Y_i) = 0$ to get

$$\mathbb{E}e^{tY_i} \leq -\frac{a_i}{b_i - a_i}e^{tb_i} + \frac{b_i}{b_i - a_i}e^{ta_i} = e^{g(u)} \qquad (4.9)$$

where $u = t(b_i - a_i)$, $g(u) = -\gamma u + \log(1 - \gamma + \gamma e^u)$ and $\gamma = -a_i/(b_i - a_i)$.

Note that $g(0) = g'(0) = 0$. Also, $g''(u) \leq 1/4$ for all $u > 0$. By Taylor's theorem, there is a $\xi \in (0, u)$ such that

$$\begin{aligned} g(u) &= g(0) + ug'(0) + \frac{u^2}{2}g''(\xi) \\ &= \frac{u^2}{2}g''(\xi) \leq \frac{u^2}{8} = \frac{t^2(b_i - a_i)^2}{8}. \end{aligned}$$

Hence,

$$\mathbb{E}e^{tY_i} \leq e^{g(u)} \leq e^{t^2(b_i - a_i)^2/8}.$$

The result follows from (4.8). ∎

PROOF of Theorem 4.5. Let $Y_i = (1/n)(X_i - p)$. Then $\mathbb{E}(Y_i) = 0$ and $a \leq Y_i \leq b$ where $a = -p/n$ and $b = (1 - p)/n$. Also, $(b - a)^2 = 1/n^2$. Applying Theorem 4.4 we get

$$\mathbb{P}(\overline{X}_n - p > \epsilon) = \mathbb{P}\left(\sum_i Y_i > \epsilon\right) \leq e^{-t\epsilon}e^{t^2/(8n)}.$$

The above holds for any $t > 0$. In particular, take $t = 4n\epsilon$ and we get $\mathbb{P}(\overline{X}_n - p > \epsilon) \leq e^{-2n\epsilon^2}$. By a similar argument we can show that $\mathbb{P}(\overline{X}_n - p < -\epsilon) \leq e^{-2n\epsilon^2}$. Putting these together we get $\mathbb{P}\left(|\overline{X}_n - p| > \epsilon\right) \leq 2e^{-2n\epsilon^2}$. ∎

4.5   Exercises

1. Let $X \sim$ Exponential(β). Find $\mathbb{P}(|X - \mu_X| \geq k\sigma_X)$ for $k > 1$. Compare this to the bound you get from Chebyshev's inequality.

2. Let $X \sim$ Poisson(λ). Use Chebyshev's inequality to show that $\mathbb{P}(X \geq 2\lambda) \leq 1/\lambda$.

3. Let $X_1, \ldots, X_n \sim$ Bernoulli(p) and $\overline{X}_n = n^{-1} \sum_{i=1}^{n} X_i$. Bound $\mathbb{P}(|\overline{X}_n - p| > \epsilon)$ using Chebyshev's inequality and using Hoeffding's inequality. Show that, when n is large, the bound from Hoeffding's inequality is smaller than the bound from Chebyshev's inequality.

4. Let $X_1, \ldots, X_n \sim$ Bernoulli(p).

   (a) Let $\alpha > 0$ be fixed and define

   $$\epsilon_n = \sqrt{\frac{1}{2n} \log\left(\frac{2}{\alpha}\right)}.$$

   Let $\widehat{p}_n = n^{-1} \sum_{i=1}^{n} X_i$. Define $C_n = (\widehat{p}_n - \epsilon_n, \ \widehat{p}_n + \epsilon_n)$. Use Hoeffding's inequality to show that

   $$\mathbb{P}(C_n \text{ contains } p) \geq 1 - \alpha.$$

   In practice, we truncate the interval so it does not go below 0 or above 1.

   (b) (Computer Experiment.) Let's examine the properties of this confidence interval. Let $\alpha = 0.05$ and $p = 0.4$. Conduct a simulation study to see how often the interval contains p (called the coverage). Do this for various values of n between 1 and 10000. Plot the coverage versus n.

   (c) Plot the length of the interval versus n. Suppose we want the length of the interval to be no more than .05. How large should n be?

5. Prove Mill's inequality, Theorem 4.7. Hint. Note that $\mathbb{P}(|Z| > t) = 2\mathbb{P}(Z > t)$. Now write out what $\mathbb{P}(Z > t)$ means and note that $x/t > 1$ whenever $x > t$.

6. Let $Z \sim N(0,1)$. Find $\mathbb{P}(|Z| > t)$ and plot this as a function of t. From Markov's inequality, we have the bound $\mathbb{P}(|Z| > t) \leq \frac{\mathbb{E}|Z|^k}{t^k}$ for any $k > 0$. Plot these bounds for $k = 1, 2, 3, 4, 5$ and compare them to the true value of $\mathbb{P}(|Z| > t)$. Also, plot the bound from Mill's inequality.

7. Let $X_1, \ldots, X_n \sim N(0,1)$. Bound $\mathbb{P}(|\overline{X}_n| > t)$ using Mill's inequality, where $\overline{X}_n = n^{-1} \sum_{i=1}^{n} X_i$. Compare to the Chebyshev bound.

5
Convergence of Random Variables

5.1 Introduction

The most important aspect of probability theory concerns the behavior of sequences of random variables. This part of probability is called **large sample theory**, or **limit theory**, or **asymptotic theory**. The basic question is this: what can we say about the limiting behavior of a sequence of random variables $X_1, X_2, X_3, \ldots$? Since statistics and data mining are all about gathering data, we will naturally be interested in what happens as we gather more and more data.

In calculus we say that a sequence of real numbers x_n converges to a limit x if, for every $\epsilon > 0$, $|x_n - x| < \epsilon$ for all large n. In probability, convergence is more subtle. Going back to calculus for a moment, suppose that $x_n = x$ for all n. Then, trivially, $\lim_{n \to \infty} x_n = x$. Consider a probabilistic version of this example. Suppose that $X_1, X_2, \ldots$ is a sequence of random variables which are independent and suppose each has a $N(0,1)$ distribution. Since these all have the same distribution, we are tempted to say that X_n "converges" to $X \sim N(0,1)$. But this can't quite be right since $\mathbb{P}(X_n = X) = 0$ for all n. (Two continuous random variables are equal with probability zero.)

Here is another example. Consider $X_1, X_2, \ldots$ where $X_i \sim N(0, 1/n)$. Intuitively, X_n is very concentrated around 0 for large n so we would like to say that X_n converges to 0. But $\mathbb{P}(X_n = 0) = 0$ for all n. Clearly, we need to develop some tools for discussing convergence in a rigorous way. This chapter develops the appropriate methods.

There are two main ideas in this chapter which we state informally here:

1. The **law of large numbers** says that the sample average $\overline{X}_n = n^{-1} \sum_{i=1}^n X_i$ **converges in probability** to the expectation $\mu = \mathbb{E}(X_i)$. This means that $\overline{X}_n$ is close to μ with high probability.

2. The **central limit theorem** says that $\sqrt{n}(\overline{X}_n - \mu)$ **converges in distribution** to a Normal distribution. This means that the sample average has approximately a Normal distribution for large n.

5.2   Types of Convergence

The two main types of convergence are defined as follows.

5.1 Definition. *Let $X_1, X_2, \ldots$ be a sequence of random variables and let X be another random variable. Let F_n denote the CDF of X_n and let F denote the CDF of X.*

1. X_n converges to X in probability, written $X_n \xrightarrow{\text{P}} X$, if, for every $\epsilon > 0$,

$$\mathbb{P}(|X_n - X| > \epsilon) \to 0 \tag{5.1}$$

as $n \to \infty$.

2. X_n converges to X in distribution, written $X_n \rightsquigarrow X$, if

$$\lim_{n \to \infty} F_n(t) = F(t) \tag{5.2}$$

at all t for which F is continuous.

When the limiting random variable is a point mass, we change the notation slightly. If $\mathbb{P}(X = c) = 1$ and $X_n \xrightarrow{\text{P}} X$ then we write $X_n \xrightarrow{\text{P}} c$. Similarly, if $X_n \rightsquigarrow X$ we write $X_n \rightsquigarrow c$.

There is another type of convergence which we introduce mainly because it is useful for proving convergence in probability.

FIGURE 5.1. Example 5.3. X_n converges in distribution to X because $F_n(t)$ converges to $F(t)$ at all points except $t = 0$. Convergence is not required at $t = 0$ because $t = 0$ is not a point of continuity for F.

5.2 Definition. X_n **converges to** X **in quadratic mean** (*also called* *convergence in L_2*), *written* $X_n \xrightarrow{\text{qm}} X$, *if*

$$\mathbb{E}(X_n - X)^2 \to 0 \qquad (5.3)$$

as $n \to \infty$.

Again, if X is a point mass at c we write $X_n \xrightarrow{\text{qm}} c$ instead of $X_n \xrightarrow{\text{qm}} X$.

5.3 Example. Let $X_n \sim N(0, 1/n)$. Intuitively, X_n is concentrating at 0 so we would like to say that X_n converges to 0. Let's see if this is true. Let F be the distribution function for a point mass at 0. Note that $\sqrt{n}X_n \sim N(0, 1)$. Let Z denote a standard normal random variable. For $t < 0$, $F_n(t) = \mathbb{P}(X_n < t) = \mathbb{P}(\sqrt{n}X_n < \sqrt{n}t) = \mathbb{P}(Z < \sqrt{n}t) \to 0$ since $\sqrt{n}t \to -\infty$. For $t > 0$, $F_n(t) = \mathbb{P}(X_n < t) = \mathbb{P}(\sqrt{n}X_n < \sqrt{n}t) = \mathbb{P}(Z < \sqrt{n}t) \to 1$ since $\sqrt{n}t \to \infty$. Hence, $F_n(t) \to F(t)$ for all $t \neq 0$ and so $X_n \rightsquigarrow 0$. Notice that $F_n(0) = 1/2 \neq F(1/2) = 1$ so convergence fails at $t = 0$. That doesn't matter because $t = 0$ is not a continuity point of F and the definition of convergence in distribution only requires convergence at continuity points. See Figure 5.1. Now consider convergence in probability. For any $\epsilon > 0$, using Markov's inequality,

$$\mathbb{P}(|X_n| > \epsilon) = \mathbb{P}(|X_n|^2 > \epsilon^2)$$
$$\leq \frac{\mathbb{E}(X_n^2)}{\epsilon^2} = \frac{\frac{1}{n}}{\epsilon^2} \to 0$$

as $n \to \infty$. Hence, $X_n \xrightarrow{\text{P}} 0$. ∎

The next theorem gives the relationship between the types of convergence. The results are summarized in Figure 5.2.

5.4 Theorem. *The following relationships hold:*
(a) $X_n \xrightarrow{\text{qm}} X$ *implies that* $X_n \xrightarrow{\text{P}} X$.
(b) $X_n \xrightarrow{\text{P}} X$ *implies that* $X_n \rightsquigarrow X$.
(c) *If* $X_n \rightsquigarrow X$ *and if* $\mathbb{P}(X = c) = 1$ *for some real number c, then* $X_n \xrightarrow{\text{P}} X$.

In general, none of the reverse implications hold except the special case in (c).

PROOF. We start by proving (a). Suppose that $X_n \xrightarrow{\text{qm}} X$. Fix $\epsilon > 0$. Then, using Markov's inequality,

$$\mathbb{P}(|X_n - X| > \epsilon) = \mathbb{P}(|X_n - X|^2 > \epsilon^2) \le \frac{\mathbb{E}|X_n - X|^2}{\epsilon^2} \to 0.$$

Proof of (b). This proof is a little more complicated. You may skip it if you wish. Fix $\epsilon > 0$ and let x be a continuity point of F. Then

$$
\begin{aligned}
F_n(x) &= \mathbb{P}(X_n \le x) = \mathbb{P}(X_n \le x, X \le x + \epsilon) + \mathbb{P}(X_n \le x, X > x + \epsilon) \\
&\le \mathbb{P}(X \le x + \epsilon) + \mathbb{P}(|X_n - X| > \epsilon) \\
&= F(x + \epsilon) + \mathbb{P}(|X_n - X| > \epsilon).
\end{aligned}
$$

Also,

$$
\begin{aligned}
F(x - \epsilon) &= \mathbb{P}(X \le x - \epsilon) = \mathbb{P}(X \le x - \epsilon, X_n \le x) + \mathbb{P}(X \le x - \epsilon, X_n > x) \\
&\le F_n(x) + \mathbb{P}(|X_n - X| > \epsilon).
\end{aligned}
$$

Hence,

$$F(x - \epsilon) - \mathbb{P}(|X_n - X| > \epsilon) \le F_n(x) \le F(x + \epsilon) + \mathbb{P}(|X_n - X| > \epsilon).$$

Take the limit as $n \to \infty$ to conclude that

$$F(x - \epsilon) \le \liminf_{n \to \infty} F_n(x) \le \limsup_{n \to \infty} F_n(x) \le F(x + \epsilon).$$

This holds for all $\epsilon > 0$. Take the limit as $\epsilon \to 0$ and use the fact that F is continuous at x and conclude that $\lim_n F_n(x) = F(x)$.

Proof of (c). Fix $\epsilon > 0$. Then,

$$
\begin{aligned}
\mathbb{P}(|X_n - c| > \epsilon) &= \mathbb{P}(X_n < c - \epsilon) + \mathbb{P}(X_n > c + \epsilon) \\
&\le \mathbb{P}(X_n \le c - \epsilon) + \mathbb{P}(X_n > c + \epsilon) \\
&= F_n(c - \epsilon) + 1 - F_n(c + \epsilon) \\
&\to F(c - \epsilon) + 1 - F(c + \epsilon) \\
&= 0 + 1 - 1 = 0.
\end{aligned}
$$

Let us now show that the reverse implications do not hold.

CONVERGENCE IN PROBABILITY DOES NOT IMPLY CONVERGENCE IN QUADRATIC MEAN. Let $U \sim \text{Unif}(0, 1)$ and let $X_n = \sqrt{n} I_{(0,1/n)}(U)$. Then $\mathbb{P}(|X_n| > \epsilon) =$

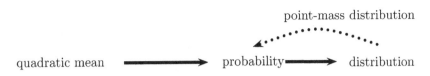

FIGURE 5.2. Relationship between types of convergence.

$\mathbb{P}(\sqrt{n}I_{(0,1/n)}(U) > \epsilon) = \mathbb{P}(0 \le U < 1/n) = 1/n \to 0$. Hence, $X_n \xrightarrow{P} 0$. But $\mathbb{E}(X_n^2) = n \int_0^{1/n} du = 1$ for all n so X_n does not converge in quadratic mean.

CONVERGENCE IN DISTRIBUTION DOES NOT IMPLY CONVERGENCE IN PROBABILITY. Let $X \sim N(0,1)$. Let $X_n = -X$ for $n = 1,2,3,\ldots$; hence $X_n \sim N(0,1)$. X_n has the same distribution function as X for all n so, trivially, $\lim_n F_n(x) = F(x)$ for all x. Therefore, $X_n \rightsquigarrow X$. But $\mathbb{P}(|X_n - X| > \epsilon). = \mathbb{P}(|2X| > \epsilon) = \mathbb{P}(|X| > \epsilon/2) \neq 0$. So X_n does not converge to X in probability. ∎

Warning! One might conjecture that if $X_n \xrightarrow{P} b$, then $\mathbb{E}(X_n) \to b$. This is not[1] true. Let X_n be a random variable defined by $\mathbb{P}(X_n = n^2) = 1/n$ and $\mathbb{P}(X_n = 0) = 1 - (1/n)$. Now, $\mathbb{P}(|X_n| < \epsilon) = \mathbb{P}(X_n = 0) = 1 - (1/n) \to 1$. Hence, $X_n \xrightarrow{P} 0$. However, $\mathbb{E}(X_n) = [n^2 \times (1/n)] + [0 \times (1 - (1/n))] = n$. Thus, $\mathbb{E}(X_n) \to \infty$.

Summary. Stare at Figure 5.2.

Some convergence properties are preserved under transformations.

5.5 Theorem. *Let X_n, X, Y_n, Y be random variables. Let g be a continuous function.*

(a) *If $X_n \xrightarrow{P} X$ and $Y_n \xrightarrow{P} Y$, then $X_n + Y_n \xrightarrow{P} X + Y$.*

(b) *If $X_n \xrightarrow{qm} X$ and $Y_n \xrightarrow{qm} Y$, then $X_n + Y_n \xrightarrow{qm} X + Y$.*

(c) *If $X_n \rightsquigarrow X$ and $Y_n \rightsquigarrow c$, then $X_n + Y_n \rightsquigarrow X + c$.*

(d) *If $X_n \xrightarrow{P} X$ and $Y_n \xrightarrow{P} Y$, then $X_n Y_n \xrightarrow{P} XY$.*

(e) *If $X_n \rightsquigarrow X$ and $Y_n \rightsquigarrow c$, then $X_n Y_n \rightsquigarrow cX$.*

(f) *If $X_n \xrightarrow{P} X$, then $g(X_n) \xrightarrow{P} g(X)$.*

(g) *If $X_n \rightsquigarrow X$, then $g(X_n) \rightsquigarrow g(X)$.*

Parts (c) and (e) are know as **Slutzky's theorem**. It is worth noting that $X_n \rightsquigarrow X$ and $Y_n \rightsquigarrow Y$ does not in general imply that $X_n + Y_n \rightsquigarrow X + Y$.

---

[1]We can conclude that $\mathbb{E}(X_n) \to b$ if X_n is uniformly integrable. See the appendix.

5.3   The Law of Large Numbers

Now we come to a crowning achievement in probability, the law of large numbers. This theorem says that the mean of a large sample is close to the mean of the distribution. For example, the proportion of heads of a large number of tosses is expected to be close to $1/2$. We now make this more precise.

Let $X_1, X_2, \ldots$ be an IID sample, let $\mu = \mathbb{E}(X_1)$ and [2] $\sigma^2 = \mathbb{V}(X_1)$. Recall that the sample mean is defined as $\overline{X}_n = n^{-1} \sum_{i=1}^{n} X_i$ and that $\mathbb{E}(\overline{X}_n) = \mu$ and $\mathbb{V}(\overline{X}_n) = \sigma^2/n$.

5.6 Theorem (The Weak Law of Large Numbers (WLLN)). [3]
If $X_1, \ldots, X_n$ are IID, then $\overline{X}_n \xrightarrow{P} \mu$.

Interpretation of the WLLN: The distribution of $\overline{X}_n$ becomes more concentrated around μ as n gets large.

PROOF. Assume that $\sigma < \infty$. This is not necessary but it simplifies the proof. Using Chebyshev's inequality,

$$\mathbb{P}\left(|\overline{X}_n - \mu| > \epsilon\right) \leq \frac{\mathbb{V}(\overline{X}_n)}{\epsilon^2} = \frac{\sigma^2}{n\epsilon^2}$$

which tends to 0 as $n \to \infty$. ∎

5.7 Example. Consider flipping a coin for which the probability of heads is p. Let X_i denote the outcome of a single toss (0 or 1). Hence, $p = P(X_i = 1) = E(X_i)$. The fraction of heads after n tosses is $\overline{X}_n$. According to the law of large numbers, $\overline{X}_n$ converges to p in probability. This does not mean that $\overline{X}_n$ will numerically equal p. It means that, when n is large, the distribution of $\overline{X}_n$ is tightly concentrated around p. Suppose that $p = 1/2$. How large should n be so that $P(.4 \leq \overline{X}_n \leq .6) \geq .7$? First, $\mathbb{E}(\overline{X}_n) = p = 1/2$ and $\mathbb{V}(\overline{X}_n) = \sigma^2/n = p(1-p)/n = 1/(4n)$. From Chebyshev's inequality,

$$
\begin{aligned}
\mathbb{P}(.4 \leq \overline{X}_n \leq .6) &= \mathbb{P}(|\overline{X}_n - \mu| \leq .1) \\
&= 1 - \mathbb{P}(|\overline{X}_n - \mu| > .1) \\
&\geq 1 - \frac{1}{4n(.1)^2} = 1 - \frac{25}{n}.
\end{aligned}
$$

The last expression will be larger than .7 if $n = 84$. ∎

---

[2]Note that $\mu = \mathbb{E}(X_i)$ is the same for all i so we can define $\mu = \mathbb{E}(X_i)$ for any i. By convention, we often write $\mu = \mathbb{E}(X_1)$.

[3]There is a stronger theorem in the appendix called the strong law of large numbers.

5.4   The Central Limit Theorem

The law of large numbers says that the distribution of $\overline{X}_n$ piles up near μ. This isn't enough to help us approximate probability statements about $\overline{X}_n$. For this we need the central limit theorem.

Suppose that $X_1, \ldots, X_n$ are IID with mean μ and variance σ^2. The central limit theorem (CLT) says that $\overline{X}_n = n^{-1} \sum_i X_i$ has a distribution which is approximately Normal with mean μ and variance σ^2/n. This is remarkable since nothing is assumed about the distribution of X_i, except the existence of the mean and variance.

5.8 Theorem (The Central Limit Theorem (CLT)). *Let $X_1, \ldots, X_n$ be IID with mean μ and variance σ^2. Let $\overline{X}_n = n^{-1} \sum_{i=1}^{n} X_i$. Then*

$$Z_n \equiv \frac{\overline{X}_n - \mu}{\sqrt{\mathbb{V}(\overline{X}_n)}} = \frac{\sqrt{n}(\overline{X}_n - \mu)}{\sigma} \rightsquigarrow Z$$

where $Z \sim N(0,1)$. In other words,

$$\lim_{n \to \infty} \mathbb{P}(Z_n \leq z) = \Phi(z) = \int_{-\infty}^{z} \frac{1}{\sqrt{2\pi}} e^{-x^2/2} dx.$$

Interpretation: Probability statements about $\overline{X}_n$ can be approximated using a Normal distribution. It's the probability statements that we are approximating, not the random variable itself.

In addition to $Z_n \rightsquigarrow N(0,1)$, there are several forms of notation to denote the fact that the distribution of Z_n is converging to a Normal. They all mean the same thing. Here they are:

$$Z_n \approx N(0,1)$$
$$\overline{X}_n \approx N\left(\mu, \frac{\sigma^2}{n}\right)$$
$$\overline{X}_n - \mu \approx N\left(0, \frac{\sigma^2}{n}\right)$$
$$\sqrt{n}(\overline{X}_n - \mu) \approx N(0, \sigma^2)$$
$$\frac{\sqrt{n}(\overline{X}_n - \mu)}{\sigma} \approx N(0,1).$$

5.9 Example. Suppose that the number of errors per computer program has a Poisson distribution with mean 5. We get 125 programs. Let $X_1, \ldots, X_{125}$ be the number of errors in the programs. We want to approximate $\mathbb{P}(\overline{X}_n < 5.5)$. Let $\mu = \mathbb{E}(X_1) = \lambda = 5$ and $\sigma^2 = \mathbb{V}(X_1) = \lambda = 5$. Then,

$$
\begin{aligned}
\mathbb{P}(\overline{X}_n < 5.5) &= \mathbb{P}\left(\frac{\sqrt{n}(\overline{X}_n - \mu)}{\sigma} < \frac{\sqrt{n}(5.5 - \mu)}{\sigma} \right) \\
&\approx \mathbb{P}(Z < 2.5) = .9938. \quad \blacksquare
\end{aligned}
$$

The central limit theorem tells us that $Z_n = \sqrt{n}(\overline{X}_n - \mu)/\sigma$ is approximately $N(0,1)$. However, we rarely know σ. Later, we will see that we can estimate σ^2 from $X_1, \ldots, X_n$ by

$$
S_n^2 = \frac{1}{n-1} \sum_{i=1}^{n} (X_i - \overline{X}_n)^2.
$$

This raises the following question: if we replace σ with S_n, is the central limit theorem still true? The answer is yes.

5.10 Theorem. *Assume the same conditions as the CLT. Then,*

$$
\frac{\sqrt{n}(\overline{X}_n - \mu)}{S_n} \rightsquigarrow N(0,1).
$$

You might wonder, how accurate the normal approximation is. The answer is given in the Berry-Essèen theorem.

5.11 Theorem (The Berry-Essèen Inequality). *Suppose that $\mathbb{E}|X_1|^3 < \infty$. Then*

$$
\sup_z |\mathbb{P}(Z_n \leq z) - \Phi(z)| \leq \frac{33}{4} \frac{\mathbb{E}|X_1 - \mu|^3}{\sqrt{n}\sigma^3}. \tag{5.4}
$$

There is also a multivariate version of the central limit theorem.

5.12 Theorem (Multivariate central limit theorem). *Let $X_1, \ldots, X_n$ be IID random vectors where*

$$
X_i = \begin{pmatrix} X_{1i} \\ X_{2i} \\ \vdots \\ X_{ki} \end{pmatrix}
$$

with mean

$$
\mu = \begin{pmatrix} \mu_1 \\ \mu_2 \\ \vdots \\ \mu_k \end{pmatrix} = \begin{pmatrix} \mathbb{E}(X_{1i}) \\ \mathbb{E}(X_{2i}) \\ \vdots \\ \mathbb{E}(X_{ki}) \end{pmatrix}
$$

and variance matrix Σ. *Let*

$$\overline{X} = \begin{pmatrix} \overline{X}_1 \\ \overline{X}_2 \\ \vdots \\ \overline{X}_k \end{pmatrix}.$$

where $\overline{X}_j = n^{-1} \sum_{i=1}^n X_{ji}$. *Then,*

$$\sqrt{n}(\overline{X} - \mu) \rightsquigarrow N(0, \Sigma).$$

5.5   The Delta Method

If Y_n has a limiting Normal distribution then the delta method allows us to find the limiting distribution of $g(Y_n)$ where g is any smooth function.

5.13 Theorem (The Delta Method). *Suppose that*

$$\frac{\sqrt{n}(Y_n - \mu)}{\sigma} \rightsquigarrow N(0, 1)$$

and that g *is a differentiable function such that* $g'(\mu) \neq 0$. *Then*

$$\frac{\sqrt{n}(g(Y_n) - g(\mu))}{|g'(\mu)|\sigma} \rightsquigarrow N(0, 1).$$

In other words,

$$Y_n \approx N\left(\mu, \frac{\sigma^2}{n}\right) \quad \text{implies that} \quad g(Y_n) \approx N\left(g(\mu), (g'(\mu))^2 \frac{\sigma^2}{n}\right).$$

5.14 Example. Let $X_1, \ldots, X_n$ be IID with finite mean μ and finite variance σ^2. By the central limit theorem, $\sqrt{n}(\overline{X}_n - \mu)/\sigma \rightsquigarrow N(0, 1)$. Let $W_n = e^{\overline{X}_n}$. Thus, $W_n = g(\overline{X}_n)$ where $g(s) = e^s$. Since $g'(s) = e^s$, the delta method implies that $W_n \approx N(e^\mu, e^{2\mu}\sigma^2/n)$. ∎

There is also a multivariate version of the delta method.

5.15 Theorem (The Multivariate Delta Method). *Suppose that* $Y_n = (Y_{n1}, \ldots, Y_{nk})$ *is a sequence of random vectors such that*

$$\sqrt{n}(Y_n - \mu) \rightsquigarrow N(0, \Sigma).$$

Let $g : \mathbb{R}^k \to \mathbb{R}$ and let

$$\nabla g(y) = \begin{pmatrix} \frac{\partial g}{\partial y_1} \\ \vdots \\ \frac{\partial g}{\partial y_k} \end{pmatrix}.$$

Let ∇_μ denote $\nabla g(y)$ evaluated at $y = \mu$ and assume that the elements of ∇_μ are nonzero. Then

$$\sqrt{n}(g(Y_n) - g(\mu)) \rightsquigarrow N\left(0, \nabla_\mu^T \Sigma \nabla_\mu\right).$$

5.16 Example. Let

$$\begin{pmatrix} X_{11} \\ X_{21} \end{pmatrix}, \begin{pmatrix} X_{12} \\ X_{22} \end{pmatrix}, \ldots, \begin{pmatrix} X_{1n} \\ X_{2n} \end{pmatrix}$$

be IID random vectors with mean $\mu = (\mu_1, \mu_2)^T$ and variance Σ. Let

$$\overline{X}_1 = \frac{1}{n} \sum_{i=1}^{n} X_{1i}, \quad \overline{X}_2 = \frac{1}{n} \sum_{i=1}^{n} X_{2i}$$

and define $Y_n = \overline{X}_1 \overline{X}_2$. Thus, $Y_n = g(\overline{X}_1, \overline{X}_2)$ where $g(s_1, s_2) = s_1 s_2$. By the central limit theorem,

$$\sqrt{n}\begin{pmatrix} \overline{X}_1 - \mu_1 \\ \overline{X}_2 - \mu_2 \end{pmatrix} \rightsquigarrow N(0, \Sigma).$$

Now

$$\nabla g(s) = \begin{pmatrix} \frac{\partial g}{\partial s_1} \\ \frac{\partial g}{\partial s_2} \end{pmatrix} = \begin{pmatrix} s_2 \\ s_1 \end{pmatrix}$$

and so

$$\nabla_\mu^T \Sigma \nabla_\mu = (\mu_2 \ \ \mu_1) \begin{pmatrix} \sigma_{11} & \sigma_{12} \\ \sigma_{12} & \sigma_{22} \end{pmatrix} \begin{pmatrix} \mu_2 \\ \mu_1 \end{pmatrix} = \mu_2^2 \sigma_{11} + 2\mu_1 \mu_2 \sigma_{12} + \mu_1^2 \sigma_{22}.$$

Therefore,

$$\sqrt{n}(\overline{X}_1 \overline{X}_2 - \mu_1 \mu_2) \rightsquigarrow N\left(0, \mu_2^2 \sigma_{11} + 2\mu_1 \mu_2 \sigma_{12} + \mu_1^2 \sigma_{22}\right). \quad \blacksquare$$

5.6   Bibliographic Remarks

Convergence plays a central role in modern probability theory. For more details, see Grimmett and Stirzaker (1982), Karr (1993), and Billingsley (1979). Advanced convergence theory is explained in great detail in van der Vaart and Wellner (1996) and and van der Vaart (1998).

5.7   Appendix

5.7.1   Almost Sure and L_1 Convergence

We say that X_n **converges almost surely** to X, written $X_n \xrightarrow{\text{as}} X$, if

$$\mathbb{P}(\{s :\ X_n(s) \to X(s)\}) = 1.$$

We say that X_n **converges in L_1** to X, written $X_n \xrightarrow{L_1} X$, if

$$\mathbb{E}|X_n - X| \to 0$$

as $n \to \infty$.

5.17 Theorem. *Let X_n and X be random variables. Then:*
  (a) $X_n \xrightarrow{\text{as}} X$ implies that $X_n \xrightarrow{\text{P}} X$.
  (b) $X_n \xrightarrow{\text{qm}} X$ implies that $X_n \xrightarrow{L_1} X$.
  (c) $X_n \xrightarrow{L_1} X$ implies that $X_n \xrightarrow{\text{P}} X$.

The weak law of large numbers says that $\overline{X}_n$ converges to $\mathbb{E}(X_1)$ in probability. The strong law asserts that this is also true almost surely.

5.18 Theorem (The Strong Law of Large Numbers). *Let $X_1, X_2, \ldots$ be IID. If $\mu = \mathbb{E}|X_1| < \infty$ then $\overline{X}_n \xrightarrow{\text{as}} \mu$.*

A sequence X_n is **asymptotically uniformly integrable** if

$$\lim_{M \to \infty} \limsup_{n \to \infty} \mathbb{E}\left(|X_n|I(|X_n| > M)\right) = 0.$$

5.19 Theorem. *If $X_n \xrightarrow{\text{P}} b$ and X_n is asymptotically uniformly integrable, then $\mathbb{E}(X_n) \to b$.*

5.7.2   Proof of the Central Limit Theorem

Recall that if X is a random variable, its moment generating function (MGF) is $\psi_X(t) = \mathbb{E}e^{tX}$. Assume in what follows that the MGF is finite in a neighborhood around $t = 0$.

5.20 Lemma. *Let $Z_1, Z_2, \ldots$ be a sequence of random variables. Let ψ_n be the MGF of Z_n. Let Z be another random variable and denote its MGF by ψ. If $\psi_n(t) \to \psi(t)$ for all t in some open interval around 0, then $Z_n \rightsquigarrow Z$.*

PROOF OF THE CENTRAL LIMIT THEOREM. Let $Y_i = (X_i - \mu)/\sigma$. Then, $Z_n = n^{-1/2}\sum_i Y_i$. Let $\psi(t)$ be the MGF of Y_i. The MGF of $\sum_i Y_i$ is $(\psi(t))^n$ and MGF of Z_n is $[\psi(t/\sqrt{n})]^n \equiv \xi_n(t)$. Now $\psi'(0) = \mathbb{E}(Y_1) = 0$, $\psi''(0) = \mathbb{E}(Y_1^2) = \mathbb{V}(Y_1) = 1$. So,

$$
\begin{aligned}
\psi(t) &= \psi(0) + t\psi'(0) + \frac{t^2}{2!}\psi''(0) + \frac{t^3}{3!}\psi'''(0) + \cdots \\
&= 1 + 0 + \frac{t^2}{2} + \frac{t^3}{3!}\psi'''(0) + \cdots \\
&= 1 + \frac{t^2}{2} + \frac{t^3}{3!}\psi'''(0) + \cdots
\end{aligned}
$$

Now,

$$
\begin{aligned}
\xi_n(t) &= \left[\psi\left(\frac{t}{\sqrt{n}}\right)\right]^n \\
&= \left[1 + \frac{t^2}{2n} + \frac{t^3}{3!n^{3/2}}\psi'''(0) + \cdots\right]^n \\
&= \left[1 + \frac{\frac{t^2}{2} + \frac{t^3}{3!n^{1/2}}\psi'''(0) + \cdots}{n}\right]^n \\
&\rightarrow e^{t^2/2}
\end{aligned}
$$

which is the MGF of a $N(0,1)$. The result follows from the previous Theorem. In the last step we used the fact that if $a_n \rightarrow a$ then

$$
\left(1 + \frac{a_n}{n}\right)^n \rightarrow e^a. \quad \blacksquare
$$

5.8   Exercises

1. Let $X_1, \ldots, X_n$ be IID with finite mean $\mu = \mathbb{E}(X_1)$ and finite variance $\sigma^2 = \mathbb{V}(X_1)$. Let $\overline{X}_n$ be the sample mean and let S_n^2 be the sample variance.

   (a) Show that $\mathbb{E}(S_n^2) = \sigma^2$.

   (b) Show that $S_n^2 \xrightarrow{P} \sigma^2$. Hint: Show that $S_n^2 = c_n n^{-1}\sum_{i=1}^n X_i^2 - d_n \overline{X}_n^2$ where $c_n \rightarrow 1$ and $d_n \rightarrow 1$. Apply the law of large numbers to $n^{-1}\sum_{i=1}^n X_i^2$ and to $\overline{X}_n$. Then use part (e) of Theorem 5.5.

2. Let $X_1, X_2, \ldots$ be a sequence of random variables. Show that $X_n \xrightarrow{qm} b$ if and only if

$$
\lim_{n\to\infty} \mathbb{E}(X_n) = b \quad \text{and} \quad \lim_{n\to\infty} \mathbb{V}(X_n) = 0.
$$

3. Let $X_1, \ldots, X_n$ be IID and let $\mu = \mathbb{E}(X_1)$. Suppose that the variance is finite. Show that $\overline{X}_n \xrightarrow{\text{qm}} \mu$.

4. Let $X_1, X_2, \ldots$ be a sequence of random variables such that

$$\mathbb{P}\left(X_n = \frac{1}{n}\right) = 1 - \frac{1}{n^2} \quad \text{and} \quad \mathbb{P}(X_n = n) = \frac{1}{n^2}.$$

Does X_n converge in probability? Does X_n converge in quadratic mean?

5. Let $X_1, \ldots, X_n \sim \text{Bernoulli}(p)$. Prove that

$$\frac{1}{n} \sum_{i=1}^{n} X_i^2 \xrightarrow{P} p \quad \text{and} \quad \frac{1}{n} \sum_{i=1}^{n} X_i^2 \xrightarrow{\text{qm}} p.$$

6. Suppose that the height of men has mean 68 inches and standard deviation 2.6 inches. We draw 100 men at random. Find (approximately) the probability that the average height of men in our sample will be at least 68 inches.

7. Let $\lambda_n = 1/n$ for $n = 1, 2, \ldots$. Let $X_n \sim \text{Poisson}(\lambda_n)$.

   (a) Show that $X_n \xrightarrow{P} 0$.

   (b) Let $Y_n = nX_n$. Show that $Y_n \xrightarrow{P} 0$.

8. Suppose we have a computer program consisting of $n = 100$ pages of code. Let X_i be the number of errors on the ith page of code. Suppose that the $X_i's$ are Poisson with mean 1 and that they are independent. Let $Y = \sum_{i=1}^{n} X_i$ be the total number of errors. Use the central limit theorem to approximate $\mathbb{P}(Y < 90)$.

9. Suppose that $\mathbb{P}(X = 1) = \mathbb{P}(X = -1) = 1/2$. Define

$$X_n = \begin{cases} X & \text{with probability } 1 - \frac{1}{n} \\ e^n & \text{with probability } \frac{1}{n}. \end{cases}$$

Does X_n converge to X in probability? Does X_n converge to X in distribution? Does $\mathbb{E}(X - X_n)^2$ converge to 0?

10. Let $Z \sim N(0, 1)$. Let $t > 0$. Show that, for any $k > 0$,

$$\mathbb{P}(|Z| > t) \leq \frac{\mathbb{E}|Z|^k}{t^k}.$$

Compare this to Mill's inequality in Chapter 4.

11. Suppose that $X_n \sim N(0, 1/n)$ and let X be a random variable with distribution $F(x) = 0$ if $x < 0$ and $F(x) = 1$ if $x \geq 0$. Does X_n converge to X in probability? (Prove or disprove). Does X_n converge to X in distribution? (Prove or disprove).

12. Let $X, X_1, X_2, X_3, \ldots$ be random variables that are positive and integer valued. Show that $X_n \rightsquigarrow X$ if and only if

$$\lim_{n \to \infty} \mathbb{P}(X_n = k) = \mathbb{P}(X = k)$$

for every integer k.

13. Let $Z_1, Z_2, \ldots$ be IID random variables with density f. Suppose that $\mathbb{P}(Z_i > 0) = 1$ and that $\lambda = \lim_{x \downarrow 0} f(x) > 0$. Let

$$X_n = n \, \min\{Z_1, \ldots, Z_n\}.$$

Show that $X_n \rightsquigarrow Z$ where Z has an exponential distribution with mean $1/\lambda$.

14. Let $X_1, \ldots, X_n \sim \text{Uniform}(0, 1)$. Let $Y_n = \overline{X}_n^2$. Find the limiting distribution of Y_n.

15. Let

$$\begin{pmatrix} X_{11} \\ X_{21} \end{pmatrix}, \begin{pmatrix} X_{12} \\ X_{22} \end{pmatrix}, \ldots, \begin{pmatrix} X_{1n} \\ X_{2n} \end{pmatrix}$$

be IID random vectors with mean $\mu = (\mu_1, \mu_2)$ and variance Σ. Let

$$\overline{X}_1 = \frac{1}{n} \sum_{i=1}^{n} X_{1i}, \quad \overline{X}_2 = \frac{1}{n} \sum_{i=1}^{n} X_{2i}$$

and define $Y_n = \overline{X}_1 / \overline{X}_2$. Find the limiting distribution of Y_n.

16. Construct an example where $X_n \rightsquigarrow X$ and $Y_n \rightsquigarrow Y$ but $X_n + Y_n$ does not converge in distribution to $X + Y$.

Part II

Statistical Inference

6
Models, Statistical Inference and Learning

6.1 Introduction

Statistical inference, or "learning" as it is called in computer science, is the process of using data to infer the distribution that generated the data. A typical statistical inference question is:

> Given a sample $X_1, \ldots, X_n \sim F$, how do we infer F?

In some cases, we may want to infer only some feature of F such as its mean.

6.2 Parametric and Nonparametric Models

A **statistical model** $\mathfrak{F}$ is a set of distributions (or densities or regression functions). A **parametric model** is a set $\mathfrak{F}$ that can be parameterized by a finite number of parameters. For example, if we assume that the data come from a Normal distribution, then the model is

$$\mathfrak{F} = \left\{ f(x; \mu, \sigma) = \frac{1}{\sigma\sqrt{2\pi}} \exp\left\{ -\frac{1}{2\sigma^2}(x - \mu)^2 \right\}, \ \mu \in \mathbb{R}, \ \sigma > 0 \right\}. \quad (6.1)$$

This is a two-parameter model. We have written the density as $f(x; \mu, \sigma)$ to show that x is a value of the random variable whereas μ and σ are parameters.

In general, a parametric model takes the form

$$\mathfrak{F} = \left\{ f(x; \theta) : \ \theta \in \Theta \right\} \tag{6.2}$$

where θ is an unknown parameter (or vector of parameters) that can take values in the **parameter space** Θ. If θ is a vector but we are only interested in one component of θ, we call the remaining parameters **nuisance parameters**. A **nonparametric model** is a set $\mathfrak{F}$ that cannot be parameterized by a finite number of parameters. For example, $\mathfrak{F}_{ALL} = \{$all CDF's$\}$ is nonparametric. [1]

6.1 Example (One-dimensional Parametric Estimation). Let $X_1, \ldots, X_n$ be independent Bernoulli(p) observations. The problem is to estimate the parameter p. ■

6.2 Example (Two-dimensional Parametric Estimation). Suppose that $X_1, \ldots, X_n \sim F$ and we assume that the PDF $f \in \mathfrak{F}$ where $\mathfrak{F}$ is given in (6.1). In this case there are two parameters, μ and σ. The goal is to estimate the parameters from the data. If we are only interested in estimating μ, then μ is the parameter of interest and σ is a nuisance parameter. ■

6.3 Example (Nonparametric estimation of the CDF). Let $X_1, \ldots, X_n$ be independent observations from a CDF F. The problem is to estimate F assuming only that $F \in \mathfrak{F}_{ALL} = \{$all CDF's$\}$. ■

6.4 Example (Nonparametric density estimation). Let $X_1, \ldots, X_n$ be independent observations from a CDF F and let $f = F'$ be the PDF. Suppose we want to estimate the PDF f. It is not possible to estimate f assuming only that $F \in \mathfrak{F}_{ALL}$. We need to assume some smoothness on f. For example, we might assume that $f \in \mathfrak{F} = \mathfrak{F}_{DENS} \cap \mathfrak{F}_{SOB}$ where $\mathfrak{F}_{DENS}$ is the set of all probability density functions and

$$\mathfrak{F}_{SOB} = \left\{ f : \ \int (f''(x))^2 dx < \infty \right\}.$$

The class $\mathfrak{F}_{SOB}$ is called a **Sobolev space**; it is the set of functions that are not "too wiggly." ■

6.5 Example (Nonparametric estimation of functionals). Let $X_1, \ldots, X_n \sim F$. Suppose we want to estimate $\mu = \mathbb{E}(X_1) = \int x \, dF(x)$ assuming only that

---

[1] The distinction between parametric and nonparametric is more subtle than this but we don't need a rigorous definition for our purposes.

μ exists. The mean μ may be thought of as a function of F: we can write $\mu = T(F) = \int x \, dF(x)$. In general, any function of F is called a **statistical functional**. Other examples of functionals are the variance $T(F) = \int x^2 dF(x) - \left(\int x dF(x) \right)^2$ and the median $T(F) = F^{-1}(1/2)$. ∎

6.6 Example (Regression, prediction, and classification). Suppose we observe pairs of data $(X_1, Y_1), \ldots (X_n, Y_n)$. Perhaps X_i is the blood pressure of subject i and Y_i is how long they live. X is called a **predictor** or **regressor** or **feature** or **independent variable**. Y is called the **outcome** or the **response variable** or the **dependent variable**. We call $r(x) = \mathbb{E}(Y|X = x)$ the **regression function**. If we assume that $r \in \mathfrak{F}$ where $\mathfrak{F}$ is finite dimensional — the set of straight lines for example — then we have a **parametric regression model**. If we assume that $r \in \mathfrak{F}$ where $\mathfrak{F}$ is not finite dimensional then we have a **nonparametric regression model**. The goal of predicting Y for a new patient based on their X value is called **prediction**. If Y is discrete (for example, live or die) then prediction is instead called **classification**. If our goal is to estimate the function r, then we call this **regression** or **curve estimation**. Regression models are sometimes written as

$$Y = r(X) + \epsilon \tag{6.3}$$

where $\mathbb{E}(\epsilon) = 0$. We can always rewrite a regression model this way. To see this, define $\epsilon = Y - r(X)$ and hence $Y = Y + r(X) - r(X) = r(X) + \epsilon$. Moreover, $\mathbb{E}(\epsilon) = \mathbb{E}\mathbb{E}(\epsilon|X) = \mathbb{E}(\mathbb{E}(Y - r(X))|X) = \mathbb{E}(\mathbb{E}(Y|X) - r(X)) = \mathbb{E}(r(X) - r(X)) = 0$. ∎

WHAT'S NEXT? It is traditional in most introductory courses to start with parametric inference. Instead, we will start with nonparametric inference and then we will cover parametric inference. In some respects, nonparametric inference is easier to understand and is more useful than parametric inference.

FREQUENTISTS AND BAYESIANS. There are many approaches to statistical inference. The two dominant approaches are called **frequentist inference** and **Bayesian inference**. We'll cover both but we will start with frequentist inference. We'll postpone a discussion of the pros and cons of these two until later.

SOME NOTATION. If $\mathfrak{F} = \{f(x; \theta) : \theta \in \Theta\}$ is a parametric model, we write $\mathbb{P}_\theta(X \in A) = \int_A f(x; \theta) dx$ and $\mathbb{E}_\theta(r(X)) = \int r(x) f(x; \theta) dx$. The subscript θ indicates that the probability or expectation is with respect to $f(x; \theta)$; it does not mean we are averaging over θ. Similarly, we write $\mathbb{V}_\theta$ for the variance.

6.3   Fundamental Concepts in Inference

Many inferential problems can be identified as being one of three types: estimation, confidence sets, or hypothesis testing. We will treat all of these problems in detail in the rest of the book. Here, we give a brief introduction to the ideas.

6.3.1   Point Estimation

Point estimation refers to providing a single "best guess" of some quantity of interest. The quantity of interest could be a parameter in a parametric model, a CDF F, a probability density function f, a regression function r, or a prediction for a future value Y of some random variable.

> By convention, we denote a point estimate of θ by $\widehat{\theta}$ or $\widehat{\theta}_n$. Remember that θ is a fixed, unknown quantity. The estimate $\widehat{\theta}$ depends on the data so $\widehat{\theta}$ is a random variable.

More formally, let $X_1, \ldots, X_n$ be n IID data points from some distribution F. A point estimator $\widehat{\theta}_n$ of a parameter θ is some function of $X_1, \ldots, X_n$:

$$\widehat{\theta}_n = g(X_1, \ldots, X_n).$$

The bias of an estimator is defined by

$$\text{bias}(\widehat{\theta}_n) = \mathbb{E}_\theta(\widehat{\theta}_n) - \theta. \tag{6.4}$$

We say that $\widehat{\theta}_n$ is **unbiased** if $\mathbb{E}(\widehat{\theta}_n) = \theta$. Unbiasedness used to receive much attention but these days is considered less important; many of the estimators we will use are biased. A reasonable requirement for an estimator is that it should converge to the true parameter value as we collect more and more data. This requirement is quantified by the following definition:

6.7 Definition. *A point estimator* $\widehat{\theta}_n$ *of a parameter* θ *is* **consistent** *if* $\widehat{\theta}_n \xrightarrow{\text{P}} \theta$.

The distribution of $\widehat{\theta}_n$ is called the **sampling distribution**. The standard deviation of $\widehat{\theta}_n$ is called the **standard error**, denoted by se:

$$\text{se} = \text{se}(\widehat{\theta}_n) = \sqrt{\mathbb{V}(\widehat{\theta}_n)}. \tag{6.5}$$

Often, the standard error depends on the unknown F. In those cases, se is an unknown quantity but we usually can estimate it. The estimated standard error is denoted by $\widehat{\text{se}}$.

6.8 Example. Let $X_1, \ldots, X_n \sim \text{Bernoulli}(p)$ and let $\widehat{p}_n = n^{-1} \sum_i X_i$. Then $\mathbb{E}(\widehat{p}_n) = n^{-1} \sum_i \mathbb{E}(X_i) = p$ so $\widehat{p}_n$ is unbiased. The standard error is $\text{se} = \sqrt{\mathbb{V}(\widehat{p}_n)} = \sqrt{p(1-p)/n}$. The estimated standard error is $\widehat{\text{se}} = \sqrt{\widehat{p}(1-\widehat{p})/n}$. ∎

The quality of a point estimate is sometimes assessed by the **mean squared error**, or MSE defined by

$$\text{MSE} = \mathbb{E}_\theta(\widehat{\theta}_n - \theta)^2. \tag{6.6}$$

Keep in mind that $\mathbb{E}_\theta(\cdot)$ refers to expectation with respect to the distribution

$$f(x_1, \ldots, x_n; \theta) = \prod_{i=1}^{n} f(x_i; \theta)$$

that generated the data. It does not mean we are averaging over a distribution for θ.

6.9 Theorem. *The* MSE *can be written as*

$$\text{MSE} = \text{bias}^2(\widehat{\theta}_n) + \mathbb{V}_\theta(\widehat{\theta}_n). \tag{6.7}$$

PROOF. Let $\overline{\theta}_n = E_\theta(\widehat{\theta}_n)$. Then

$$
\begin{aligned}
\mathbb{E}_\theta(\widehat{\theta}_n - \theta)^2 &= \mathbb{E}_\theta(\widehat{\theta}_n - \overline{\theta}_n + \overline{\theta}_n - \theta)^2 \\
&= \mathbb{E}_\theta(\widehat{\theta}_n - \overline{\theta}_n)^2 + 2(\overline{\theta}_n - \theta)\mathbb{E}_\theta(\widehat{\theta}_n - \overline{\theta}_n) + \mathbb{E}_\theta(\overline{\theta}_n - \theta)^2 \\
&= (\overline{\theta}_n - \theta)^2 + \mathbb{E}_\theta(\widehat{\theta}_n - \overline{\theta}_n)^2 \\
&= \text{bias}^2(\widehat{\theta}_n) + \mathbb{V}(\widehat{\theta}_n)
\end{aligned}
$$

where we have used the fact that $\mathbb{E}_\theta(\widehat{\theta}_n - \overline{\theta}_n) = \overline{\theta}_n - \overline{\theta}_n = 0$. ∎

6.10 Theorem. *If* bias $\to 0$ *and* se $\to 0$ *as* $n \to \infty$ *then* $\widehat{\theta}_n$ *is consistent, that is,* $\widehat{\theta}_n \xrightarrow{P} \theta$.

PROOF. If bias $\to 0$ and se $\to 0$ then, by Theorem 6.9, MSE $\to 0$. It follows that $\widehat{\theta}_n \xrightarrow{\text{qm}} \theta$. (Recall Definition 5.2.) The result follows from part (b) of Theorem 5.4. ∎

6.11 Example. Returning to the coin flipping example, we have that $\mathbb{E}_p(\widehat{p}_n) = p$ so the bias $= p - p = 0$ and se $= \sqrt{p(1-p)/n} \to 0$. Hence, $\widehat{p}_n \xrightarrow{P} p$, that is, $\widehat{p}_n$ is a consistent estimator. ∎

Many of the estimators we will encounter will turn out to have, approximately, a Normal distribution.

6.12 Definition. *An estimator is* **asymptotically Normal** *if*

$$\frac{\widehat{\theta}_n - \theta}{\text{se}} \rightsquigarrow N(0,1). \tag{6.8}$$

6.3.2 Confidence Sets

A $1 - \alpha$ **confidence interval** for a parameter θ is an interval $C_n = (a,b)$ where $a = a(X_1, \ldots, X_n)$ and $b = b(X_1, \ldots, X_n)$ are functions of the data such that

$$\mathbb{P}_\theta(\theta \in C_n) \geq 1 - \alpha, \quad \text{for all } \theta \in \Theta. \tag{6.9}$$

In words, (a,b) traps θ with probability $1 - \alpha$. We call $1 - \alpha$ the **coverage** of the confidence interval.

Warning! C_n is random and θ is fixed.

Commonly, people use 95 percent confidence intervals, which corresponds to choosing $\alpha = 0.05$. If θ is a vector then we use a **confidence set** (such as a sphere or an ellipse) instead of an interval.

Warning! There is much confusion about how to interpret a confidence interval. A confidence interval is not a probability statement about θ since θ is a fixed quantity, not a random variable. Some texts interpret confidence intervals as follows: if I repeat the experiment over and over, the interval will contain the parameter 95 percent of the time. This is correct but useless since we rarely repeat the same experiment over and over. A better interpretation is this:

> On day 1, you collect data and construct a 95 percent confidence interval for a parameter θ_1. On day 2, you collect new data and construct a 95 percent confidence interval for an unrelated parameter θ_2. On day 3, you collect new data and construct a 95 percent confidence interval for an unrelated parameter θ_3. You continue this way constructing confidence intervals for a sequence of unrelated parameters $\theta_1, \theta_2, \ldots$ Then 95 percent of your intervals will trap the true parameter value. There is no need to introduce the idea of repeating the same experiment over and over.

6.13 Example. Every day, newspapers report opinion polls. For example, they might say that "83 percent of the population favor arming pilots with guns." Usually, you will see a statement like "this poll is accurate to within 4 points

95 percent of the time." They are saying that 83 ± 4 is a 95 percent confidence interval for the true but unknown proportion p of people who favor arming pilots with guns. If you form a confidence interval this way every day for the rest of your life, 95 percent of your intervals will contain the true parameter. This is true even though you are estimating a different quantity (a different poll question) every day. ∎

6.14 Example. The fact that a confidence interval is not a probability statement about θ is confusing. Consider this example from Berger and Wolpert (1984). Let θ be a fixed, known real number and let X_1, X_2 be independent random variables such that $\mathbb{P}(X_i = 1) = \mathbb{P}(X_i = -1) = 1/2$. Now define $Y_i = \theta + X_i$ and suppose that you only observe Y_1 and Y_2. Define the following "confidence interval" which actually only contains one point:

$$C = \begin{cases} \{Y_1 - 1\} & \text{if } Y_1 = Y_2 \\ \{(Y_1 + Y_2)/2\} & \text{if } Y_1 \neq Y_2. \end{cases}$$

You can check that, no matter what θ is, we have $\mathbb{P}_\theta(\theta \in C) = 3/4$ so this is a 75 percent confidence interval. Suppose we now do the experiment and we get $Y_1 = 15$ and $Y_2 = 17$. Then our 75 percent confidence interval is $\{16\}$. However, we are certain that $\theta = 16$. If you wanted to make a probability statement about θ you would probably say that $\mathbb{P}(\theta \in C | Y_1, Y_2) = 1$. There is nothing wrong with saying that $\{16\}$ is a 75 percent confidence interval. But is it not a probability statement about θ. ∎

In Chapter 11 we will discuss Bayesian methods in which we treat θ as if it were a random variable and we do make probability statements about θ. In particular, we will make statements like "the probability that θ is in C_n, given the data, is 95 percent." However, these Bayesian intervals refer to degree-of-belief probabilities. These Bayesian intervals will not, in general, trap the parameter 95 percent of the time.

6.15 Example. In the coin flipping setting, let $C_n = (\hat{p}_n - \epsilon_n, \hat{p}_n + \epsilon_n)$ where $\epsilon_n^2 = \log(2/\alpha)/(2n)$. From Hoeffding's inequality (4.4) it follows that

$$\mathbb{P}(p \in C_n) \geq 1 - \alpha$$

for every p. Hence, C_n is a $1 - \alpha$ confidence interval. ∎

As mentioned earlier, point estimators often have a limiting Normal distribution, meaning that equation (6.8) holds, that is, $\hat{\theta}_n \approx N(\theta, \hat{se}^2)$. In this case we can construct (approximate) confidence intervals as follows.

6.16 Theorem (Normal-based Confidence Interval). *Suppose that $\widehat{\theta}_n \approx N(\theta, \widehat{se}^2)$. Let Φ be the CDF of a standard Normal and let $z_{\alpha/2} = \Phi^{-1}(1 - (\alpha/2))$, that is, $\mathbb{P}(Z > z_{\alpha/2}) = \alpha/2$ and $\mathbb{P}(-z_{\alpha/2} < Z < z_{\alpha/2}) = 1 - \alpha$ where $Z \sim N(0,1)$. Let*

$$C_n = (\widehat{\theta}_n - z_{\alpha/2}\,\widehat{se}, \ \widehat{\theta}_n + z_{\alpha/2}\,\widehat{se}). \tag{6.10}$$

Then

$$\mathbb{P}_\theta(\theta \in C_n) \to 1 - \alpha. \tag{6.11}$$

PROOF. Let $Z_n = (\widehat{\theta}_n - \theta)/\widehat{se}$. By assumption $Z_n \rightsquigarrow Z$ where $Z \sim N(0,1)$. Hence,

$$
\begin{aligned}
\mathbb{P}_\theta(\theta \in C_n) &= \mathbb{P}_\theta\left(\widehat{\theta}_n - z_{\alpha/2}\,\widehat{se} < \theta < \widehat{\theta}_n + z_{\alpha/2}\,\widehat{se}\right) \\
&= \mathbb{P}_\theta\left(-z_{\alpha/2} < \frac{\widehat{\theta}_n - \theta}{\widehat{se}} < z_{\alpha/2}\right) \\
&\to \mathbb{P}\left(-z_{\alpha/2} < Z < z_{\alpha/2}\right) \\
&= 1 - \alpha. \quad \blacksquare
\end{aligned}
$$

For 95 percent confidence intervals, $\alpha = 0.05$ and $z_{\alpha/2} = 1.96 \approx 2$ leading to the approximate 95 percent confidence interval $\widehat{\theta}_n \pm 2\,\widehat{se}$.

6.17 Example. Let $X_1, \ldots, X_n \sim$ Bernoulli(p) and let $\widehat{p}_n = n^{-1}\sum_{i=1}^n X_i$. Then $\mathbb{V}(\widehat{p}_n) = n^{-2}\sum_{i=1}^n \mathbb{V}(X_i) = n^{-2}\sum_{i=1}^n p(1-p) = n^{-2}np(1-p) = p(1-p)/n$. Hence, $se = \sqrt{p(1-p)/n}$ and $\widehat{se} = \sqrt{\widehat{p}_n(1-\widehat{p}_n)/n}$. By the Central Limit Theorem, $\widehat{p}_n \approx N(p, \widehat{se}^2)$. Therefore, an approximate $1 - \alpha$ confidence interval is

$$\widehat{p}_n \pm z_{\alpha/2}\widehat{se} = \widehat{p}_n \pm z_{\alpha/2}\sqrt{\frac{\widehat{p}_n(1-\widehat{p}_n)}{n}}.$$

Compare this with the confidence interval in example 6.15. The Normal-based interval is shorter but it only has approximately (large sample) correct coverage. $\blacksquare$

6.3.3 Hypothesis Testing

In **hypothesis testing**, we start with some default theory — called a **null hypothesis** — and we ask if the data provide sufficient evidence to reject the theory. If not we retain the null hypothesis. [2]

---

[2] The term "retaining the null hypothesis" is due to Chris Genovese. Other terminology is "accepting the null" or "failing to reject the null."

6.18 Example (Testing if a Coin is Fair). Let

$$X_1, \ldots, X_n \sim \text{Bernoulli}(p)$$

be n independent coin flips. Suppose we want to test if the coin is fair. Let H_0 denote the hypothesis that the coin is fair and let H_1 denote the hypothesis that the coin is not fair. H_0 is called the **null hypothesis** and H_1 is called the **alternative hypothesis**. We can write the hypotheses as

$$H_0 : p = 1/2 \quad \text{versus} \quad H_1 : p \neq 1/2.$$

It seems reasonable to reject H_0 if $T - |\widehat{p}_n - (1/2)|$ is large. When we discuss hypothesis testing in detail, we will be more precise about how large T should be to reject H_0. ∎

6.4 Bibliographic Remarks

Statistical inference is covered in many texts. Elementary texts include DeGroot and Schervish (2002) and Larsen and Marx (1986). At the intermediate level I recommend Casella and Berger (2002), Bickel and Doksum (2000), and Rice (1995). At the advanced level, Cox and Hinkley (2000), Lehmann and Casella (1998), Lehmann (1986), and van der Vaart (1998).

6.5 Appendix

Our definition of confidence interval requires that $\mathbb{P}_\theta(\theta \in C_n) \geq 1 - \alpha$ for all $\theta \in \Theta$. A **pointwise asymptotic** confidence interval requires that $\liminf_{n \to \infty} \mathbb{P}_\theta(\theta \in C_n) \geq 1 - \alpha$ for all $\theta \in \Theta$. A **uniform asymptotic** confidence interval requires that $\liminf_{n \to \infty} \inf_{\theta \in \Theta} \mathbb{P}_\theta(\theta \in C_n) \geq 1 - \alpha$. The approximate Normal-based interval is a pointwise asymptotic confidence interval.

6.6 Exercises

1. Let $X_1, \ldots, X_n \sim \text{Poisson}(\lambda)$ and let $\widehat{\lambda} = n^{-1} \sum_{i=1}^{n} X_i$. Find the bias, se, and MSE of this estimator.

2. Let $X_1, \ldots, X_n \sim \text{Uniform}(0, \theta)$ and let $\widehat{\theta} = \max\{X_1, \ldots, X_n\}$. Find the bias, se, and MSE of this estimator.

3. Let $X_1, \ldots, X_n \sim \text{Uniform}(0, \theta)$ and let $\widehat{\theta} = 2\overline{X}_n$. Find the bias, se, and MSE of this estimator.

7
Estimating the CDF and Statistical Functionals

The first inference problem we will consider is nonparametric estimation of the CDF F. Then we will estimate statistical functionals, which are functions of CDF, such as the mean, the variance, and the correlation. The nonparametric method for estimating functionals is called the plug-in method.

7.1 The Empirical Distribution Function

Let $X_1, \ldots, X_n \sim F$ be an IID sample where F is a distribution function on the real line. We will estimate F with the empirical distribution function, which is defined as follows.

7.1 Definition. *The* **empirical distribution function** $\widehat{F}_n$ *is the* CDF *that puts mass* $1/n$ *at each data point* X_i. *Formally,*

$$\widehat{F}_n(x) = \frac{\sum_{i=1}^{n} I(X_i \leq x)}{n} \tag{7.1}$$

where

$$I(X_i \leq x) = \begin{cases} 1 & \text{if } X_i \leq x \\ 0 & \text{if } X_i > x. \end{cases}$$

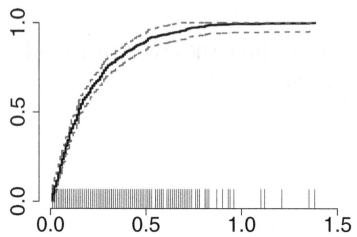

FIGURE 7.1. Nerve data. Each vertical line represents one data point. The solid line is the empirical distribution function. The lines above and below the middle line are a 95 percent confidence band.

7.2 Example (Nerve Data). Cox and Lewis (1966) reported 799 waiting times between successive pulses along a nerve fiber. Figure 7.1 shows the empirical CDF $\widehat{F}_n$. The data points are shown as small vertical lines at the bottom of the plot. Suppose we want to estimate the fraction of waiting times between .4 and .6 seconds. The estimate is $\widehat{F}_n(.6) - \widehat{F}_n(.4) = .93 - .84 = .09$. ∎

7.3 Theorem. *At any fixed value of x,*

$$\mathbb{E}\left(\widehat{F}_n(x)\right) = F(x),$$

$$\mathbb{V}\left(\widehat{F}_n(x)\right) = \frac{F(x)(1 - F(x))}{n},$$

$$\text{MSE} = \frac{F(x)(1 - F(x))}{n} \to 0,$$

$$\widehat{F}_n(x) \xrightarrow{\text{P}} F(x).$$

7.4 Theorem (The Glivenko-Cantelli Theorem). *Let $X_1, \ldots, X_n \sim F$. Then* [1]

$$\sup_x |\widehat{F}_n(x) - F(x)| \xrightarrow{\text{P}} 0.$$

Now we give an inequality that will be used to construct a confidence band.

---

[1] More precisely, $\sup_x |\widehat{F}_n(x) - F(x)|$ converges to 0 almost surely.

7.5 Theorem (The Dvoretzky-Kiefer-Wolfowitz (DKW) Inequality). *Let* $X_1, \ldots,$
$X_n \sim F$. *Then, for any* $\epsilon > 0$,

$$\mathbb{P}\left(\sup_x |F(x) - \widehat{F}_n(x)| > \epsilon \right) \leq 2e^{-2n\epsilon^2}. \tag{7.2}$$

From the DKW inequality, we can construct a confidence set as follows:

A Nonparametric $1 - \alpha$ Confidence Band for F

Define,

$$
\begin{aligned}
L(x) &= \max\{\widehat{F}_n(x) - \epsilon_n, 0\} \\
U(x) &= \min\{\widehat{F}_n(x) + \epsilon_n, 1\}
\end{aligned}
$$

$$\text{where} \quad \epsilon_n = \sqrt{\frac{1}{2n} \log\left(\frac{2}{\alpha}\right)}.$$

It follows from (7.2) that for any F,

$$\mathbb{P}\left(L(x) \leq F(x) \leq U(x) \text{ for all } x \right) \geq 1 - \alpha. \tag{7.3}$$

7.6 Example. The dashed lines in Figure 7.1 give a 95 percent confidence band using $\epsilon_n = \sqrt{\frac{1}{2n} \log\left(\frac{2}{.05}\right)} = .048$. ∎

7.2   Statistical Functionals

A **statistical functional** $T(F)$ is any function of F. Examples are the mean $\mu = \int x \, dF(x)$, the variance $\sigma^2 = \int (x - \mu)^2 \, dF(x)$ and the median $m = F^{-1}(1/2)$.

7.7 Definition. *The **plug-in estimator** of* $\theta = T(F)$ *is defined by*

$$\widehat{\theta}_n = T(\widehat{F}_n).$$

In other words, just plug in $\widehat{F}_n$ *for the unknown* F.

7.8 Definition. *If* $T(F) = \int r(x) dF(x)$ *for some function* $r(x)$ *then* T *is called a **linear functional**.*

The reason $T(F) = \int r(x)dF(x)$ is called a linear functional is because T satisfies

$$T(aF + bG) = aT(F) + bT(G),$$

hence T is linear in its arguments. Recall that $\int r(x)dF(x)$ is defined to be $\int r(x)f(x)dx$ in the continuous case and $\sum_j r(x_j)f(x_j)$ in the discrete. The empirical cdf $\widehat{F}_n(x)$ is discrete, putting mass $1/n$ at each X_i. Hence, if $T(F) = \int r(x)dF(x)$ is a linear functional then we have:

7.9 Theorem. *The plug-in estimator for linear functional* $T(F) = \int r(x)dF(x)$ *is:*

$$T(\widehat{F}_n) = \int r(x)d\widehat{F}_n(x) = \frac{1}{n}\sum_{i=1}^{n} r(X_i). \qquad (7.4)$$

Sometimes we can find the estimated standard error se of $T(\widehat{F}_n)$ by doing some calculations. However, in other cases it is not obvious how to estimate the standard error. In the next chapter, we will discuss a general method for finding $\widehat{\text{se}}$. For now, let us just assume that somehow we can find $\widehat{\text{se}}$.

In many cases, it turns out that

$$T(\widehat{F}_n) \approx N(T(F), \widehat{\text{se}}^2). \qquad (7.5)$$

By equation (6.11), an approximate $1 - \alpha$ confidence interval for $T(F)$ is then

$$T(\widehat{F}_n) \pm z_{\alpha/2}\,\widehat{\text{se}}. \qquad (7.6)$$

We will call this the **Normal-based interval.** For a 95 percent confidence interval, $z_{\alpha/2} = z_{.05/2} = 1.96 \approx 2$ so the interval is

$$T(\widehat{F}_n) \pm 2\,\widehat{\text{se}}.$$

7.10 Example (The mean). Let $\mu = T(F) = \int x\,dF(x)$. The plug-in estimator is $\widehat{\mu} = \int x\,d\widehat{F}_n(x) = \overline{X}_n$. The standard error is se $= \sqrt{\mathbb{V}(\overline{X}_n)} = \sigma/\sqrt{n}$. If $\widehat{\sigma}$ denotes an estimate of σ, then the estimated standard error is $\widehat{\sigma}/\sqrt{n}$. (In the next example, we shall see how to estimate σ.) A Normal-based confidence interval for μ is $\overline{X}_n \pm z_{\alpha/2}\,\widehat{\text{se}}$. ∎

7.11 Example (The Variance). Let $\sigma^2 = T(F) = \mathbb{V}(X) = \int x^2 dF(x) - \left(\int x\,dF(x)\right)^2$. The plug-in estimator is

$$\widehat{\sigma}^2 = \int x^2 d\widehat{F}_n(x) - \left(\int x\,d\widehat{F}_n(x)\right)^2$$

$$= \frac{1}{n}\sum_{i=1}^{n}X_i^2 - \left(\frac{1}{n}\sum_{i=1}^{n}X_i\right)^2$$

$$= \frac{1}{n}\sum_{i=1}^{n}(X_i - \overline{X}_n)^2.$$

Another reasonable estimator of σ^2 is the sample variance

$$S_n^2 = \frac{1}{n-1}\sum_{i=1}^{n}(X_i - \overline{X}_n)^2.$$

In practice, there is little difference between $\widehat{\sigma}^2$ and S_n^2 and you can use either one. Returning to the last example, we now see that the estimated standard error of the estimate of the mean is $\widehat{se} = \widehat{\sigma}/\sqrt{n}$. ∎

7.12 Example (The Skewness). Let μ and σ^2 denote the mean and variance of a random variable X. The skewness is defined to be

$$\kappa = \frac{\mathbb{E}(X-\mu)^3}{\sigma^3} = \frac{\int(x-\mu)^3 dF(x)}{\left\{\int(x-\mu)^2 dF(x)\right\}^{3/2}}.$$

The skewness measures the lack of symmetry of a distribution. To find the plug-in estimate, first recall that $\widehat{\mu} = n^{-1}\sum_i X_i$ and $\widehat{\sigma}^2 = n^{-1}\sum_i(X_i - \widehat{\mu})^2$. The plug-in estimate of κ is

$$\widehat{\kappa} = \frac{\int(x-\mu)^3 d\widehat{F}_n(x)}{\left\{\int(x-\mu)^2 d\widehat{F}_n(x)\right\}^{3/2}} = \frac{\frac{1}{n}\sum_i(X_i - \widehat{\mu})^3}{\widehat{\sigma}^3}. \quad ∎$$

7.13 Example (Correlation). Let $Z = (X,Y)$ and let $\rho = T(F) = \mathbb{E}(X - \mu_X)(Y - \mu_Y)/(\sigma_x\sigma_y)$ denote the correlation between X and Y, where $F(x,y)$ is bivariate. We can write

$$T(F) = a(T_1(F), T_2(F), T_3(F), T_4(F), T_5(F))$$

where

$$T_1(F) = \int x\, dF(z), \quad T_2(F) = \int y\, dF(z), \quad T_3(F) = \int xy\, dF(z),$$
$$T_4(F) = \int x^2\, dF(z), \quad T_5(F) = \int y^2\, dF(z),$$

and

$$a(t_1,\ldots,t_5) = \frac{t_3 - t_1 t_2}{\sqrt{(t_4 - t_1^2)(t_5 - t_2^2)}}.$$

Replace F with $\widehat{F}_n$ in $T_1(F), \ldots, T_5(F)$, and take

$$\widehat{\rho} = a(T_1(\widehat{F}_n), T_2(\widehat{F}_n), T_3(\widehat{F}_n), T_4(\widehat{F}_n), T_5(\widehat{F}_n)).$$

We get

$$\widehat{\rho} = \frac{\sum_i (X_i - \overline{X}_n)(Y_i - \overline{Y}_n)}{\sqrt{\sum_i (X_i - \overline{X}_n)^2}\sqrt{\sum_i (Y_i - \overline{Y}_n)^2}}$$

which is called the **sample correlation**. ∎

7.14 Example (Quantiles). Let F be strictly increasing with density f. For $0 < p < 1$, the pth quantile is defined by $T(F) = F^{-1}(p)$. The estimate if $T(F)$ is $\widehat{F}_n^{-1}(p)$. We have to be a bit careful since $\widehat{F}_n$ is not invertible. To avoid ambiguity we define

$$\widehat{F}_n^{-1}(p) = \inf\{x : \ \widehat{F}_n(x) \geq p\}.$$

We call $T(\widehat{F}_n) = \widehat{F}_n^{-1}(p)$ the pth **sample quantile**. ∎

Only in the first example did we compute a standard error or a confidence interval. How shall we handle the other examples? When we discuss parametric methods, we will develop formulas for standard errors and confidence intervals. But in our nonparametric setting we need something else. In the next chapter, we will introduce the bootstrap for getting standard errors and confidence intervals.

7.15 Example (Plasma Cholesterol). Figure 7.2 shows histograms for plasma cholesterol (in mg/dl) for 371 patients with chest pain (Scott et al. (1978)). The histograms show the percentage of patients in 10 bins. The first histogram is for 51 patients who had no evidence of heart disease while the second histogram is for 320 patients who had narrowing of the arteries. Is the mean cholesterol different in the two groups? Let us regard these data as samples from two distributions F_1 and F_2. Let $\mu_1 = \int x dF_1(x)$ and $\mu_2 = \int x dF_2(x)$ denote the means of the two populations. The plug-in estimates are $\widehat{\mu}_1 = \int x d\widehat{F}_{n,1}(x) = \overline{X}_{n,1} = 195.27$ and $\widehat{\mu}_2 = \int x d\widehat{F}_{n,2}(x) = \overline{X}_{n,2} = 216.19$. Recall that the standard error of the sample mean $\widehat{\mu} = \frac{1}{n}\sum_{i=1}^{n} X_i$ is

$$\mathsf{se}(\widehat{\mu}) = \sqrt{\mathbb{V}\left(\frac{1}{n}\sum_{i=1}^{n} X_i\right)} = \sqrt{\frac{1}{n^2}\sum_{i=1}^{n} \mathbb{V}(X_i)} = \sqrt{\frac{n\sigma^2}{n^2}} = \frac{\sigma}{\sqrt{n}}$$

which we estimate by

$$\widehat{\mathsf{se}}(\widehat{\mu}) = \frac{\widehat{\sigma}}{\sqrt{n}}$$

where

$$\widehat{\sigma} = \sqrt{\frac{1}{n}\sum_{i=1}^{n}(X_i - \overline{X})^2}.$$

For the two groups this yields $\widehat{se}(\widehat{\mu}_1) = 5.0$ and $\widehat{se}(\widehat{\mu}_2) = 2.4$. Approximate 95 percent confidence intervals for μ_1 and μ_2 are $\widehat{\mu}_1 \pm 2\widehat{se}(\widehat{\mu}_1) = (185, 205)$ and $\widehat{\mu}_2 \pm 2\widehat{se}(\widehat{\mu}_2) = (211, 221)$.

Now, consider the functional $\theta = T(F_2) - T(F_1)$ whose plug-in estimate is $\widehat{\theta} = \widehat{\mu}_2 - \widehat{\mu}_1 = 216.19 - 195.27 = 20.92$. The standard error of $\widehat{\theta}$ is

$$se = \sqrt{\mathbb{V}(\widehat{\mu}_2 - \widehat{\mu}_1)} = \sqrt{\mathbb{V}(\widehat{\mu}_2) + \mathbb{V}(\widehat{\mu}_1)} = \sqrt{(se(\widehat{\mu}_1))^2 + (se(\widehat{\mu}_2))^2}$$

and we estimate this by

$$\widehat{se} = \sqrt{(\widehat{se}(\widehat{\mu}_1))^2 + (\widehat{se}(\widehat{\mu}_2))^2} = 5.55.$$

An approximate 95 percent confidence interval for θ is $\widehat{\theta} \pm 2\,\widehat{se}(\widehat{\theta}_n) = (9.8, 32.0)$. This suggests that cholesterol is higher among those with narrowed arteries. We should not jump to the conclusion (from these data) that cholesterol causes heart disease. The leap from statistical evidence to causation is very subtle and is discussed in Chapter 16. ■

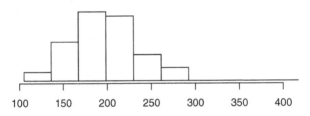

plasma cholesterol for patients without heart disease

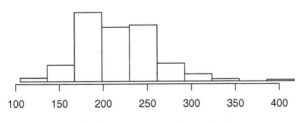

plasma cholesterol for patients with heart disease

FIGURE 7.2. Plasma cholesterol for 51 patients with no heart disease and 320 patients with narrowing of the arteries.

7.3 Bibliographic Remarks

The Glivenko-Cantelli theorem is the tip of the iceberg. The theory of distribution functions is a special case of what are called empirical processes which underlie much of modern statistical theory. Some references on empirical processes are Shorack and Wellner (1986) and van der Vaart and Wellner (1996).

7.4 Exercises

1. Prove Theorem 7.3.

2. Let $X_1, \ldots, X_n \sim$ Bernoulli(p) and let $Y_1, \ldots, Y_m \sim$ Bernoulli(q). Find the plug-in estimator and estimated standard error for p. Find an approximate 90 percent confidence interval for p. Find the plug-in estimator and estimated standard error for $p - q$. Find an approximate 90 percent confidence interval for $p - q$.

3. (Computer Experiment.) Generate 100 observations from a N(0,1) distribution. Compute a 95 percent confidence band for the CDF F (as described in the appendix). Repeat this 1000 times and see how often the confidence band contains the true distribution function. Repeat using data from a Cauchy distribution.

4. Let $X_1, \ldots, X_n \sim F$ and let $\widehat{F}_n(x)$ be the empirical distribution function. For a fixed x, use the central limit theorem to find the limiting distribution of $\widehat{F}_n(x)$.

5. Let x and y be two distinct points. Find $\text{Cov}(\widehat{F}_n(x), \widehat{F}_n(y))$.

6. Let $X_1, \ldots, X_n \sim F$ and let $\widehat{F}$ be the empirical distribution function. Let $a < b$ be fixed numbers and define $\theta = T(F) = F(b) - F(a)$. Let $\widehat{\theta} = T(\widehat{F}_n) = \widehat{F}_n(b) - \widehat{F}_n(a)$. Find the estimated standard error of $\widehat{\theta}$. Find an expression for an approximate $1 - \alpha$ confidence interval for θ.

7. Data on the magnitudes of earthquakes near Fiji are available on the website for this book. Estimate the CDF $F(x)$. Compute and plot a 95 percent confidence envelope for F (as described in the appendix). Find an approximate 95 percent confidence interval for $F(4.9) - F(4.3)$.

8. Get the data on eruption times and waiting times between eruptions of the Old Faithful geyser from the website. Estimate the mean waiting time and give a standard error for the estimate. Also, give a 90 percent confidence interval for the mean waiting time. Now estimate the median waiting time. In the next chapter we will see how to get the standard error for the median.

9. 100 people are given a standard antibiotic to treat an infection and another 100 are given a new antibiotic. In the first group, 90 people recover; in the second group, 85 people recover. Let p_1 be the probability of recovery under the standard treatment and let p_2 be the probability of recovery under the new treatment. We are interested in estimating $\theta = p_1 - p_2$. Provide an estimate, standard error, an 80 percent confidence interval, and a 95 percent confidence interval for θ.

10. In 1975, an experiment was conducted to see if cloud seeding produced rainfall. 26 clouds were seeded with silver nitrate and 26 were not. The decision to seed or not was made at random. Get the data from

    http://lib.stat.cmu.edu/DASL/Stories/CloudSeeding.html

    Let θ be the difference in the mean precipitation from the two groups. Estimate θ. Estimate the standard error of the estimate and produce a 95 percent confidence interval.

8
The Bootstrap

The **bootstrap** is a method for estimating standard errors and computing confidence intervals. Let $T_n = g(X_1, \ldots, X_n)$ be a **statistic**, that is, T_n is any function of the data. Suppose we want to know $\mathbb{V}_F(T_n)$, the variance of T_n. We have written $\mathbb{V}_F$ to emphasize that the variance usually depends on the unknown distribution function F. For example, if $T_n = \overline{X}_n$ then $\mathbb{V}_F(T_n) = \sigma^2/n$ where $\sigma^2 = \int (x - \mu)^2 dF(x)$ and $\mu = \int x dF(x)$. Thus the variance of T_n is a function of F. The bootstrap idea has two steps:

Step 1: Estimate $\mathbb{V}_F(T_n)$ with $\mathbb{V}_{\widehat{F}_n}(T_n)$.

Step 2: Approximate $\mathbb{V}_{\widehat{F}_n}(T_n)$ using simulation.

For $T_n = \overline{X}_n$, we have for Step 1 that $\mathbb{V}_{\widehat{F}_n}(T_n) = \widehat{\sigma}^2/n$ where $\widehat{\sigma}^2 = n^{-1} \sum_{i=1}^n (X_i - \overline{X}_n)$. In this case, Step 1 is enough. However, in more complicated cases we cannot write down a simple formula for $\mathbb{V}_{\widehat{F}_n}(T_n)$ which is why we need Step 2. Before proceeding, let us discuss the idea of simulation.

8.1   Simulation

Suppose we draw an IID sample $Y_1, \ldots, Y_B$ from a distribution G. By the law of large numbers,

$$\overline{Y}_n = \frac{1}{B} \sum_{j=1}^{B} Y_j \xrightarrow{\text{P}} \int y \, dG(y) = \mathbb{E}(Y)$$

as $B \to \infty$. So if we draw a large sample from G, we can use the sample mean $\overline{Y}_n$ to approximate $\mathbb{E}(Y)$. In a simulation, we can make B as large as we like, in which case, the difference between $\overline{Y}_n$ and $\mathbb{E}(Y)$ is negligible. More generally, if h is any function with finite mean then

$$\frac{1}{B} \sum_{j=1}^{B} h(Y_j) \xrightarrow{\text{P}} \int h(y) dG(y) = \mathbb{E}(h(Y))$$

as $B \to \infty$. In particular,

$$\frac{1}{B} \sum_{j=1}^{B} (Y_j - \overline{Y})^2 = \frac{1}{B} \sum_{j=1}^{B} Y_j^2 - \left(\frac{1}{B} \sum_{j=1}^{B} Y_j \right)^2$$

$$\xrightarrow{\text{P}} \int y^2 dF(y) - \left(\int y \, dF(y) \right)^2 = \mathbb{V}(Y).$$

Hence, we can use the sample variance of the simulated values to approximate $\mathbb{V}(Y)$.

8.2   Bootstrap Variance Estimation

According to what we just learned, we can approximate $\mathbb{V}_{\widehat{F}_n}(T_n)$ by simulation. Now $\mathbb{V}_{\widehat{F}_n}(T_n)$ means "the variance of T_n if the distribution of the data is $\widehat{F}_n$." How can we simulate from the distribution of T_n when the data are assumed to have distribution $\widehat{F}_n$? The answer is to simulate $X_1^*, \ldots, X_n^*$ from $\widehat{F}_n$ and then compute $T_n^* = g(X_1^*, \ldots, X_n^*)$. This constitutes one draw from the distribution of T_n. The idea is illustrated in the following diagram:

$$
\begin{array}{lllll}
\text{Real world} & F & \Longrightarrow & X_1, \ldots, X_n & \Longrightarrow & T_n = g(X_1, \ldots, X_n) \\
\text{Bootstrap world} & \widehat{F}_n & \Longrightarrow & X_1^*, \ldots, X_n^* & \Longrightarrow & T_n^* = g(X_1^*, \ldots, X_n^*)
\end{array}
$$

How do we simulate $X_1^*, \ldots, X_n^*$ from $\widehat{F}_n$? Notice that $\widehat{F}_n$ puts mass $1/n$ at each data point $X_1, \ldots, X_n$. Therefore,

> drawing an observation from $\widehat{F}_n$ is equivalent to drawing
> one point at random from the original data set.

Thus, to simulate $X_1^*, \ldots, X_n^* \sim \widehat{F}_n$ it suffices to draw n observations with replacement from $X_1, \ldots, X_n$. Here is a summary:

Bootstrap Variance Estimation

1. Draw $X_1^*, \ldots, X_n^* \sim \widehat{F}_n$.

2. Compute $T_n^* = g(X_1^*, \ldots, X_n^*)$.

3. Repeat steps 1 and 2, B times, to get $T_{n,1}^*, \ldots, T_{n,B}^*$.

4. Let

$$v_{\text{boot}} = \frac{1}{B} \sum_{b=1}^{B} \left(T_{n,b}^* - \frac{1}{B} \sum_{r=1}^{B} T_{n,r}^* \right)^2. \tag{8.1}$$

8.1 Example. The following pseudocode shows how to use the bootstrap to estimate the standard error of the median.

Bootstrap for The Median

```
Given data X = (X(1), ..., X(n)):


T <- median(X)
Tboot <- vector of length B
for(i in 1:B){
    Xstar <- sample of size n from X (with replacement)
    Tboot[i] <- median(Xstar)
    }
se <- sqrt(variance(Tboot))
```

The following schematic diagram will remind you that we are using two approximations:

$$\mathbb{V}_F(T_n) \overset{\overset{\text{not so small}}{\frown}}{\approx} \mathbb{V}_{\widehat{F}_n}(T_n) \overset{\overset{\text{small}}{\frown}}{\approx} v_{\text{boot}}.$$

8.2 Example. Consider the nerve data. Let $\theta = T(F) = \int (x-\mu)^3 dF(x)/\sigma^3$ be the skewness. The skewness is a measure of asymmetry. A Normal distribution, for example, has skewness 0. The plug-in estimate of the skewness is

$$\widehat{\theta} = T(\widehat{F}_n) = \frac{\int (x - \mu)^3 d\widehat{F}_n(x)}{\widehat{\sigma}^3} = \frac{\frac{1}{n}\sum_{i=1}^{n}(X_i - \overline{X}_n)^3}{\widehat{\sigma}^3} = 1.76.$$

To estimate the standard error with the bootstrap we follow the same steps as with the median example except we compute the skewness from each bootstrap sample. When applied to the nerve data, the bootstrap, based on $B = 1,000$ replications, yields a standard error for the estimated skewness of .16. ∎

8.3 Bootstrap Confidence Intervals

There are several ways to construct bootstrap confidence intervals. Here we discuss three methods.

Method 1: The Normal Interval. The simplest method is the Normal interval

$$T_n \pm z_{\alpha/2}\, \widehat{se}_{\text{boot}} \tag{8.2}$$

where $\widehat{se}_{\text{boot}} = \sqrt{v_{\text{boot}}}$ is the bootstrap estimate of the standard error. This interval is not accurate unless the distribution of T_n is close to Normal.

Method 2: Pivotal Intervals. Let $\theta = T(F)$ and $\widehat{\theta}_n = T(\widehat{F}_n)$ and define the **pivot** $R_n = \widehat{\theta}_n - \theta$. Let $\widehat{\theta}_{n,1}^*, \ldots, \widehat{\theta}_{n,B}^*$ denote bootstrap replications of $\widehat{\theta}_n$. Let $H(r)$ denote the CDF of the pivot:

$$H(r) = \mathbb{P}_F(R_n \le r). \tag{8.3}$$

Define $C_n^\star = (a, b)$ where

$$a = \widehat{\theta}_n - H^{-1}\left(1 - \frac{\alpha}{2}\right) \quad \text{and} \quad b = \widehat{\theta}_n - H^{-1}\left(\frac{\alpha}{2}\right). \tag{8.4}$$

It follows that

$$
\begin{aligned}
\mathbb{P}(a \le \theta \le b) &= \mathbb{P}(a - \widehat{\theta}_n \le \theta - \widehat{\theta}_n \le b - \widehat{\theta}_n) \\
&= \mathbb{P}(\widehat{\theta}_n - b \le \widehat{\theta}_n - \theta \le \widehat{\theta}_n - a) \\
&= \mathbb{P}(\widehat{\theta}_n - b \le R_n \le \widehat{\theta}_n - a) \\
&= H(\widehat{\theta}_n - a) - H(\widehat{\theta}_n - b) \\
&= H\left(H^{-1}\left(1 - \frac{\alpha}{2}\right)\right) - H\left(H^{-1}\left(\frac{\alpha}{2}\right)\right) \\
&= 1 - \frac{\alpha}{2} - \frac{\alpha}{2} = 1 - \alpha.
\end{aligned}
$$

Hence, C_n^* is an exact $1 - \alpha$ confidence interval for θ. Unfortunately, a and b depend on the unknown distribution H but we can form a bootstrap estimate of H:

$$\widehat{H}(r) = \frac{1}{B} \sum_{b=1}^{B} I(R_{n,b}^* \leq r) \tag{8.5}$$

where $R_{n,b}^* = \widehat{\theta}_{n,b}^* - \widehat{\theta}_n$. Let r_β^* denote the β sample quantile of $(R_{n,1}^*, \ldots, R_{n,B}^*)$ and let θ_β^* denote the β sample quantile of $(\widehat{\theta}_{n,1}^*, \ldots, \widehat{\theta}_{n,B}^*)$. Note that $r_\beta^* = \theta_\beta^* - \widehat{\theta}_n$. It follows that an approximate $1 - \alpha$ confidence interval is $C_n = (\widehat{a}, \widehat{b})$ where

$$
\begin{aligned}
\widehat{a} &= \widehat{\theta}_n - \widehat{H}^{-1}\left(1 - \frac{\alpha}{2}\right) = \widehat{\theta}_n - r_{1-\alpha/2}^* = 2\widehat{\theta}_n - \theta_{1-\alpha/2}^* \\
\widehat{b} &= \widehat{\theta}_n - \widehat{H}^{-1}\left(\frac{\alpha}{2}\right) \quad = \widehat{\theta}_n - r_{\alpha/2}^* = 2\widehat{\theta}_n - \theta_{\alpha/2}^*.
\end{aligned}
$$

In summary, the $1 - \alpha$ **bootstrap pivotal confidence** interval is

$$C_n = \left(2\widehat{\theta}_n - \widehat{\theta}_{1-\alpha/2}^*, \ 2\widehat{\theta}_n - \widehat{\theta}_{\alpha/2}^*\right). \tag{8.6}$$

8.3 Theorem. *Under weak conditions on $T(F)$,*

$$\mathbb{P}_F\left(T(F) \in C_n\right) \to 1 - \alpha$$

as $n \to \infty$, where C_n is given in (8.6).

Method 3: Percentile Intervals. The **bootstrap percentile interval** is defined by

$$C_n = \left(\theta_{\alpha/2}^*, \ \theta_{1-\alpha/2}^*\right).$$

The justification for this interval is given in the appendix.

8.4 Example. For estimating the skewness of the nerve data, here are the various confidence intervals.

Method	95% Interval
Normal	(1.44, 2.09)
Pivotal	(1.48, 2.11)
Percentile	(1.42, 2.03)

All these confidence intervals are approximate. The probability that $T(F)$ is in the interval is not exactly $1 - \alpha$. All three intervals have the same level of accuracy. There are more accurate bootstrap confidence intervals but they are more complicated and we will not discuss them here.

8.5 Example (The Plasma Cholesterol Data). Let us return to the cholesterol data. Suppose we are interested in the difference of the medians. Pseudocode for the bootstrap analysis is as follows:

```
x1 <- first sample
x2 <- second sample
n1 <- length(x1)
n2 <- length(x2)
th.hat <- median(x2) - median(x1)
B <- 1000
Tboot <- vector of length B
for(i in 1:B){
    xx1 <- sample of size n1 with replacement from x1
    xx2 <- sample of size n2 with replacement from x2
    Tboot[i] <- median(xx2) - median(xx1)
    }
se <- sqrt(variance(Tboot))
Normal      <- (th.hat - 2*se, th.hat + 2*se)
percentile <- (quantile(Tboot,.025), quantile(Tboot,.975))
pivotal     <- ( 2*th.hat-quantile(Tboot,.975),
                2*th.hat-quantile(Tboot,.025) )
```

The point estimate is 18.5, the bootstrap standard error is 7.42 and the resulting approximate 95 percent confidence intervals are as follows:

Method	95% Interval
Normal	(3.7, 33.3)
Pivotal	(5.0, 34.0)
Percentile	(5.0, 33.3)

Since these intervals exclude 0, it appears that the second group has higher cholesterol although there is considerable uncertainty about how much higher as reflected in the width of the intervals. ■

The next two examples are based on small sample sizes. In practice, statistical methods based on very small sample sizes might not be reliable. We include the examples for their pedagogical value but we do want to sound a note of caution about interpreting the results with some skepticism.

8.6 Example. Here is an example that was one of the first used to illustrate the bootstrap by Bradley Efron, the inventor of the bootstrap. The data are LSAT scores (for entrance to law school) and GPA.

LSAT	576	635	558	578	666	580	555	661
	651	605	653	575	545	572	594	

GPA	3.39	3.30	2.81	3.03	3.44	3.07	3.00	3.43
	3.36	3.13	3.12	2.74	2.76	2.88	3.96	

Each data point is of the form $X_i = (Y_i, Z_i)$ where $Y_i = \text{LSAT}_i$ and $Z_i = \text{GPA}_i$. The law school is interested in the correlation

$$\theta = \frac{\int \int (y - \mu_Y)(z - \mu_Z) dF(y, z)}{\sqrt{\int (y - \mu_Y)^2 dF(y) \int (z - \mu_Z)^2 dF(z)}}.$$

The plug-in estimate is the sample correlation

$$\widehat{\theta} = \frac{\sum_i (Y_i - \overline{Y})(Z_i - \overline{Z})}{\sqrt{\sum_i (Y_i - \overline{Y})^2 \sum_i (Z_i - \overline{Z})^2}}.$$

The estimated correlation is $\widehat{\theta} = .776$. The bootstrap based on $B = 1000$ gives $\widehat{se} = .137$. Figure 8.1 shows the data and a histogram of the bootstrap replications $\widehat{\theta}_1^*, \ldots, \widehat{\theta}_B^*$. This histogram is an approximation to the sampling distribution of $\widehat{\theta}$. The Normal-based 95 percent confidence interval is $.78 \pm 2\widehat{se} = (.51, 1.00)$ while the percentile interval is $(.46, .96)$. In large samples, the two methods will show closer agreement. ∎

8.7 Example. This example is from Efron and Tibshirani (1993). When drug companies introduce new medications, they are sometimes required to show bioequivalence. This means that the new drug is not substantially different than the current treatment. Here are data on eight subjects who used medical patches to infuse a hormone into the blood. Each subject received three treatments: placebo, old-patch, new-patch.

subject	placebo	old	new	old − placebo	new − old
1	9243	17649	16449	8406	-1200
2	9671	12013	14614	2342	2601
3	11792	19979	17274	8187	-2705
4	13357	21816	23798	8459	1982
5	9055	13850	12560	4795	-1290
6	6290	9806	10157	3516	351
7	12412	17208	16570	4796	-638
8	18806	29044	26325	10238	-2719

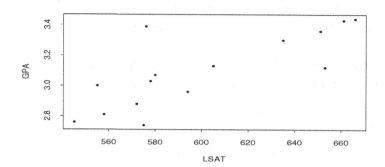

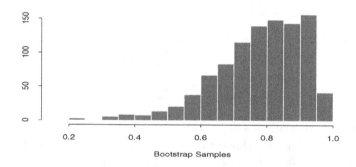

FIGURE 8.1. Law school data. The top panel shows the raw data. The bottom panel is a histogram of the correlations computed from each bootstrap sample.

Let Z = old − placebo and Y = new − old. The Food and Drug Administration (FDA) requirement for bioequivalence is that $|\theta| \le .20$ where

$$\theta = \frac{\mathbb{E}_F(Y)}{\mathbb{E}_F(Z)}.$$

The plug-in estimate of θ is

$$\widehat{\theta} = \frac{\overline{Y}}{\overline{Z}} = \frac{-452.3}{6342} = -0.0713.$$

The bootstrap standard error is $\widehat{se} = 0.105$. To answer the bioequivalence question, we compute a confidence interval. From $B = 1000$ bootstrap replications we get the 95 percent interval (-0.24,0.15). This is not quite contained

in (-0.20,0.20) so at the 95 percent level we have not demonstrated bioequivalence. Figure 8.2 shows the histogram of the bootstrap values. ∎

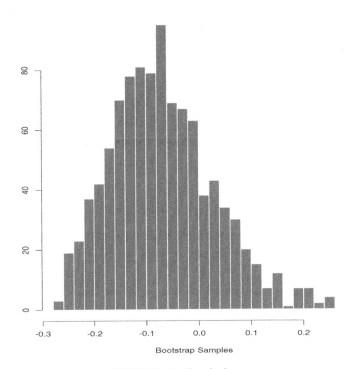

FIGURE 8.2. Patch data.

8.4 Bibliographic Remarks

The bootstrap was invented by Efron (1979). There are several books on these
topics including Efron and Tibshirani (1993), Davison and Hinkley (1997),
Hall (1992) and Shao and Tu (1995). Also, see section 3.6 of van der Vaart
and Wellner (1996).

8.5 Appendix

8.5.1 The Jackknife

There is another method for computing standard errors called the **jackknife**,
due to Quenouille (1949). It is less computationally expensive than the boot-

strap but is less general. Let $T_n = T(X_1, \ldots, X_n)$ be a statistic and $T_{(-i)}$ denote the statistic with the ith observation removed. Let $\overline{T}_n = n^{-1} \sum_{i=1}^{n} T_{(-i)}$. The jackknife estimate of $\text{var}(T_n)$ is

$$v_{\text{jack}} = \frac{n-1}{n} \sum_{i=1}^{n} (T_{(-i)} - \overline{T}_n)^2$$

and the jackknife estimate of the standard error is $\widehat{\text{se}}_{jack} = \sqrt{v_{jack}}$. Under suitable conditions on T, it can be shown that v_{jack} consistently estimates $\text{var}(T_n)$ in the sense that $v_{\text{jack}}/\text{var}(T_n) \xrightarrow{P} 1$. However, unlike the bootstrap, the jackknife does not produce consistent estimates of the standard error of sample quantiles.

8.5.2 Justification For The Percentile Interval

Suppose there exists a monotone transformation $U = m(T)$ such that $U \sim N(\phi, c^2)$ where $\phi = m(\theta)$. We do not suppose we know the transformation, only that one exists. Let $U_b^* = m(\theta_{n,b}^*)$. Let u_β^* be the β sample quantile of the U_b^*'s. Since a monotone transformation preserves quantiles, we have that $u_{\alpha/2}^* = m(\theta_{\alpha/2}^*)$. Also, since $U \sim N(\phi, c^2)$, the $\alpha/2$ quantile of U is $\phi - z_{\alpha/2}c$. Hence $u_{\alpha/2}^* = \phi - z_{\alpha/2}c$. Similarly, $u_{1-\alpha/2}^* = \phi + z_{\alpha/2}c$. Therefore,

$$
\begin{aligned}
\mathbb{P}(\theta_{\alpha/2}^* \le \theta \le \theta_{1-\alpha/2}^*) &= \mathbb{P}(m(\theta_{\alpha/2}^*) \le m(\theta) \le m(\theta_{1-\alpha/2}^*)) \\
&= \mathbb{P}(u_{\alpha/2}^* \le \phi \le u_{1-\alpha/2}^*) \\
&= \mathbb{P}(U - cz_{\alpha/2} \le \phi \le U + cz_{\alpha/2}) \\
&= \mathbb{P}\left(-z_{\alpha/2} \le \frac{U - \phi}{c} \le z_{\alpha/2}\right) \\
&= 1 - \alpha.
\end{aligned}
$$

An exact normalizing transformation will rarely exist but there may exist approximate normalizing transformations.

8.6 Exercises

1. Consider the data in Example 8.6. Find the plug-in estimate of the correlation coefficient. Estimate the standard error using the bootstrap. Find a 95 percent confidence interval using the Normal, pivotal, and percentile methods.

2. (Computer Experiment.) Conduct a simulation to compare the various bootstrap confidence interval methods. Let $n = 50$ and let $T(F) = \int (x - \mu)^3 dF(x)/\sigma^3$ be the skewness. Draw $Y_1, \ldots, Y_n \sim N(0, 1)$ and set $X_i = e^{Y_i}$, $i = 1, \ldots, n$. Construct the three types of bootstrap 95 percent intervals for $T(F)$ from the data $X_1, \ldots, X_n$. Repeat this whole thing many times and estimate the true coverage of the three intervals.

3. Let

$$X_1, \ldots, X_n \sim t_3$$

where $n = 25$. Let $\theta = T(F) = (q_{.75} - q_{.25})/1.34$ where q_p denotes the pth quantile. Do a simulation to compare the coverage and length of the following confidence intervals for θ: (i) Normal interval with standard error from the bootstrap, (ii) bootstrap percentile interval, and (iii) pivotal bootstrap interval.

4. Let $X_1, \ldots, X_n$ be distinct observations (no ties). Show that there are

$$\binom{2n - 1}{n}$$

distinct bootstrap samples.

Hint: Imagine putting n balls into n buckets.

5. Let $X_1, \ldots, X_n$ be distinct observations (no ties). Let $X_1^*, \ldots, X_n^*$ denote a bootstrap sample and let $\overline{X}_n^* = n^{-1} \sum_{i=1}^n X_i^*$. Find: $\mathbb{E}(\overline{X}_n^* | X_1, \ldots, X_n)$, $\mathbb{V}(\overline{X}_n^* | X_1, \ldots, X_n)$, $\mathbb{E}(\overline{X}_n^*)$ and $\mathbb{V}(\overline{X}_n^*)$.

6. (Computer Experiment.) Let $X_1, \ldots, X_n$ Normal$(\mu, 1)$. Let $\theta = e^\mu$ and let $\widehat{\theta} = e^{\overline{X}}$. Create a data set (using $\mu = 5$) consisting of n=100 observations.

(a) Use the bootstrap to get the se and 95 percent confidence interval for θ.

(b) Plot a histogram of the bootstrap replications. This is an estimate of the distribution of $\widehat{\theta}$. Compare this to the true sampling distribution of $\widehat{\theta}$.

7. Let $X_1, \ldots, X_n \sim$ Uniform$(0, \theta)$. Let $\widehat{\theta} = X_{max} = \max\{X_1, \ldots, X_n\}$. Generate a data set of size 50 with $\theta = 1$.

(a) Find the distribution of $\widehat{\theta}$. Compare the true distribution of $\widehat{\theta}$ to the histograms from the bootstrap.

(b) This is a case where the bootstrap does very poorly. In fact, we can prove that this is the case. Show that $P(\widehat{\theta} = \widehat{\theta}) = 0$ and yet $P(\widehat{\theta}^* = \widehat{\theta}) \approx .632$. Hint: show that, $P(\widehat{\theta}^* = \widehat{\theta}) = 1 - (1 - (1/n))^n$ then take the limit as n gets large.

8. Let $T_n = \overline{X}_n^2$, $\mu = \mathbb{E}(X_1)$, $\alpha_k = \int |x-\mu|^k dF(x)$ and $\widehat{\alpha}_k = n^{-1}\sum_{i=1}^n |X_i - \overline{X}_n|^k$. Show that

$$v_{\text{boot}} = \frac{4\overline{X}_n^2 \widehat{\alpha}_2}{n} + \frac{4\overline{X}_n \widehat{\alpha}_3}{n^2} + \frac{\widehat{\alpha}_4}{n^3}.$$

9
Parametric Inference

We now turn our attention to parametric models, that is, models of the form

$$\mathfrak{F} = \left\{ f(x; \theta) : \; \theta \in \Theta \right\} \tag{9.1}$$

where the $\Theta \subset \mathbb{R}^k$ is the parameter space and $\theta = (\theta_1, \ldots, \theta_k)$ is the parameter. The problem of inference then reduces to the problem of estimating the parameter θ.

Students learning statistics often ask: how would we ever know that the distribution that generated the data is in some parametric model? This is an excellent question. Indeed, we would rarely have such knowledge which is why nonparametric methods are preferable. Still, studying methods for parametric models is useful for two reasons. First, there are some cases where background knowledge suggests that a parametric model provides a reasonable approximation. For example, counts of traffic accidents are known from prior experience to follow approximately a Poisson model. Second, the inferential concepts for parametric models provide background for understanding certain nonparametric methods.

We begin with a brief discussion about parameters of interest and nuisance parameters in the next section, then we will discuss two methods for estimating θ, the method of moments and the method of maximum likelihood.

9.1   Parameter of Interest

Often, we are only interested in some function $T(\theta)$. For example, if $X \sim N(\mu, \sigma^2)$ then the parameter is $\theta = (\mu, \sigma)$. If our goal is to estimate μ then $\mu = T(\theta)$ is called the **parameter of interest** and σ is called a **nuisance parameter**. The parameter of interest might be a complicated function of θ as in the following example.

9.1 Example. Let $X_1, \ldots, X_n \sim \text{Normal}(\mu, \sigma^2)$. The parameter is $\theta = (\mu, \sigma)$ and the parameter space is $\Theta = \{(\mu, \sigma) : \mu \in \mathbb{R}, \ \sigma > 0\}$. Suppose that X_i is the outcome of a blood test and suppose we are interested in τ, the fraction of the population whose test score is larger than 1. Let Z denote a standard Normal random variable. Then

$$
\begin{aligned}
\tau &= \mathbb{P}(X > 1) = 1 - \mathbb{P}(X < 1) = 1 - \mathbb{P}\left(\frac{X - \mu}{\sigma} < \frac{1 - \mu}{\sigma}\right) \\
&= 1 - \mathbb{P}\left(Z < \frac{1 - \mu}{\sigma}\right) = 1 - \Phi\left(\frac{1 - \mu}{\sigma}\right).
\end{aligned}
$$

The parameter of interest is $\tau = T(\mu, \sigma) = 1 - \Phi((1 - \mu)/\sigma)$. ∎

9.2 Example. Recall that X has a Gamma(α, β) distribution if

$$
f(x; \alpha, \beta) = \frac{1}{\beta^\alpha \Gamma(\alpha)} x^{\alpha - 1} e^{-x/\beta}, \quad x > 0
$$

where $\alpha, \beta > 0$ and

$$
\Gamma(\alpha) = \int_0^\infty y^{\alpha - 1} e^{-y} dy
$$

is the Gamma function. The parameter is $\theta = (\alpha, \beta)$. The Gamma distribution is sometimes used to model lifetimes of people, animals, and electronic equipment. Suppose we want to estimate the mean lifetime. Then $T(\alpha, \beta) = \mathbb{E}_\theta(X_1) = \alpha\beta$. ∎

9.2   The Method of Moments

The first method for generating parametric estimators that we will study is called the method of moments. We will see that these estimators are not optimal but they are often easy to compute. They are are also useful as starting values for other methods that require iterative numerical routines.

Suppose that the parameter $\theta = (\theta_1, \ldots, \theta_k)$ has k components. For $1 \leq j \leq k$, define the jth **moment**

$$\alpha_j \equiv \alpha_j(\theta) = \mathbb{E}_\theta(X^j) = \int x^j dF_\theta(x) \tag{9.2}$$

and the jth **sample moment**

$$\widehat{\alpha}_j = \frac{1}{n} \sum_{i=1}^n X_i^j. \tag{9.3}$$

9.3 Definition. *The* **method of moments estimator** $\widehat{\theta}_n$ *is defined to be the value of θ such that*

$$\begin{aligned}
\alpha_1(\widehat{\theta}_n) &= \widehat{\alpha}_1 \\
\alpha_2(\widehat{\theta}_n) &= \widehat{\alpha}_2 \\
\vdots \quad \vdots \quad &\vdots \\
\alpha_k(\widehat{\theta}_n) &= \widehat{\alpha}_k.
\end{aligned} \tag{9.4}$$

Formula (9.4) defines a system of k equations with k unknowns.

9.4 Example. Let $X_1, \ldots, X_n \sim$ Bernoulli(p). Then $\alpha_1 = \mathbb{E}_p(X) = p$ and $\widehat{\alpha}_1 = n^{-1} \sum_{i=1}^n X_i$. By equating these we get the estimator

$$\widehat{p}_n = \frac{1}{n} \sum_{i=1}^n X_i. \quad \blacksquare$$

9.5 Example. Let $X_1, \ldots, X_n \sim$ Normal(μ, σ^2). Then, $\alpha_1 = \mathbb{E}_\theta(X_1) = \mu$ and $\alpha_2 = \mathbb{E}_\theta(X_1^2) = \mathbb{V}_\theta(X_1) + (\mathbb{E}_\theta(X_1))^2 = \sigma^2 + \mu^2$. We need to solve the equations[1]

$$\begin{aligned}
\widehat{\mu} &= \frac{1}{n} \sum_{i=1}^n X_i \\
\widehat{\sigma}^2 + \widehat{\mu}^2 &= \frac{1}{n} \sum_{i=1}^n X_i^2.
\end{aligned}$$

This is a system of 2 equations with 2 unknowns. The solution is

$$\widehat{\mu} = \overline{X}_n$$

---

[1] Recall that $\mathbb{V}(X) = \mathbb{E}(X^2) - (\mathbb{E}(X))^2$. Hence, $\mathbb{E}(X^2) = \mathbb{V}(X) + (\mathbb{E}(X))^2$.

$$\widehat{\sigma}^2 \;=\; \frac{1}{n}\sum_{i=1}^{n}(X_i - \overline{X}_n)^2. \quad \blacksquare$$

9.6 Theorem. *Let $\widehat{\theta}_n$ denote the method of moments estimator. Under appropriate conditions on the model, the following statements hold:*

1. *The estimate $\widehat{\theta}_n$ exists with probability tending to 1.*

2. *The estimate is consistent: $\widehat{\theta}_n \xrightarrow{\text{P}} \theta$.*

3. *The estimate is asymptotically Normal:*

$$\sqrt{n}(\widehat{\theta}_n - \theta) \rightsquigarrow N(0, \Sigma)$$

where

$$\Sigma = g\mathbb{E}_\theta(YY^T)g^T,$$

$Y = (X, X^2, \ldots, X^k)^T$, $g = (g_1, \ldots, g_k)$ *and* $g_j = \partial\alpha_j^{-1}(\theta)/\partial\theta$.

The last statement in the theorem above can be used to find standard errors and confidence intervals. However, there is an easier way: the bootstrap. We defer discussion of this until the end of the chapter.

9.3   Maximum Likelihood

The most common method for estimating parameters in a parametric model is the **maximum likelihood method**. Let $X_1, \ldots, X_n$ be IID with PDF $f(x; \theta)$.

9.7 Definition. *The* **likelihood function** *is defined by*

$$\mathcal{L}_n(\theta) = \prod_{i=1}^{n} f(X_i; \theta). \tag{9.5}$$

The **log-likelihood function** *is defined by $\ell_n(\theta) = \log \mathcal{L}_n(\theta)$.*

The likelihood function is just the joint density of the data, except that we **treat it is a function of the parameter** θ. Thus, $\mathcal{L}_n : \Theta \to [0, \infty)$. The likelihood function is not a density function: in general, it is **not** true that $\mathcal{L}_n(\theta)$ integrates to 1 (with respect to θ).

9.8 Definition. *The* **maximum likelihood estimator** MLE, *denoted by $\widehat{\theta}_n$, is the value of θ that maximizes $\mathcal{L}_n(\theta)$.*

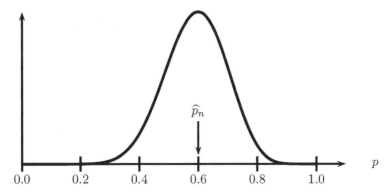

FIGURE 9.1. Likelihood function for Bernoulli with $n = 20$ and $\sum_{i=1}^{n} X_i = 12$. The MLE is $\widehat{p}_n = 12/20 = 0.6$.

The maximum of $\ell_n(\theta)$ occurs at the same place as the maximum of $\mathcal{L}_n(\theta)$, so maximizing the log-likelihood leads to the same answer as maximizing the likelihood. Often, it is easier to work with the log-likelihood.

9.9 Remark. If we multiply $\mathcal{L}_n(\theta)$ by any positive constant c (not depending on θ) then this will not change the MLE. Hence, we shall often drop constants in the likelihood function.

9.10 Example. Suppose that $X_1, \ldots, X_n \sim \text{Bernoulli}(p)$. The probability function is $f(x; p) = p^x (1-p)^{1-x}$ for $x = 0, 1$. The unknown parameter is p. Then,

$$\mathcal{L}_n(p) = \prod_{i=1}^{n} f(X_i; p) = \prod_{i=1}^{n} p^{X_i} (1-p)^{1-X_i} = p^S (1-p)^{n-S}$$

where $S = \sum_i X_i$. Hence,

$$\ell_n(p) = S \log p + (n - S) \log(1 - p).$$

Take the derivative of $\ell_n(p)$, set it equal to 0 to find that the MLE is $\widehat{p}_n = S/n$. See Figure 9.1. ∎

9.11 Example. Let $X_1, \ldots, X_n \sim N(\mu, \sigma^2)$. The parameter is $\theta = (\mu, \sigma)$ and the likelihood function (ignoring some constants) is:

$$
\begin{aligned}
\mathcal{L}_n(\mu, \sigma) &= \prod_i \frac{1}{\sigma} \exp\left\{ -\frac{1}{2\sigma^2}(X_i - \mu)^2 \right\} \\
&= \sigma^{-n} \exp\left\{ -\frac{1}{2\sigma^2} \sum_i (X_i - \mu)^2 \right\}
\end{aligned}
$$

$$= \sigma^{-n} \exp\left\{-\frac{nS^2}{2\sigma^2}\right\} \exp\left\{-\frac{n(\overline{X} - \mu)^2}{2\sigma^2}\right\}$$

where $\overline{X} = n^{-1}\sum_i X_i$ is the sample mean and $S^2 = n^{-1}\sum_i (X_i - \overline{X})^2$. The last equality above follows from the fact that $\sum_i (X_i - \mu)^2 = nS^2 + n(\overline{X} - \mu)^2$ which can be verified by writing $\sum_i (X_i - \mu)^2 = \sum_i (X_i - \overline{X} + \overline{X} - \mu)^2$ and then expanding the square. The log-likelihood is

$$\ell(\mu, \sigma) = -n \log \sigma - \frac{nS^2}{2\sigma^2} - \frac{n(\overline{X} - \mu)^2}{2\sigma^2}.$$

Solving the equations

$$\frac{\partial \ell(\mu, \sigma)}{\partial \mu} = 0 \quad \text{and} \quad \frac{\partial \ell(\mu, \sigma)}{\partial \sigma} = 0,$$

we conclude that $\widehat{\mu} = \overline{X}$ and $\widehat{\sigma} = S$. It can be verified that these are indeed global maxima of the likelihood. ∎

9.12 Example (A Hard Example). Here is an example that many people find confusing. Let $X_1, \ldots, X_n \sim Unif(0, \theta)$. Recall that

$$f(x; \theta) = \begin{cases} 1/\theta & 0 \le x \le \theta \\ 0 & \text{otherwise.} \end{cases}$$

Consider a fixed value of θ. Suppose $\theta < X_i$ for some i. Then, $f(X_i; \theta) = 0$ and hence $\mathcal{L}_n(\theta) = \prod_i f(X_i; \theta) = 0$. It follows that $\mathcal{L}_n(\theta) = 0$ if any $X_i > \theta$. Therefore, $\mathcal{L}_n(\theta) = 0$ if $\theta < X_{(n)}$ where $X_{(n)} = \max\{X_1, \ldots, X_n\}$. Now consider any $\theta \ge X_{(n)}$. For every X_i we then have that $f(X_i; \theta) = 1/\theta$ so that $\mathcal{L}_n(\theta) = \prod_i f(X_i; \theta) = \theta^{-n}$. In conclusion,

$$\mathcal{L}_n(\theta) = \begin{cases} \left(\frac{1}{\theta}\right)^n & \theta \ge X_{(n)} \\ 0 & \theta < X_{(n)}. \end{cases}$$

See Figure 9.2. Now $\mathcal{L}_n(\theta)$ is strictly decreasing over the interval $[X_{(n)}, \infty)$. Hence, $\widehat{\theta}_n = X_{(n)}$. ∎

The maximum likelihood estimators for the multivariate Normal and the multinomial can be found in Theorems 14.5 and 14.3.

9.4   Properties of Maximum Likelihood Estimators

Under certain conditions on the model, the maximum likelihood estimator $\widehat{\theta}_n$ possesses many properties that make it an appealing choice of estimator. The main properties of the MLE are:

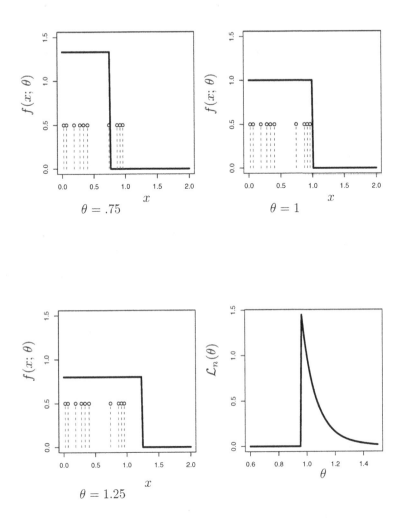

FIGURE 9.2. Likelihood function for Uniform $(0, \theta)$. The vertical lines show the observed data. The first three plots show $f(x; \theta)$ for three different values of θ. When $\theta < X_{(n)} = \max\{X_1, \ldots, X_n\}$, as in the first plot, $f(X_{(n)}; \theta) = 0$ and hence $\mathcal{L}_n(\theta) = \prod_{i=1}^{n} f(X_i; \theta) = 0$. Otherwise $f(X_i; \theta) = 1/\theta$ for each i and hence $\mathcal{L}_n(\theta) = \prod_{i=1}^{n} f(X_i; \theta) = (1/\theta)^n$. The last plot shows the likelihood function.

1. The MLE is **consistent**: $\widehat{\theta}_n \xrightarrow{P} \theta_\star$ where $\theta_\star$ denotes the true value of the parameter θ;

2. The MLE is **equivariant**: if $\widehat{\theta}_n$ is the MLE of θ then $g(\widehat{\theta}_n)$ is the MLE of $g(\theta)$;

3. The MLE is **asymptotically Normal**: $(\widehat{\theta} - \theta_\star)/\widehat{se} \rightsquigarrow N(0,1)$; also, the estimated standard error $\widehat{se}$ can often be computed analytically;

4. The MLE is **asymptotically optimal** or **efficient**: roughly, this means that among all well-behaved estimators, the MLE has the smallest variance, at least for large samples;

5. The MLE is approximately the Bayes estimator. (This point will be explained later.)

We will spend some time explaining what these properties mean and why they are good things. In sufficiently complicated problems, these properties will no longer hold and the MLE will no longer be a good estimator. For now we focus on the simpler situations where the MLE works well. The properties we discuss only hold if the model satisfies certain **regularity conditions**. These are essentially smoothness conditions on $f(x; \theta)$. **Unless otherwise stated, we shall tacitly assume that these conditions hold.**

9.5   Consistency of Maximum Likelihood Estimators

Consistency means that the MLE converges in probability to the true value. To proceed, we need a definition. If f and g are PDF's, define the **Kullback-Leibler distance** [2] between f and g to be

$$D(f,g) = \int f(x) \log\left(\frac{f(x)}{g(x)}\right) dx. \tag{9.6}$$

It can be shown that $D(f,g) \geq 0$ and $D(f,f) = 0$. For any $\theta, \psi \in \Theta$ write $D(\theta, \psi)$ to mean $D(f(x; \theta), f(x; \psi))$.

We will say that the model $\mathfrak{F}$ is **identifiable** if $\theta \neq \psi$ implies that $D(\theta, \psi) > 0$. This means that different values of the parameter correspond to different distributions. We will assume from now on the the model is identifiable.

---

[2] This is not a distance in the formal sense because $D(f,g)$ is not symmetric.

Let $\theta_\star$ denote the true value of θ. Maximizing $\ell_n(\theta)$ is equivalent to maximizing

$$M_n(\theta) = \frac{1}{n} \sum_i \log \frac{f(X_i; \theta)}{f(X_i; \theta_\star)}.$$

This follows since $M_n(\theta) = n^{-1}(\ell_n(\theta) - \ell_n(\theta_\star))$ and $\ell_n(\theta_\star)$ is a constant (with respect to θ). By the law of large numbers, $M_n(\theta)$ converges to

$$\begin{aligned}
\mathbb{E}_{\theta_\star}\left(\log \frac{f(X_i; \theta)}{f(X_i; \theta_\star)}\right) &= \int \log\left(\frac{f(x; \theta)}{f(x; \theta_\star)}\right) f(x; \theta_\star) dx \\
&= -\int \log\left(\frac{f(x; \theta_\star)}{f(x; \theta)}\right) f(x; \theta_\star) dx \\
&= -D(\theta_\star, \theta).
\end{aligned}$$

Hence, $M_n(\theta) \approx -D(\theta_\star, \theta)$ which is maximized at $\theta_\star$ since $-D(\theta_\star, \theta_\star) = 0$ and $-D(\theta_\star, \theta) < 0$ for $\theta \neq \theta_\star$. Therefore, we expect that the maximizer will tend to $\theta_\star$. To prove this formally, we need more than $M_n(\theta) \xrightarrow{P} -D(\theta_\star, \theta)$. We need this convergence to be uniform over θ. We also have to make sure that the function $D(\theta_\star, \theta)$ is well behaved. Here are the formal details.

9.13 Theorem. *Let $\theta_\star$ denote the true value of θ. Define*

$$M_n(\theta) = \frac{1}{n} \sum_i \log \frac{f(X_i; \theta)}{f(X_i; \theta_\star)}$$

and $M(\theta) = -D(\theta_\star, \theta)$. Suppose that

$$\sup_{\theta \in \Theta} |M_n(\theta) - M(\theta)| \xrightarrow{P} 0 \tag{9.7}$$

and that, for every $\epsilon > 0$,

$$\sup_{\theta : |\theta - \theta_\star| \geq \epsilon} M(\theta) < M(\theta_\star). \tag{9.8}$$

Let $\widehat{\theta}_n$ denote the MLE. Then $\widehat{\theta}_n \xrightarrow{P} \theta_\star$.

The proof is in the appendix.

9.6   Equivariance of the MLE

9.14 Theorem. *Let $\tau = g(\theta)$ be a function of θ. Let $\widehat{\theta}_n$ be the MLE of θ. Then $\widehat{\tau}_n = g(\widehat{\theta}_n)$ is the MLE of τ.*

PROOF. Let $h = g^{-1}$ denote the inverse of g. Then $\widehat{\theta}_n = h(\widehat{\tau}_n)$. For any τ, $\mathcal{L}(\tau) = \prod_i f(x_i; h(\tau)) = \prod_i f(x_i; \theta) = \mathcal{L}(\theta)$ where $\theta = h(\tau)$. Hence, for any τ, $\mathcal{L}_n(\tau) = \mathcal{L}(\theta) \leq \mathcal{L}(\widehat{\theta}) = \mathcal{L}_n(\widehat{\tau})$. ∎

9.15 Example. Let $X_1, \ldots, X_n \sim N(\theta, 1)$. The MLE for θ is $\widehat{\theta}_n = \overline{X}_n$. Let $\tau = e^\theta$. Then, the MLE for τ is $\widehat{\tau} = e^{\widehat{\theta}} = e^{\overline{X}}$. ∎

9.7   Asymptotic Normality

It turns out that the distribution of $\widehat{\theta}_n$ is approximately Normal and we can compute its approximate variance analytically. To explore this, we first need a few definitions.

9.16 Definition. *The* **score function** *is defined to be*

$$s(X; \theta) = \frac{\partial \log f(X; \theta)}{\partial \theta}. \tag{9.9}$$

The **Fisher information** *is defined to be*

$$I_n(\theta) = \mathbb{V}_\theta \left(\sum_{i=1}^n s(X_i; \theta) \right)$$

$$= \sum_{i=1}^n \mathbb{V}_\theta \left(s(X_i; \theta) \right). \tag{9.10}$$

For $n = 1$ we will sometimes write $I(\theta)$ instead of $I_1(\theta)$. It can be shown that $\mathbb{E}_\theta(s(X; \theta)) = 0$. It then follows that $\mathbb{V}_\theta(s(X; \theta)) = \mathbb{E}_\theta(s^2(X; \theta))$. In fact, a further simplification of $I_n(\theta)$ is given in the next result.

9.17 Theorem. $I_n(\theta) = nI(\theta)$. Also,

$$I(\theta) = -\mathbb{E}_\theta \left(\frac{\partial^2 \log f(X; \theta)}{\partial \theta^2} \right)$$

$$= -\int \left(\frac{\partial^2 \log f(x; \theta)}{\partial \theta^2} \right) f(x; \theta) dx. \tag{9.11}$$

9.18 Theorem (Asymptotic Normality of the MLE). *Let* $\mathrm{se} = \sqrt{\mathbb{V}(\widehat{\theta}_n)}$. *Under appropriate regularity conditions, the following hold:*

1. $\mathrm{se} \approx \sqrt{1/I_n(\theta)}$ *and*

$$\frac{(\widehat{\theta}_n - \theta)}{\mathrm{se}} \rightsquigarrow N(0,1). \tag{9.12}$$

2. *Let* $\widehat{\mathrm{se}} = \sqrt{1/I_n(\widehat{\theta}_n)}$. *Then,*

$$\frac{(\widehat{\theta}_n - \theta)}{\widehat{\mathrm{se}}} \rightsquigarrow N(0,1). \tag{9.13}$$

The proof is in the appendix. The first statement says that $\widehat{\theta}_n \approx N(\theta, \mathrm{se})$ where the approximate standard error of $\widehat{\theta}_n$ is $\mathrm{se} = \sqrt{1/I_n(\theta)}$. The second statement says that this is still true even if we replace the standard error by its estimated standard error $\widehat{\mathrm{se}} = \sqrt{1/I_n(\widehat{\theta}_n)}$.

Informally, the theorem says that the distribution of the MLE can be approximated with $N(\theta, \widehat{\mathrm{se}}^2)$. From this fact we can construct an (asymptotic) confidence interval.

9.19 Theorem. *Let*

$$C_n = \left(\widehat{\theta}_n - z_{\alpha/2}\,\widehat{\mathrm{se}}, \ \widehat{\theta}_n + z_{\alpha/2}\,\widehat{\mathrm{se}} \right).$$

Then, $\mathbb{P}_\theta(\theta \in C_n) \to 1 - \alpha$ *as* $n \to \infty$.

PROOF. Let Z denote a standard normal random variable. Then,

$$
\begin{aligned}
\mathbb{P}_\theta(\theta \in C_n) &= \mathbb{P}_\theta\left(\widehat{\theta}_n - z_{\alpha/2}\,\widehat{\mathrm{se}} \le \theta \le \widehat{\theta}_n + z_{\alpha/2}\,\widehat{\mathrm{se}} \right) \\
&= \mathbb{P}_\theta\left(-z_{\alpha/2} \le \frac{\widehat{\theta}_n - \theta}{\widehat{\mathrm{se}}} \le z_{\alpha/2} \right) \\
&\to \mathbb{P}(-z_{\alpha/2} < Z < z_{\alpha/2}) = 1 - \alpha. \quad \blacksquare
\end{aligned}
$$

For $\alpha = .05$, $z_{\alpha/2} = 1.96 \approx 2$, so:

$$\widehat{\theta}_n \pm 2\,\widehat{\mathrm{se}} \tag{9.14}$$

is an approximate 95 percent confidence interval.

When you read an opinion poll in the newspaper, you often see a statement like: the poll is accurate to within one point, 95 percent of the time. They are simply giving a 95 percent confidence interval of the form $\widehat{\theta}_n \pm 2\,\widehat{se}$.

9.20 Example. Let $X_1, \ldots, X_n \sim$ Bernoulli(p). The MLE is $\widehat{p}_n = \sum_i X_i/n$ and $f(x;p) = p^x(1-p)^{1-x}$, $\log f(x;p) = x\log p + (1-x)\log(1-p)$,

$$s(X;p) = \frac{X}{p} - \frac{1-X}{1-p},$$

and

$$-s'(X;p) = \frac{X}{p^2} + \frac{1-X}{(1-p)^2}.$$

Thus,

$$I(p) = \mathbb{E}_p(-s'(X;p)) = \frac{p}{p^2} + \frac{(1-p)}{(1-p)^2} = \frac{1}{p(1-p)}.$$

Hence,

$$\widehat{se} = \frac{1}{\sqrt{I_n(\widehat{p}_n)}} = \frac{1}{\sqrt{nI(\widehat{p}_n)}} = \left\{\frac{\widehat{p}(1-\widehat{p})}{n}\right\}^{1/2}.$$

An approximate 95 percent confidence interval is

$$\widehat{p}_n \pm 2\left\{\frac{\widehat{p}_n(1-\widehat{p}_n)}{n}\right\}^{1/2}. \quad \blacksquare$$

9.21 Example. Let $X_1, \ldots, X_n \sim N(\theta, \sigma^2)$ where σ^2 is known. The score function is $s(X;\theta) = (X-\theta)/\sigma^2$ and $s'(X;\theta) = -1/\sigma^2$ so that $I_1(\theta) = 1/\sigma^2$. The MLE is $\widehat{\theta}_n = \overline{X}_n$. According to Theorem 9.18, $\overline{X}_n \approx N(\theta, \sigma^2/n)$. In this case, the Normal approximation is actually exact. $\blacksquare$

9.22 Example. Let $X_1, \ldots, X_n \sim$ Poisson(λ). Then $\widehat{\lambda}_n = \overline{X}_n$ and some calculations show that $I_1(\lambda) = 1/\lambda$, so

$$\widehat{se} = \frac{1}{\sqrt{nI(\widehat{\lambda}_n)}} = \sqrt{\frac{\widehat{\lambda}_n}{n}}.$$

Therefore, an approximate $1-\alpha$ confidence interval for λ is $\widehat{\lambda}_n \pm z_{\alpha/2}\sqrt{\widehat{\lambda}_n/n}$. $\blacksquare$

9.8 Optimality

Suppose that $X_1, \ldots, X_n \sim N(\theta, \sigma^2)$. The MLE is $\widehat{\theta}_n = \overline{X}_n$. Another reasonable estimator of θ is the sample median $\widetilde{\theta}_n$. The MLE satisfies

$$\sqrt{n}(\widehat{\theta}_n - \theta) \rightsquigarrow N(0, \sigma^2).$$

It can be proved that the median satisfies

$$\sqrt{n}(\widetilde{\theta}_n - \theta) \rightsquigarrow N\left(0, \sigma^2 \frac{\pi}{2}\right).$$

This means that the median converges to the right value but has a larger variance than the MLE.

More generally, consider two estimators T_n and U_n and suppose that

$$\sqrt{n}(T_n - \theta) \rightsquigarrow N(0, t^2),$$

and that

$$\sqrt{n}(U_n - \theta) \rightsquigarrow N(0, u^2).$$

We define the asymptotic relative efficiency of U to T by $\mathrm{ARE}(U, T) = t^2/u^2$. In the Normal example, $\mathrm{ARE}(\widetilde{\theta}_n, \widehat{\theta}_n) = 2/\pi = .63$. The interpretation is that if you use the median, you are effectively using only a fraction of the data.

9.23 Theorem. *If $\widehat{\theta}_n$ is the* MLE *and $\widetilde{\theta}_n$ is any other estimator then* [3]

$$\mathrm{ARE}(\widetilde{\theta}_n, \widehat{\theta}_n) \le 1.$$

Thus, the MLE *has the smallest (asymptotic) variance and we say that the* MLE *is* **efficient** *or* **asymptotically optimal.**

This result is predicated upon the assumed model being correct. If the model is wrong, the MLE may no longer be optimal. We will discuss optimality in more generality when we discuss decision theory in Chapter 12.

9.9 The Delta Method

Let $\tau = g(\theta)$ where g is a smooth function. The maximum likelihood estimator of τ is $\widehat{\tau} = g(\widehat{\theta})$. Now we address the following question: what is the distribution of $\widehat{\tau}$?

9.24 Theorem (The Delta Method). *If $\tau = g(\theta)$ where g is differentiable and $g'(\theta) \ne 0$ then*

$$\frac{(\widehat{\tau}_n - \tau)}{\widehat{\mathrm{se}}(\widehat{\tau})} \rightsquigarrow N(0, 1) \qquad (9.15)$$

---

[3] The result is actually more subtle than this but the details are too complicated to consider here.

where $\widehat{\tau}_n = g(\widehat{\theta}_n)$ and

$$\widehat{se}(\widehat{\tau}_n) = |g'(\widehat{\theta})| \widehat{se}(\widehat{\theta}_n) \qquad (9.16)$$

Hence, if

$$C_n = \left(\widehat{\tau}_n - z_{\alpha/2}\, \widehat{se}(\widehat{\tau}_n),\ \widehat{\tau}_n + z_{\alpha/2}\, \widehat{se}(\widehat{\tau}_n) \right) \qquad (9.17)$$

then $\mathbb{P}_\theta(\tau \in C_n) \to 1 - \alpha$ as $n \to \infty$.

9.25 Example. Let $X_1, \ldots, X_n \sim \text{Bernoulli}(p)$ and let $\psi = g(p) = \log(p/(1 - p))$. The Fisher information function is $I(p) = 1/(p(1 - p))$ so the estimated standard error of the MLE $\widehat{p}_n$ is

$$\widehat{se} = \sqrt{\frac{\widehat{p}_n(1 - \widehat{p}_n)}{n}}.$$

The MLE of ψ is $\widehat{\psi} = \log \widehat{p}/(1 - \widehat{p})$. Since, $g'(p) = 1/(p(1 - p))$, according to the delta method

$$\widehat{se}(\widehat{\psi}_n) = |g'(\widehat{p}_n)|\widehat{se}(\widehat{p}_n) = \frac{1}{\sqrt{n\widehat{p}_n(1 - \widehat{p}_n)}}.$$

An approximate 95 percent confidence interval is

$$\widehat{\psi}_n \pm \frac{2}{\sqrt{n\widehat{p}_n(1 - \widehat{p}_n)}}. \quad \blacksquare$$

9.26 Example. Let $X_1, \ldots, X_n \sim N(\mu, \sigma^2)$. Suppose that μ is known, σ is unknown and that we want to estimate $\psi = \log \sigma$. The log-likelihood is $\ell(\sigma) = -n \log \sigma - \frac{1}{2\sigma^2} \sum_i (x_i - \mu)^2$. Differentiate and set equal to 0 and conclude that

$$\widehat{\sigma}_n = \sqrt{\frac{\sum_i (X_i - \mu)^2}{n}}.$$

To get the standard error we need the Fisher information. First,

$$\log f(X; \sigma) = -\log \sigma - \frac{(X - \mu)^2}{2\sigma^2}$$

with second derivative

$$\frac{1}{\sigma^2} - \frac{3(X - \mu)^2}{\sigma^4},$$

and hence

$$I(\sigma) = -\frac{1}{\sigma^2} + \frac{3\sigma^2}{\sigma^4} = \frac{2}{\sigma^2}.$$

Therefore, $\widehat{se} = \widehat{\sigma}_n/\sqrt{2n}$. Let $\psi = g(\sigma) = \log \sigma$. Then, $\widehat{\psi}_n = \log \widehat{\sigma}_n$. Since $g' = 1/\sigma$,

$$\widehat{se}(\widehat{\psi}_n) = \frac{1}{\widehat{\sigma}_n} \frac{\widehat{\sigma}_n}{\sqrt{2n}} = \frac{1}{\sqrt{2n}},$$

and an approximate 95 percent confidence interval is $\widehat{\psi}_n \pm 2/\sqrt{2n}$. ∎

9.10   Multiparameter Models

These ideas can directly be extended to models with several parameters. Let $\theta = (\theta_1, \ldots, \theta_k)$ and let $\widehat{\theta} = (\widehat{\theta}_1, \ldots, \widehat{\theta}_k)$ be the MLE. Let $\ell_n = \sum_{i=1}^{n} \log f(X_i; \theta)$,

$$H_{jj} = \frac{\partial^2 \ell_n}{\partial \theta_j^2} \quad \text{and} \quad H_{jk} = \frac{\partial^2 \ell_n}{\partial \theta_j \partial \theta_k}.$$

Define the **Fisher Information Matrix** by

$$I_n(\theta) = -\begin{bmatrix} \mathbb{E}_\theta(H_{11}) & \mathbb{E}_\theta(H_{12}) & \cdots & \mathbb{E}_\theta(H_{1k}) \\ \mathbb{E}_\theta(H_{21}) & \mathbb{E}_\theta(H_{22}) & \cdots & \mathbb{E}_\theta(H_{2k}) \\ \vdots & \vdots & \vdots & \vdots \\ \mathbb{E}_\theta(H_{k1}) & \mathbb{E}_\theta(H_{k2}) & \cdots & \mathbb{E}_\theta(H_{kk}) \end{bmatrix}. \tag{9.18}$$

Let $J_n(\theta) = I_n^{-1}(\theta)$ be the inverse of I_n.

9.27 Theorem. *Under appropriate regularity conditions,*

$$(\widehat{\theta} - \theta) \approx N(0, J_n).$$

Also, if $\widehat{\theta}_j$ is the jth component of $\widehat{\theta}$, then

$$\frac{(\widehat{\theta}_j - \theta_j)}{\widehat{se}_j} \rightsquigarrow N(0, 1) \tag{9.19}$$

where $\widehat{se}_j^2 = J_n(j, j)$ is the jth diagonal element of J_n. The approximate covariance of $\widehat{\theta}_j$ and $\widehat{\theta}_k$ is $\text{Cov}(\widehat{\theta}_j, \widehat{\theta}_k) \approx J_n(j, k)$.

There is also a multiparameter delta method. Let $\tau = g(\theta_1, \ldots, \theta_k)$ be a function and let

$$\nabla g = \begin{pmatrix} \dfrac{\partial g}{\partial \theta_1} \\ \vdots \\ \dfrac{\partial g}{\partial \theta_k} \end{pmatrix}$$

be the gradient of g.

9.28 Theorem (Multiparameter delta method). *Suppose that ∇g evaluated at $\widehat{\theta}$ is not 0. Let $\widehat{\tau} = g(\widehat{\theta})$. Then*

$$\frac{(\widehat{\tau} - \tau)}{\widehat{\text{se}}(\widehat{\tau})} \leadsto N(0, 1)$$

where

$$\widehat{\text{se}}(\widehat{\tau}) = \sqrt{(\widehat{\nabla}g)^T \widehat{J}_n (\widehat{\nabla}g)}, \tag{9.20}$$

$\widehat{J}_n = J_n(\widehat{\theta}_n)$ *and* $\widehat{\nabla}g$ *is* ∇g *evaluated at* $\theta = \widehat{\theta}$.

9.29 Example. Let $X_1, \ldots, X_n \sim N(\mu, \sigma^2)$. Let $\tau = g(\mu, \sigma) = \sigma/\mu$. In Excercise 8 you will show that

$$I_n(\mu, \sigma) = \begin{bmatrix} \frac{n}{\sigma^2} & 0 \\ 0 & \frac{2n}{\sigma^2} \end{bmatrix}.$$

Hence,

$$J_n = I_n^{-1}(\mu, \sigma) = \frac{1}{n} \begin{bmatrix} \sigma^2 & 0 \\ 0 & \frac{\sigma^2}{2} \end{bmatrix}.$$

The gradient of g is

$$\nabla g = \begin{pmatrix} -\frac{\sigma}{\mu^2} \\ \frac{1}{\mu} \end{pmatrix}.$$

Thus,

$$\widehat{\text{se}}(\widehat{\tau}) = \sqrt{(\widehat{\nabla}g)^T \widehat{J}_n (\widehat{\nabla}g)} = \frac{1}{\sqrt{n}} \sqrt{\frac{1}{\widehat{\mu}^4} + \frac{\widehat{\sigma}^2}{2\widehat{\mu}^2}}. \quad \blacksquare$$

9.11 The Parametric Bootstrap

For parametric models, standard errors and confidence intervals may also be estimated using the bootstrap. There is only one change. In the nonparametric bootstrap, we sampled $X_1^*, \ldots, X_n^*$ from the empirical distribution $\widehat{F}_n$. In the parametric bootstrap we sample instead from $f(x; \widehat{\theta}_n)$. Here, $\widehat{\theta}_n$ could be the MLE or the method of moments estimator.

9.30 Example. Consider example 9.29. To get the bootstrap standard error, simulate $X_1, \ldots, X_n^* \sim N(\widehat{\mu}, \widehat{\sigma}^2)$, compute $\widehat{\mu}^* = n^{-1} \sum_i X_i^*$ and $\widehat{\sigma}^{2*} = n^{-1} \sum_i (X_i^* - \widehat{\mu}^*)^2$. Then compute $\widehat{\tau}^* = g(\widehat{\mu}^*, \widehat{\sigma}^*) = \widehat{\sigma}^*/\widehat{\mu}^*$. Repeating this B times yields bootstrap replications

$$\widehat{\tau}_1^*, \ldots, \widehat{\tau}_B^*$$

and the estimated standard error is

$$\widehat{se}_{\text{boot}} = \sqrt{\frac{\sum_{b=1}^{B}(\widehat{\tau}_b^* - \widehat{\tau})^2}{B}}. \quad \blacksquare$$

The bootstrap is much easier than the delta method. On the other hand, the delta method has the advantage that it gives a closed form expression for the standard error.

9.12 Checking Assumptions

If we assume the data come from a parametric model, then it is a good idea to check that assumption. One possibility is to check the assumptions informally by inspecting plots of the data. For example, if a histogram of the data looks very bimodal, then the assumption of Normality might be questionable. A formal way to test a parametric model is to use a **goodness-of-fit test.** See Section 10.8.

9.13 Appendix

9.13.1 Proofs

PROOF OF THEOREM 9.13. Since $\widehat{\theta}_n$ maximizes $M_n(\theta)$, we have $M_n(\widehat{\theta}_n) \geq M_n(\theta_\star)$. Hence,

$$
\begin{aligned}
M(\theta_\star) - M(\widehat{\theta}_n) &= M_n(\theta_\star) - M(\widehat{\theta}_n) + M(\theta_\star) - M_n(\theta_\star) \\
&\leq M_n(\widehat{\theta}_n) - M(\widehat{\theta}_n) + M(\theta_\star) - M_n(\theta_\star) \\
&\leq \sup_\theta |M_n(\theta) - M(\theta)| + M(\theta_\star) - M_n(\theta_\star) \\
&\xrightarrow{\text{P}} 0
\end{aligned}
$$

where the last line follows from (9.7). It follows that, for any $\delta > 0$,

$$\mathbb{P}\left(M(\widehat{\theta}_n) < M(\theta_\star) - \delta\right) \to 0.$$

Pick any $\epsilon > 0$. By (9.8), there exists $\delta > 0$ such that $|\theta - \theta_\star| \geq \epsilon$ implies that $M(\theta) < M(\theta_\star) - \delta$. Hence,

$$\mathbb{P}(|\widehat{\theta}_n - \theta_\star| > \epsilon) \leq \mathbb{P}\left(M(\widehat{\theta}_n) < M(\theta_\star) - \delta\right) \to 0. \quad \blacksquare$$

Next we want to prove Theorem 9.18. First we need a lemma.

9.31 Lemma. *The score function satisfies*

$$\mathbb{E}_\theta\left[s(X;\theta)\right] = 0.$$

PROOF. Note that $1 = \int f(x;\theta)dx$. Differentiate both sides of this equation to conclude that

$$
\begin{aligned}
0 &= \frac{\partial}{\partial\theta}\int f(x;\theta)dx = \int \frac{\partial}{\partial\theta}f(x;\theta)dx \\
&= \int \frac{\frac{\partial f(x;\theta)}{\partial\theta}}{f(x;\theta)}f(x;\theta)dx = \int \frac{\partial\log f(x;\theta)}{\partial\theta}f(x;\theta)dx \\
&= \int s(x;\theta)f(x;\theta)dx = \mathbb{E}_\theta s(X;\theta). \quad\blacksquare
\end{aligned}
$$

PROOF OF THEOREM 9.18. Let $\ell(\theta) = \log\mathcal{L}(\theta)$. Then,

$$0 = \ell'(\widehat{\theta}) \approx \ell'(\theta) + (\widehat{\theta} - \theta)\ell''(\theta).$$

Rearrange the above equation to get $\widehat{\theta} - \theta = -\ell'(\theta)/\ell''(\theta)$ or, in other words,

$$\sqrt{n}(\widehat{\theta} - \theta) = \frac{\frac{1}{\sqrt{n}}\ell'(\theta)}{-\frac{1}{n}\ell''(\theta)} \equiv \frac{\text{TOP}}{\text{BOTTOM}}.$$

Let $Y_i = \partial\log f(X_i;\theta)/\partial\theta$. Recall that $\mathbb{E}(Y_i) = 0$ from the previous lemma and also $\mathbb{V}(Y_i) = I(\theta)$. Hence,

$$\text{TOP} = n^{-1/2}\sum_i Y_i = \sqrt{n}\,\overline{Y} = \sqrt{n}(\overline{Y} - 0) \rightsquigarrow W \sim N(0, I(\theta))$$

by the central limit theorem. Let $A_i = -\partial^2\log f(X_i;\theta)/\partial\theta^2$. Then $\mathbb{E}(A_i) = I(\theta)$ and

$$\text{BOTTOM} = \overline{A} \xrightarrow{\text{P}} I(\theta)$$

by the law of large numbers. Apply Theorem 5.5 part (e), to conclude that

$$\sqrt{n}(\widehat{\theta} - \theta) \rightsquigarrow \frac{W}{I(\theta)} \stackrel{d}{=} N\left(0, \frac{1}{I(\theta)}\right).$$

Assuming that $I(\theta)$ is a continuous function of θ, it follows that $I(\widehat{\theta}_n) \xrightarrow{\text{P}} I(\theta)$. Now

$$
\begin{aligned}
\frac{\widehat{\theta}_n - \theta}{\widehat{se}} &= \sqrt{n}I^{1/2}(\widehat{\theta}_n)(\widehat{\theta}_n - \theta) \\
&= \left\{\sqrt{n}I^{1/2}(\theta)(\widehat{\theta}_n - \theta)\right\}\sqrt{\frac{I(\widehat{\theta}_n)}{I(\theta)}}.
\end{aligned}
$$

The first term tends in distribution to N(0,1). The second term tends in probability to 1. The result follows from Theorem 5.5 part (e). ∎

OUTLINE OF PROOF OF THEOREM 9.24. Write

$$\widehat{\tau}_n = g(\widehat{\theta}_n) \approx g(\theta) + (\widehat{\theta}_n - \theta)g'(\theta) = \tau + (\widehat{\theta}_n - \theta)g'(\theta).$$

Thus,

$$\sqrt{n}(\widehat{\tau}_n - \tau) \approx \sqrt{n}(\widehat{\theta}_n - \theta)g'(\theta),$$

and hence

$$\frac{\sqrt{nI(\theta)}(\widehat{\tau}_n - \tau)}{g'(\theta)} \approx \sqrt{nI(\theta)}(\widehat{\theta}_n - \theta).$$

Theorem 9.18 tells us that the right-hand side tends in distribution to a N(0,1). Hence,

$$\frac{\sqrt{nI(\theta)}(\widehat{\tau}_n - \tau)}{g'(\theta)} \rightsquigarrow N(0,1)$$

or, in other words,

$$\widehat{\tau}_n \approx N\left(\tau, \mathsf{se}^2(\widehat{\tau}_n)\right),$$

where

$$\mathsf{se}^2(\widehat{\tau}_n) = \frac{(g'(\theta))^2}{nI(\theta)}.$$

The result remains true if we substitute $\widehat{\theta}_n$ for θ by Theorem 5.5 part (e). ∎

9.13.2 Sufficiency

A **statistic** is a function $T(X^n)$ of the data. A sufficient statistic is a statistic that contains all the information in the data. To make this more formal, we need some definitions.

9.32 Definition. *Write $x^n \leftrightarrow y^n$ if $f(x^n; \theta) = c f(y^n; \theta)$ for some constant c that might depend on x^n and y^n but not θ. A statistic $T(x^n)$ is* **sufficient** *if $T(x^n) \leftrightarrow T(y^n)$ implies that $x^n \leftrightarrow y^n$.*

Notice that if $x^n \leftrightarrow y^n$, then the likelihood function based on x^n has the same shape as the likelihood function based on y^n. Roughly speaking, a statistic is sufficient if we can calculate the likelihood function knowing only $T(X^n)$.

9.33 Example. Let $X_1, \ldots, X_n \sim$ Bernoulli(p). Then $\mathcal{L}(p) = p^S(1-p)^{n-S}$ where $S = \sum_i X_i$, so S is sufficient. ∎

9.34 Example. Let $X_1, \ldots, X_n \sim N(\mu, \sigma)$ and let $T = (\overline{X}, S)$. Then

$$f(X^n; \mu, \sigma) = \left(\frac{1}{\sigma\sqrt{2\pi}}\right)^n \exp\left\{-\frac{nS^2}{2\sigma^2}\right\} \exp\left\{-\frac{n(\overline{X} - \mu)^2}{2\sigma^2}\right\}$$

where S^2 is the sample variance. The last expression depends on the data only through T and therefore, $T = (\overline{X}, S)$ is a sufficient statistic. Note that $U = (17\overline{X}, S)$ is also a sufficient statistic. If I tell you the value of U then you can easily figure out T and then compute the likelihood. Sufficient statistics are far from unique. Consider the following statistics for the $N(\mu, \sigma^2)$ model:

$$\begin{aligned}
T_1(X^n) &= (X_1, \ldots, X_n) \\
T_2(X^n) &= (\overline{X}, S) \\
T_3(X^n) &= \overline{X} \\
T_4(X^n) &= (\overline{X}, S, X_3).
\end{aligned}$$

The first statistic is just the whole data set. This is sufficient. The second is also sufficient as we proved above. The third is not sufficient: you can't compute $\mathcal{L}(\mu, \sigma)$ if I only tell you $\overline{X}$. The fourth statistic T_4 is sufficient. The statistics T_1 and T_4 are sufficient but they contain redundant information. Intuitively, there is a sense in which T_2 is a "more concise" sufficient statistic than either T_1 or T_4. We can express this formally by noting that T_2 is a function of T_1 and similarly, T_2 is a function of T_4. For example, $T_2 = g(T_4)$ where $g(a_1, a_2, a_3) = (a_1, a_2)$. ∎

9.35 Definition. *A statistic T is* **minimal sufficient** *if (i) it is sufficient; and (ii) it is a function of every other sufficient statistic.*

9.36 Theorem. *T is minimal sufficient if the following is true:*

$$T(x^n) = T(y^n) \text{ if and only if } x^n \leftrightarrow y^n.$$

A statistic induces a partition on the set of outcomes. We can think of sufficiency in terms of these partitions.

9.37 Example. Let $X_1, X_2 \sim Bernoulli(\theta)$. Let $V = X_1$, $T = \sum_i X_i$ and $U = (T, X_1)$. Here is the set of outcomes and the statistics:

X_1	X_2	V	T	U
0	0	0	0	(0,0)
0	1	0	1	(1,0)
1	0	1	1	(1,1)
1	1	1	2	(2,1)

The partitions induced by these statistics are:

$$V \longrightarrow \{(0,0), (0,1)\}, \ \{(1,0), (1,1)\}$$
$$T \longrightarrow \{(0,0)\}, \ \{(0,1), (1,0)\}, \ \{(1,1)\}$$
$$U \longrightarrow \{(0,0)\}, \ \{(0,1)\}, \ \{(1,0)\}, \ \{(1,1)\}.$$

Then V is not sufficient but T and U are sufficient. T is minimal sufficient; U is not minimal since if $x^n = (1,0)$ and $y^n = (0,1)$, then $x^n \leftrightarrow y^n$ yet $U(x^n) \neq U(y^n)$. The statistic $W = 17\,T$ generates the same partition as T. It is also minimal sufficient. ∎

9.38 Example. For a $N(\mu, \sigma^2)$ model, $T = (\overline{X}, S)$ is a minimal sufficient statistic. For the Bernoulli model, $T = \sum_i X_i$ is a minimal sufficient statistic. For the Poisson model, $T = \sum_i X_i$ is a minimal sufficient statistic. Check that $T = (\sum_i X_i, X_1)$ is sufficient but not minimal sufficient. Check that $T = X_1$ is not sufficient. ∎

I did not give the usual definition of sufficiency. The usual definition is this: T is sufficient if the distribution of X^n given $T(X^n) = t$ does not depend on θ. In other words, T is sufficient if $f(x_1, \ldots, x_n | t; \theta) = h(x_1, \ldots, x_n, t)$ where h is some function that does not depend on θ.

9.39 Example. Two coin flips. Let $X = (X_1, X_2) \sim$ Bernoulli(p). Then $T = X_1 + X_2$ is sufficient. To see this, we need the distribution of (X_1, X_2) given $T = t$. Since T can take 3 possible values, there are 3 conditional distributions to check. They are: (i) the distribution of (X_1, X_2) given $T = 0$:

$$P(X_1 = 0, X_2 = 0 | t = 0) = 1, P(X_1 = 0, X_2 = 1 | t = 0) = 0,$$

$$P(X_1 = 1, X_2 = 0 | t = 0) = 0, P(X_1 = 1, X_2 = 1 | t = 0) = 0;$$

(ii) the distribution of (X_1, X_2) given $T = 1$:

$$P(X_1 = 0, X_2 = 0 | t = 1) = 0, P(X_1 = 0, X_2 = 1 | t = 1) = \frac{1}{2},$$

$$P(X_1 = 1, X_2 = 0 | t = 1) = \frac{1}{2}, P(X_1 = 1, X_2 = 1 | t = 1) = 0; \text{ and}$$

(iii) the distribution of (X_1, X_2) given $T = 2$:

$$P(X_1 = 0, X_2 = 0 | t = 2) = 0, P(X_1 = 0, X_2 = 1 | t = 2) = 0,$$

$$P(X_1 = 1, X_2 = 0 | t = 2) = 0, P(X_1 = 1, X_2 = 1 | t = 2) = 1.$$

None of these depend on the parameter p. Thus, the distribution of $X_1, X_2 | T$ does not depend on θ, so T is sufficient. ∎

9.40 Theorem (Factorization Theorem). *T is sufficient if and only if there are functions $g(t, \theta)$ and $h(x)$ such that $f(x^n; \theta) = g(t(x^n), \theta)h(x^n)$.*

9.41 Example. Return to the two coin flips. Let $t = x_1 + x_2$. Then

$$
\begin{aligned}
f(x_1, x_2; \theta) &= f(x_1; \theta)f(x_2; \theta) \\
&= \theta^{x_1}(1 - \theta)^{1-x_1}\theta^{x_2}(1 - \theta)^{1-x_2} \\
&= g(t, \theta)h(x_1, x_2)
\end{aligned}
$$

where $g(t, \theta) = \theta^t(1 - \theta)^{2-t}$ and $h(x_1, x_2) = 1$. Therefore, $T = X_1 + X_2$ is sufficient. ∎

Now we discuss an implication of sufficiency in point estimation. Let $\widehat{\theta}$ be an estimator of θ. The Rao-Blackwell theorem says that an estimator should only depend on the sufficient statistic, otherwise it can be improved. Let $R(\theta, \widehat{\theta}) = \mathbb{E}_\theta(\theta - \widehat{\theta})^2$ denote the MSE of the estimator.

9.42 Theorem (Rao-Blackwell). *Let $\widehat{\theta}$ be an estimator and let T be a sufficient statistic. Define a new estimator by*

$$
\widetilde{\theta} = \mathbb{E}(\widehat{\theta}|T).
$$

Then, for every θ, $R(\theta, \widetilde{\theta}) \le R(\theta, \widehat{\theta})$.

9.43 Example. Consider flipping a coin twice. Let $\widehat{\theta} = X_1$. This is a well-defined (and unbiased) estimator. But it is not a function of the sufficient statistic $T = X_1 + X_2$. However, note that $\widetilde{\theta} = \mathbb{E}(X_1|T) = (X_1 + X_2)/2$. By the Rao-Blackwell Theorem, $\widetilde{\theta}$ has MSE at least as small as $\widehat{\theta} = X_1$. The same applies with n coin flips. Again define $\widehat{\theta} = X_1$ and $T = \sum_i X_i$. Then $\widetilde{\theta} = \mathbb{E}(X_1|T) = n^{-1}\sum_i X_i$ has improved MSE. ∎

9.13.3 Exponential Families

Most of the parametric models we have studied so far are special cases of a general class of models called exponential families. We say that $\{f(x; \theta) : \theta \in \Theta\}$ is a **one-parameter exponential family** if there are functions $\eta(\theta)$, $B(\theta)$, $T(x)$ and $h(x)$ such that

$$
f(x; \theta) = h(x)e^{\eta(\theta)T(x)-B(\theta)}.
$$

It is easy to see that $T(X)$ is sufficient. We call T the **natural sufficient statistic**.

9.44 Example. Let $X \sim \text{Poisson}(\theta)$. Then

$$f(x; \theta) = \frac{\theta^x e^{-\theta}}{x!} = \frac{1}{x!} e^{x \log \theta - \theta}$$

and hence, this is an exponential family with $\eta(\theta) = \log \theta$, $B(\theta) = \theta$, $T(x) = x$, $h(x) = 1/x!$. ∎

9.45 Example. Let $X \sim \text{Binomial}(n, \theta)$. Then

$$f(x; \theta) = \binom{n}{x} \theta^x (1 - \theta)^{n-x} = \binom{n}{x} \exp\left\{ x \log\left(\frac{\theta}{1 - \theta} \right) + n \log(1 - \theta) \right\}.$$

In this case,

$$\eta(\theta) = \log\left(\frac{\theta}{1 - \theta} \right), B(\theta) = -n \log(\theta)$$

and

$$T(x) = x, h(x) = \binom{n}{x}.$$

∎

We can rewrite an exponential family as

$$f(x; \eta) = h(x) e^{\eta T(x) - A(\eta)}$$

where $\eta = \eta(\theta)$ is called the **natural parameter** and

$$A(\eta) = \log \int h(x) e^{\eta T(x)} dx.$$

For example a Poisson can be written as $f(x; \eta) = e^{\eta x - e^{\eta}}/x!$ where the natural parameter is $\eta = \log \theta$.

Let $X_1, \ldots, X_n$ be IID from an exponential family. Then $f(x^n; \theta)$ is an exponential family:

$$f(x^n; \theta) = h_n(x^n) h_n(x^n) e^{\eta(\theta) T_n(x^n) - B_n(\theta)}$$

where $h_n(x^n) = \prod_i h(x_i)$, $T_n(x^n) = \sum_i T(x_i)$ and $B_n(\theta) = nB(\theta)$. This implies that $\sum_i T(X_i)$ is sufficient.

9.46 Example. Let $X_1, \ldots, X_n \sim \text{Uniform}(0, \theta)$. Then

$$f(x^n; \theta) = \frac{1}{\theta^n} I(x_{(n)} \le \theta)$$

where I is 1 if the term inside the brackets is true and 0 otherwise, and $x_{(n)} = max\{x_1, \ldots, x_n\}$. Thus $T(X^n) = max\{X_1, \ldots, X_n\}$ is sufficient. But since $T(X^n) \ne \sum_i T(X_i)$, this cannot be an exponential family. ∎

9.47 Theorem. *Let X have density in an exponential family. Then,*

$$\mathbb{E}(T(X)) = A'(\eta), \quad \mathbb{V}(T(X)) = A''(\eta).$$

If $\theta = (\theta_1, \ldots, \theta_k)$ is a vector, then we say that $f(x;\theta)$ has exponential family form if

$$f(x;\theta) = h(x) \exp\left\{ \sum_{j=1}^{k} \eta_j(\theta) T_j(x) - B(\theta) \right\}.$$

Again, $T = (T_1, \ldots, T_k)$ is sufficient. An IID sample of size n also has exponential form with sufficient statistic $(\sum_i T_1(X_i), \ldots, \sum_i T_k(X_i))$.

9.48 Example. Consider the normal family with $\theta = (\mu, \sigma)$. Now,

$$f(x;\theta) = \exp\left\{ \frac{\mu}{\sigma^2} x - \frac{x^2}{2\sigma^2} - \frac{1}{2}\left(\frac{\mu^2}{\sigma^2} + \log(2\pi\sigma^2) \right) \right\}.$$

This is exponential with

$$\eta_1(\theta) = \frac{\mu}{\sigma^2}, \quad T_1(x) = x$$

$$\eta_2(\theta) = -\frac{1}{2\sigma^2}, \quad T_2(x) = x^2$$

$$B(\theta) = \frac{1}{2}\left(\frac{\mu^2}{\sigma^2} + \log(2\pi\sigma^2) \right), \quad h(x) = 1.$$

Hence, with n IID samples, $(\sum_i X_i, \sum_i X_i^2)$ is sufficient. ∎

As before we can write an exponential family as

$$f(x;\eta) = h(x) \exp\left\{ T^T(x)\eta - A(\eta) \right\},$$

where $A(\eta) = \log \int h(x) e^{T^T(x)\eta} dx$. It can be shown that

$$\mathbb{E}(T(X)) = \dot{A}(\eta) \quad \mathbb{V}(T(X)) = \ddot{A}(\eta),$$

where the first expression is the vector of partial derivatives and the second is the matrix of second derivatives.

9.13.4 Computing Maximum Likelihood Estimates

In some cases we can find the MLE $\hat{\theta}$ analytically. More often, we need to find the MLE by numerical methods. We will briefly discuss two commonly used methods: (i) Newton-Raphson, and (ii) the EM algorithm. Both are iterative methods that produce a sequence of values $\theta^0, \theta^1, \ldots$ that, under ideal conditions, converge to the MLE $\widehat{\theta}$. In each case, it is helpful to use a good starting value θ^0. Often, the method of moments estimator is a good starting value.

NEWTON-RAPHSON. To motivate Newton-Raphson, let's expand the derivative of the log-likelihood around θ^j:

$$0 = \ell'(\widehat{\theta}) \approx \ell'(\theta^j) + (\widehat{\theta} - \theta^j)\ell''(\theta^j).$$

Solving for $\widehat{\theta}$ gives

$$\widehat{\theta} \approx \theta^j - \frac{\ell'(\theta^j)}{\ell''(\theta^j)}.$$

This suggests the following iterative scheme:

$$\widehat{\theta}^{j+1} = \theta^j - \frac{\ell'(\theta^j)}{\ell''(\theta^j)}.$$

In the multiparameter case, the mle $\widehat{\theta} = (\widehat{\theta}_1, \ldots, \widehat{\theta}_k)$ is a vector and the method becomes

$$\widehat{\theta}^{j+1} = \theta^j - H^{-1}\ell'(\theta^j)$$

where $\ell'(\theta^j)$ is the vector of first derivatives and H is the matrix of second derivatives of the log-likelihood.

THE EM ALGORITHM. The letters EM stand for Expectation-Maximization. The idea is to iterate between taking an expectation then maximizing. Suppose we have data Y whose density $f(y; \theta)$ leads to a log-likelihood that is hard to maximize. But suppose we can find another random variable Z such that $f(y; \theta) = \int f(y, z; \theta)\, dz$ and such that the likelihood based on $f(y, z; \theta)$ is easy to maximize. In other words, the model of interest is the marginal of a model with a simpler likelihood. In this case, we call Y the observed data and Z the hidden (or latent or missing) data. If we could just "fill in" the missing data, we would have an easy problem. Conceptually, the EM algorithm works by filling in the missing data, maximizing the log-likelihood, and iterating.

9.49 Example (Mixture of Normals). Sometimes it is reasonable to assume that the distribution of the data is a mixture of two normals. Think of heights of people being a mixture of men and women's heights. Let $\phi(y; \mu, \sigma)$ denote a normal density with mean μ and standard deviation σ. The density of a mixture of two Normals is

$$f(y; \theta) = (1 - p)\phi(y; \mu_0, \sigma_0) + p\phi(y; \mu_1, \sigma_1).$$

The idea is that an observation is drawn from the first normal with probability p and the second with probability $1-p$. However, we don't know which Normal it was drawn from. The parameters are $\theta = (\mu_0, \sigma_0, \mu_1, \sigma_1, p)$. The likelihood function is

$$\mathcal{L}(\theta) = \prod_{i=1}^{n} \left[(1-p)\phi(y_i; \mu_0, \sigma_0) + p\phi(y_i; \mu_1, \sigma_1) \right].$$

Maximizing this function over the five parameters is hard. Imaging that we were given extra information telling us which of the two normals every observation came from. These "complete" data are of the form $(Y_1, Z_1), \ldots, (Y_n, Z_n)$, where $Z_i = 0$ represents the first normal and $Z_i = 1$ represents the second. Note that $\mathbb{P}(Z_i = 1) = p$. We shall soon see that the likelihood for the complete data $(Y_1, Z_1), \ldots, (Y_n, Z_n)$ is much simpler than the likelihood for the observed data $Y_1, \ldots, Y_n$. ∎

Now we describe the EM algorithm.

The EM Algorithm

(0) Pick a starting value θ^0. Now for $j = 1, 2, \ldots$, repeat steps 1 and 2 below:

(1) (The E-step): Calculate

$$J(\theta|\theta^j) = \mathbb{E}_{\theta^j} \left(\log \frac{f(Y^n, Z^n; \theta)}{f(Y^n, Z^n; \theta^j)} \,\middle|\, Y^n = y^n \right).$$

The expectation is over the missing data Z^n treating θ^i and the observed data Y^n as fixed.

(2) Find θ^{j+1} to maximize $J(\theta|\theta^j)$.

We now show that the EM algorithm always increases the likelihood, that is, $\mathcal{L}(\theta^{j+1}) \geq \mathcal{L}(\theta^j)$. Note that

$$
\begin{aligned}
J(\theta^{j+1}|\theta^j) &= \mathbb{E}_{\theta^j} \left(\log \frac{f(Y^n, Z^n; \theta^{j+1})}{f(Y^n, Z^n; \theta^j)} \,\middle|\, Y^n = y^n \right) \\
&= \log \frac{f(y^n; \theta^{j+1})}{f(y^n; \theta^j)} + \mathbb{E}_{\theta^j} \left(\log \frac{f(Z^n|Y^n; \theta^{j+1})}{f(Z^n|Y^n; \theta^j)} \,\middle|\, Y^n = y^n \right)
\end{aligned}
$$

and hence

$$
\frac{\mathcal{L}(\theta^{j+1})}{\mathcal{L}(\theta^j)} = \log \frac{f(y^n; \theta^{j+1})}{f(y^n; \theta^j)}
$$

$$= J(\theta^{j+1}|\theta^j) - \mathbb{E}_{\theta^j}\left(\log\frac{f(Z^n|Y^n;\theta^{j+1})}{f(Z^n|Y^n;\theta^j)}\,\bigg|\,Y^n = y^n\right)$$

$$= J(\theta^{j+1}|\theta^j) + K(f_j, f_{j+1})$$

where $f_j = f(y^n;\theta^j)$ and $f_{j+1} = f(y^n;\theta^{j+1})$ and $K(f,g) = \int f(x)\log(f(x)/g(x))\,dx$ is the Kullback-Leibler distance. Now, θ^{j+1} was chosen to maximize $J(\theta|\theta^j)$. Hence, $J(\theta^{j+1}|\theta^j) \geq J(\theta^j|\theta^j) = 0$. Also, by the properties of Kullback-Leibler divergence, $K(f_j, f_{j+1}) \geq 0$. Hence, $\mathcal{L}(\theta^{j+1}) \geq \mathcal{L}(\theta^j)$ as claimed.

9.50 Example (Continuation of Example 9.49). Consider again the mixture of two normals but, for simplicity assume that $p = 1/2$, $\sigma_1 = \sigma_2 = 1$. The density is

$$f(y;\mu_1,\mu_2) = \frac{1}{2}\phi(y;\mu_0,1) + \frac{1}{2}\phi(y;\mu_1,1).$$

Directly maximizing the likelihood is hard. Introduce latent variables $Z_1,\ldots,Z_n$ where $Z_i = 0$ if Y_i is from $\phi(y;\mu_0,1)$, and $Z_i = 1$ if Y_i is from $\phi(y;\mu_1,1)$, $\mathbb{P}(Z_i = 1) = P(Z_i = 0) = 1/2$, $f(y_i|Z_i = 0) = \phi(y;\mu_0,1)$ and $f(y_i|Z_i = 1) = \phi(y;\mu_1,1)$. So $f(y) = \sum_{z=0}^{1} f(y,z)$ where we have dropped the parameters from the density to avoid notational overload. We can write

$$f(z,y) = f(z)f(y|z) = \frac{1}{2}\phi(y;\mu_0,1)^{1-z}\phi(y;\mu_1,1)^z.$$

Hence, the complete likelihood is

$$\prod_{i=1}^{n}\phi(y_i;\mu_0,1)^{1-z_i}\phi(y_i;\mu_1,1)^{z_i}.$$

The complete log-likelihood is then

$$\widetilde{\ell} = -\frac{1}{2}\sum_{i=1}^{n}(1-z_i)(y_i-\mu_0) - \frac{1}{2}\sum_{i=1}^{n}z_i(y_i-\mu_1).$$

And so

$$J(\theta|\theta^j) = -\frac{1}{2}\sum_{i=1}^{n}(1-\mathbb{E}(Z_i|y^n,\theta^j))(y_i-\mu_0) - \frac{1}{2}\sum_{i=1}^{n}\mathbb{E}(Z_i|y^n,\theta^j))(y_i-\mu_1).$$

Since Z_i is binary, $\mathbb{E}(Z_i|y^n,\theta^j) = \mathbb{P}(Z_i = 1|y^n,\theta^j)$ and, by Bayes' theorem,

$$\mathbb{P}(Z_i = 1|y^n,\theta^i) = \frac{f(y^n|Z_i = 1;\theta^j)\mathbb{P}(Z_i = 1)}{f(y^n|Z_i = 1;\theta^j)\mathbb{P}(Z_i = 1) + f(y^n|Z_i = 0;\theta^j)\mathbb{P}(Z_i = 0)}$$

$$= \frac{\phi(y_i;\mu_1^j,1)\frac{1}{2}}{\phi(y_i;\mu_1^j,1)\frac{1}{2} + \phi(y_i;\mu_0^j,1)\frac{1}{2}}$$

$$= \frac{\phi(y_i;\mu_1^j,1)}{\phi(y_i;\mu_1^j,1) + \phi(y_i;\mu_0^j,1)}$$

$$= \tau(i).$$

Take the derivative of $J(\theta|\theta^j)$ with respect to μ_1 and μ_2, set them equal to 0 to get

$$\widehat{\mu}_1^{j+1} = \frac{\sum_{i=1}^n \tau_i y_i}{\sum_{i=1}^n \tau_i}$$

and

$$\widehat{\mu}_0^{j+1} = \frac{\sum_{i=1}^n (1 - \tau_i) y_i}{\sum_{i=1}^n (1 - \tau_i)}.$$

We then recompute τ_i using $\widehat{\mu}_1^{j+1}$ and $\widehat{\mu}_0^{j+1}$ and iterate. ∎

9.14 Exercises

1. Let $X_1, \ldots, X_n \sim$ Gamma(α, β). Find the method of moments estimator for α and β.

2. Let $X_1, \ldots, X_n \sim$ Uniform(a, b) where a and b are unknown parameters and $a < b$.

   (a) Find the method of moments estimators for a and b.

   (b) Find the MLE $\widehat{a}$ and $\widehat{b}$.

   (c) Let $\tau = \int x \, dF(x)$. Find the MLE of τ.

   (d) Let $\widehat{\tau}$ be the MLE of τ. Let $\widetilde{\tau}$ be the nonparametric plug-in estimator of $\tau = \int x \, dF(x)$. Suppose that $a = 1$, $b = 3$, and $n = 10$. Find the MSE of $\widehat{\tau}$ by simulation. Find the MSE of $\widetilde{\tau}$ analytically. Compare.

3. Let $X_1, \ldots, X_n \sim N(\mu, \sigma^2)$. Let τ be the .95 percentile, i.e. $\mathbb{P}(X < \tau) = .95$.

   (a) Find the MLE of τ.

   (b) Find an expression for an approximate $1 - \alpha$ confidence interval for τ.

   (c) Suppose the data are:

   ```
   3.23 -2.50  1.88 -0.68  4.43  0.17
   1.03 -0.07 -0.01  0.76  1.76  3.18
   0.33 -0.31  0.30 -0.61  1.52  5.43
   1.54  2.28  0.42  2.33 -1.03  4.00
   0.39
   ```

   Find the MLE $\widehat{\tau}$. Find the standard error using the delta method. Find the standard error using the parametric bootstrap.

4. Let $X_1, \ldots, X_n \sim \text{Uniform}(0, \theta)$. Show that the MLE is consistent. Hint: Let $Y = \max\{X_1, \ldots, X_n\}$. For any c, $\mathbb{P}(Y < c) = \mathbb{P}(X_1 < c, X_2 < c, \ldots, X_n < c) = \mathbb{P}(X_1 < c)\mathbb{P}(X_2 < c)\ldots\mathbb{P}(X_n < c)$.

5. Let $X_1, \ldots, X_n \sim \text{Poisson}(\lambda)$. Find the method of moments estimator, the maximum likelihood estimator and the Fisher information $I(\lambda)$.

6. Let $X_1, \ldots, X_n \sim N(\theta, 1)$. Define

$$Y_i = \begin{cases} 1 & \text{if } X_i > 0 \\ 0 & \text{if } X_i \le 0. \end{cases}$$

Let $\psi = \mathbb{P}(Y_1 = 1)$.

(a) Find the maximum likelihood estimator $\widehat{\psi}$ of ψ.

(b) Find an approximate 95 percent confidence interval for ψ.

(c) Define $\widetilde{\psi} = (1/n) \sum_i Y_i$. Show that $\widetilde{\psi}$ is a consistent estimator of ψ.

(d) Compute the asymptotic relative efficiency of $\widetilde{\psi}$ to $\widehat{\psi}$. Hint: Use the delta method to get the standard error of the MLE. Then compute the standard error (i.e. the standard deviation) of $\widetilde{\psi}$.

(e) Suppose that the data are not really normal. Show that $\widehat{\psi}$ is not consistent. What, if anything, does $\widehat{\psi}$ converge to?

7. (Comparing two treatments.) n_1 people are given treatment 1 and n_2 people are given treatment 2. Let X_1 be the number of people on treatment 1 who respond favorably to the treatment and let X_2 be the number of people on treatment 2 who respond favorably. Assume that $X_1 \sim \text{Binomial}(n_1, p_1)$ $X_2 \sim \text{Binomial}(n_2, p_2)$. Let $\psi = p_1 - p_2$.

(a) Find the MLE $\widehat{\psi}$ for ψ.

(b) Find the Fisher information matrix $I(p_1, p_2)$.

(c) Use the multiparameter delta method to find the asymptotic standard error of $\widehat{\psi}$.

(d) Suppose that $n_1 = n_2 = 200$, $X_1 = 160$ and $X_2 = 148$. Find $\widehat{\psi}$. Find an approximate 90 percent confidence interval for ψ using (i) the delta method and (ii) the parametric bootstrap.

8. Find the Fisher information matrix for Example 9.29.

9. Let $X_1, \ldots, X_n \sim \text{Normal}(\mu, 1)$. Let $\theta = e^\mu$ and let $\widehat{\theta} = e^{\overline{X}}$ be the MLE. Create a data set (using $\mu = 5$) consisting of $n=100$ observations.

(a) Use the delta method to get $\widehat{se}$ and a 95 percent confidence interval for θ. Use the parametric bootstrap to get $\widehat{se}$ and 95 percent confidence interval for θ. Use the nonparametric bootstrap to get $\widehat{se}$ and 95 percent confidence interval for θ. Compare your answers.

(b) Plot a histogram of the bootstrap replications for the parametric and nonparametric bootstraps. These are estimates of the distribution of $\widehat{\theta}$. The delta method also gives an approximation to this distribution namely, Normal$(\widehat{\theta}, \widehat{se}^2)$. Compare these to the true sampling distribution of $\widehat{\theta}$ (which you can get by simulation). Which approximation — parametric bootstrap, bootstrap, or delta method — is closer to the true distribution?

10. Let $X_1, ..., X_n \sim$ Uniform$(0, \theta)$. The MLE is $\widehat{\theta} = X_{(n)} = \max\{X_1, ..., X_n\}$. Generate a dataset of size 50 with $\theta = 1$.

(a) Find the distribution of $\widehat{\theta}$ analytically. Compare the true distribution of $\widehat{\theta}$ to the histograms from the parametric and nonparametric bootstraps.

(b) This is a case where the nonparametric bootstrap does very poorly. Show that for the parametric bootstrap $\mathbb{P}(\widehat{\theta}^* = \widehat{\theta}) = 0$, but for the nonparametric bootstrap $\mathbb{P}(\widehat{\theta}^* = \widehat{\theta}) \approx .632$. Hint: show that, $\mathbb{P}(\widehat{\theta}^* = \widehat{\theta}) = 1 - (1 - (1/n))^n$ then take the limit as n gets large. What is the implication of this?

10
Hypothesis Testing and p-values

Suppose we want to know if exposure to asbestos is associated with lung disease. We take some rats and randomly divide them into two groups. We expose one group to asbestos and leave the second group unexposed. Then we compare the disease rate in the two groups. Consider the following two hypotheses:

The Null Hypothesis: The disease rate is the same in the two groups.

The Alternative Hypothesis: The disease rate is not the same in the two groups.

If the exposed group has a much higher rate of disease than the unexposed group then we will reject the null hypothesis and conclude that the evidence favors the alternative hypothesis. This is an example of hypothesis testing.

More formally, suppose that we partition the parameter space Θ into two disjoint sets Θ_0 and Θ_1 and that we wish to test

$$H_0 : \theta \in \Theta_0 \quad \text{versus} \quad H_1 : \theta \in \Theta_1. \tag{10.1}$$

We call H_0 the **null hypothesis** and H_1 the **alternative hypothesis**.

Let X be a random variable and let $\mathcal{X}$ be the range of X. We test a hypothesis by finding an appropriate subset of outcomes $R \subset \mathcal{X}$ called the **rejection**

	Retain Null	Reject Null
H_0 true	$\checkmark$	type I error
H_1 true	type II error	$\checkmark$

TABLE 10.1. Summary of outcomes of hypothesis testing.

region. If $X \in R$ we reject the null hypothesis, otherwise, we do not reject the null hypothesis:

$$X \in R \implies \text{reject } H_0$$
$$X \notin R \implies \text{retain (do not reject) } H_0$$

Usually, the rejection region R is of the form

$$R = \left\{ x : T(x) > c \right\} \tag{10.2}$$

where T is a **test statistic** and c is a **critical value**. The problem in hypothesis testing is to find an appropriate test statistic T and an appropriate critical value c.

Warning! There is a tendency to use hypothesis testing methods even when they are not appropriate. Often, estimation and confidence intervals are better tools. Use hypothesis testing only when you want to test a well-defined hypothesis.

Hypothesis testing is like a legal trial. We assume someone is innocent unless the evidence strongly suggests that he is guilty. Similarly, we retain H_0 unless there is strong evidence to reject H_0. There are two types of errors we can make. Rejecting H_0 when H_0 is true is called a **type I error**. Retaining H_0 when H_1 is true is called a **type II error**. The possible outcomes for hypothesis testing are summarized in Tab. 10.1.

10.1 Definition. *The* **power function** *of a test with rejection region R is defined by*

$$\beta(\theta) = \mathbb{P}_\theta(X \in R). \tag{10.3}$$

The **size** *of a test is defined to be*

$$\alpha = \sup_{\theta \in \Theta_0} \beta(\theta). \tag{10.4}$$

A test is said to have **level** α *if its size is less than or equal to α.*

A hypothesis of the form $\theta = \theta_0$ is called a **simple hypothesis**. A hypothesis of the form $\theta > \theta_0$ or $\theta < \theta_0$ is called a **composite hypothesis**. A test of the form

$$H_0 : \theta = \theta_0 \quad \text{versus} \quad H_1 : \theta \neq \theta_0$$

is called a **two-sided test**. A test of the form

$$H_0 : \theta \leq \theta_0 \quad \text{versus} \quad H_1 : \theta > \theta_0$$

or

$$H_0 : \theta \geq \theta_0 \quad \text{versus} \quad H_1 : \theta < \theta_0$$

is called a **one-sided test**. The most common tests are two-sided.

10.2 Example. Let $X_1, \ldots, X_n \sim N(\mu, \sigma)$ where σ is known. We want to test $H_0 : \mu \leq 0$ versus $H_1 : \mu > 0$. Hence, $\Theta_0 = (-\infty, 0]$ and $\Theta_1 = (0, \infty)$. Consider the test:

$$\text{reject } H_0 \text{ if } T > c$$

where $T = \overline{X}$. The rejection region is

$$R = \left\{ (x_1, \ldots, x_n) : T(x_1, \ldots, x_n) > c \right\}.$$

Let Z denote a standard Normal random variable. The power function is

$$
\begin{aligned}
\beta(\mu) &= \mathbb{P}_\mu \left(\overline{X} > c \right) \\
&= \mathbb{P}_\mu \left(\frac{\sqrt{n}(\overline{X} - \mu)}{\sigma} > \frac{\sqrt{n}(c - \mu)}{\sigma} \right) \\
&= \mathbb{P} \left(Z > \frac{\sqrt{n}(c - \mu)}{\sigma} \right) \\
&= 1 - \Phi \left(\frac{\sqrt{n}(c - \mu)}{\sigma} \right).
\end{aligned}
$$

This function is increasing in μ. See Figure 10.1. Hence

$$\text{size} = \sup_{\mu \leq 0} \beta(\mu) = \beta(0) = 1 - \Phi \left(\frac{\sqrt{n}\,c}{\sigma} \right).$$

For a size α test, we set this equal to α and solve for c to get

$$c = \frac{\sigma\, \Phi^{-1}(1 - \alpha)}{\sqrt{n}}.$$

We reject when $\overline{X} > \sigma\, \Phi^{-1}(1 - \alpha)/\sqrt{n}$. Equivalently, we reject when

$$\frac{\sqrt{n}\,(\overline{X} - 0)}{\sigma} > z_\alpha.$$

where $z_\alpha = \Phi^{-1}(1 - \alpha)$. ∎

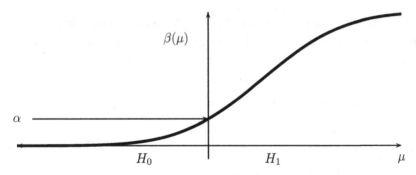

FIGURE 10.1. The power function for Example 10.2. The size of the test is the largest probability of rejecting H_0 when H_0 is true. This occurs at $\mu = 0$ hence the size is $\beta(0)$. We choose the critical value c so that $\beta(0) = \alpha$.

It would be desirable to find the test with highest power under H_1, among all size α tests. Such a test, if it exists, is called **most powerful**. Finding most powerful tests is hard and, in many cases, most powerful tests don't even exist. Instead of going into detail about when most powerful tests exist, we'll just consider four widely used tests: the Wald test,[1] the χ^2 test, the permutation test, and the likelihood ratio test.

10.1   The Wald Test

Let θ be a scalar parameter, let $\widehat{\theta}$ be an estimate of θ and let $\widehat{se}$ be the estimated standard error of $\widehat{\theta}$.

---

[1] The test is named after Abraham Wald (1902–1950), who was a very influential mathematical statistician. Wald died in a plane crash in India in 1950.

10.3 Definition. The Wald Test

Consider testing

$$H_0 : \theta = \theta_0 \quad \text{versus} \quad H_1 : \theta \neq \theta_0.$$

Assume that $\widehat{\theta}$ is asymptotically Normal:

$$\frac{(\widehat{\theta} - \theta_0)}{\widehat{\text{se}}} \rightsquigarrow N(0, 1).$$

The size α Wald test is: reject H_0 when $|W| > z_{\alpha/2}$ where

$$W = \frac{\widehat{\theta} - \theta_0}{\widehat{\text{se}}}. \tag{10.5}$$

10.4 Theorem. *Asymptotically, the Wald test has size α, that is,*

$$\mathbb{P}_{\theta_0} \left(|W| > z_{\alpha/2} \right) \to \alpha$$

as $n \to \infty$.

PROOF. Under $\theta = \theta_0$, $(\widehat{\theta} - \theta_0)/\widehat{\text{se}} \rightsquigarrow N(0, 1)$. Hence, the probability of rejecting when the null $\theta = \theta_0$ is true is

$$
\begin{aligned}
\mathbb{P}_{\theta_0} \left(|W| > z_{\alpha/2} \right) &= \mathbb{P}_{\theta_0} \left(\frac{|\widehat{\theta} - \theta_0|}{\widehat{\text{se}}} > z_{\alpha/2} \right) \\
&\to \mathbb{P} \left(|Z| > z_{\alpha/2} \right) \\
&= \alpha
\end{aligned}
$$

where $Z \sim N(0, 1)$. ∎

10.5 Remark. An alternative version of the Wald test statistic is $W = (\widehat{\theta} - \theta_0)/\text{se}_0$ where se_0 is the standard error computed at $\theta = \theta_0$. Both versions of the test are valid.

Let us consider the power of the Wald test when the null hypothesis is false.

10.6 Theorem. *Suppose the true value of θ is $\theta_\star \neq \theta_0$. The power $\beta(\theta_\star)$ — the probability of correctly rejecting the null hypothesis — is given (approximately) by*

$$1 - \Phi \left(\frac{\theta_0 - \theta_\star}{\widehat{\text{se}}} + z_{\alpha/2} \right) + \Phi \left(\frac{\theta_0 - \theta_\star}{\widehat{\text{se}}} - z_{\alpha/2} \right). \tag{10.6}$$

Recall that $\widehat{se}$ tends to 0 as the sample size increases. Inspecting (10.6) closely we note that: (i) the power is large if θ_* is far from θ_0, and (ii) the power is large if the sample size is large.

10.7 Example (Comparing Two Prediction Algorithms). We test a prediction algorithm on a test set of size m and we test a second prediction algorithm on a second test set of size n. Let X be the number of incorrect predictions for algorithm 1 and let Y be the number of incorrect predictions for algorithm 2. Then $X \sim \text{Binomial}(m, p_1)$ and $Y \sim \text{Binomial}(n, p_2)$. To test the null hypothesis that $p_1 = p_2$ write

$$H_0 : \delta = 0 \quad \text{versus} \quad H_1 : \delta \neq 0$$

where $\delta = p_1 - p_2$. The MLE is $\widehat{\delta} = \widehat{p}_1 - \widehat{p}_2$ with estimated standard error

$$\widehat{se} = \sqrt{\frac{\widehat{p}_1(1 - \widehat{p}_1)}{m} + \frac{\widehat{p}_2(1 - \widehat{p}_2)}{n}}.$$

The size α Wald test is to reject H_0 when $|W| > z_{\alpha/2}$ where

$$W = \frac{\widehat{\delta} - 0}{\widehat{se}} = \frac{\widehat{p}_1 - \widehat{p}_2}{\sqrt{\frac{\widehat{p}_1(1 - \widehat{p}_1)}{m} + \frac{\widehat{p}_2(1 - \widehat{p}_2)}{n}}}.$$

The power of this test will be largest when p_1 is far from p_2 and when the sample sizes are large.

What if we used the same test set to test both algorithms? The two samples are no longer independent. Instead we use the following strategy. Let $X_i = 1$ if algorithm 1 is correct on test case i and $X_i = 0$ otherwise. Let $Y_i = 1$ if algorithm 2 is correct on test case i, and $Y_i = 0$ otherwise. Define $D_i = X_i - Y_i$. A typical dataset will look something like this:

Test Case	X_i	Y_i	$D_i = X_i - Y_i$
1	1	0	1
2	1	1	0
3	1	1	0
4	0	1	-1
5	0	0	0
$\vdots$	$\vdots$	$\vdots$	$\vdots$
n	0	1	-1

Let

$$\delta = \mathbb{E}(D_i) = \mathbb{E}(X_i) - \mathbb{E}(Y_i) = \mathbb{P}(X_i = 1) - \mathbb{P}(Y_i = 1).$$

The nonparametric plug-in estimate of δ is $\widehat{\delta} = \overline{D} = n^{-1} \sum_{i=1}^{n} D_i$ and $\widehat{se}(\widehat{\delta}) = S/\sqrt{n}$, where $S^2 = n^{-1} \sum_{i=1}^{n} (D_i - \overline{D})^2$. To test $H_0 : \delta = 0$ versus $H_1 : \delta \neq 0$

we use $W = \hat{\delta}/\hat{se}$ and reject H_0 if $|W| > z_{\alpha/2}$. This is called a **paired comparison.** ∎

10.8 Example (Comparing Two Means). Let $X_1, \ldots, X_m$ and $Y_1, \ldots, Y_n$ be two independent samples from populations with means μ_1 and μ_2, respectively. Let's test the null hypothesis that $\mu_1 = \mu_2$. Write this as $H_0 : \delta = 0$ versus $H_1 : \delta \neq 0$ where $\delta = \mu_1 - \mu_2$. Recall that the nonparametric plug-in estimate of δ is $\hat{\delta} = \overline{X} - \overline{Y}$ with estimated standard error

$$\hat{se} = \sqrt{\frac{s_1^2}{m} + \frac{s_2^2}{n}}$$

where s_1^2 and s_2^2 are the sample variances. The size α Wald test rejects H_0 when $|W| > z_{\alpha/2}$ where

$$W = \frac{\hat{\delta} - 0}{\hat{se}} = \frac{\overline{X} - \overline{Y}}{\sqrt{\frac{s_1^2}{m} + \frac{s_2^2}{n}}}. \quad ∎$$

10.9 Example (Comparing Two Medians). Consider the previous example again but let us test whether the medians of the two distributions are the same. Thus, $H_0 : \delta = 0$ versus $H_1 : \delta \neq 0$ where $\delta = \nu_1 - \nu_2$ where ν_1 and ν_2 are the medians. The nonparametric plug-in estimate of δ is $\hat{\delta} = \hat{\nu}_1 - \hat{\nu}_2$ where $\hat{\nu}_1$ and $\hat{\nu}_2$ are the sample medians. The estimated standard error $\hat{se}$ of $\hat{\delta}$ can be obtained from the bootstrap. The Wald test statistic is $W = \hat{\delta}/\hat{se}$. ∎

There is a relationship between the Wald test and the $1 - \alpha$ asymptotic confidence interval $\hat{\theta} \pm \hat{se}\, z_{\alpha/2}$.

10.10 Theorem. *The size α Wald test rejects $H_0 : \theta = \theta_0$ versus $H_1 : \theta \neq \theta_0$ if and only if $\theta_0 \notin C$ where*

$$C = (\hat{\theta} - \hat{se}\, z_{\alpha/2}, \ \hat{\theta} + \hat{se}\, z_{\alpha/2}).$$

Thus, testing the hypothesis is equivalent to checking whether the null value is in the confidence interval.

Warning! When we reject H_0 we often say that the result is **statistically significant.** A result might be statistically significant and yet the size of the effect might be small. In such a case we have a result that is statistically significant but not scientifically or practically significant. The difference between statistical significance and scientific significance is easy to understand in light of Theorem 10.10. Any confidence interval that excludes θ_0 corresponds to rejecting H_0. But the values in the interval could be close to θ_0 (not scientifically significant) or far from θ_0 (scientifically significant). See Figure 10.2.

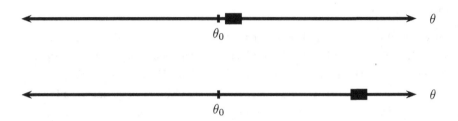

FIGURE 10.2. Scientific significance versus statistical significance. A level α test rejects $H_0 : \theta = \theta_0$ if and only if the $1 - \alpha$ confidence interval does not include θ_0. Here are two different confidence intervals. Both exclude θ_0 so in both cases the test would reject H_0. But in the first case, the estimated value of θ is close to θ_0 so the finding is probably of little scientific or practical value. In the second case, the estimated value of θ is far from θ_0 so the finding is of scientific value. This shows two things. First, statistical significance does not imply that a finding is of scientific importance. Second, confidence intervals are often more informative than tests.

10.2   p-values

Reporting "reject H_0" or "retain H_0" is not very informative. Instead, we could ask, for every α, whether the test rejects at that level. Generally, if the test rejects at level α it will also reject at level $\alpha' > \alpha$. Hence, there is a smallest α at which the test rejects and we call this number the p-value. See Figure 10.3.

10.11 Definition. *Suppose that for every $\alpha \in (0,1)$ we have a size α test with rejection region R_α. Then,*

$$\text{p-value} = \inf\left\{\alpha : \ T(X^n) \in R_\alpha\right\}.$$

That is, the p-value is the smallest level at which we can reject H_0.

Informally, the p-value is a measure of the evidence against H_0: the smaller the p-value, the stronger the evidence against H_0. Typically, researchers use the following evidence scale:

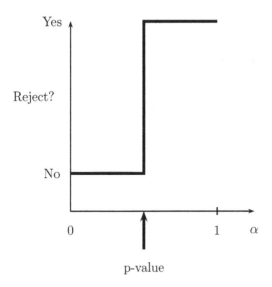

FIGURE 10.3. p-values explained. For each α we can ask: does our test reject H_0 at level α? The p-value is the smallest α at which we do reject H_0. If the evidence against H_0 is strong, the p-value will be small.

p-value	evidence
$< .01$	very strong evidence against H_0
$.01 - .05$	strong evidence against H_0
$.05 - .10$	weak evidence against H_0
$> .1$	little or no evidence against H_0

Warning! A large p-value is not strong evidence in favor of H_0. A large p-value can occur for two reasons: (i) H_0 is true or (ii) H_0 is false but the test has low power.

Warning! Do not confuse the p-value with $\mathbb{P}(H_0|\text{Data})$. [2] **The p-value is not the probability that the null hypothesis is true.**

The following result explains how to compute the p-value.

---

[2]We discuss quantities like $\mathbb{P}(H_0|\text{Data})$ in the chapter on Bayesian inference.

10.12 Theorem. *Suppose that the size α test is of the form*

$$\text{reject } H_0 \text{ if and only if } T(X^n) \geq c_\alpha.$$

Then,

$$\text{p-value} = \sup_{\theta \in \Theta_0} \mathbb{P}_\theta(T(X^n) \geq T(x^n))$$

where x^n is the observed value of X^n. If $\Theta_0 = \{\theta_0\}$ then

$$\text{p-value} = \mathbb{P}_{\theta_0}(T(X^n) \geq T(x^n)).$$

We can express Theorem 10.12 as follows:

The p-value is the probability (under H_0) of observing a value of the test statistic the same as or more extreme than what was actually observed.

10.13 Theorem. *Let $w = (\hat{\theta} - \theta_0)/\hat{se}$ denote the observed value of the Wald statistic W. The p-value is given by*

$$\text{p} - \text{value} = \mathbb{P}_{\theta_0}(|W| > |w|) \approx \mathbb{P}(|Z| > |w|) = 2\Phi(-|w|) \qquad (10.7)$$

where $Z \sim N(0,1)$.

To understand this last theorem, look at Figure 10.4.

Here is an important property of p-values.

10.14 Theorem. *If the test statistic has a continuous distribution, then under $H_0 : \theta = \theta_0$, the p-value has a Uniform $(0,1)$ distribution. Therefore, if we reject H_0 when the p-value is less than α, the probability of a type I error is α.*

In other words, if H_0 is true, the p-value is like a random draw from a Unif$(0,1)$ distribution. If H_1 is true, the distribution of the p-value will tend to concentrate closer to 0.

10.15 Example. Recall the cholesterol data from Example 7.15. To test if the means are different we compute

$$W = \frac{\hat{\delta} - 0}{\hat{se}} = \frac{\overline{X} - \overline{Y}}{\sqrt{\frac{s_1^2}{m} + \frac{s_2^2}{n}}} = \frac{216.2 - 195.3}{\sqrt{5^2 + 2.4^2}} = 3.78.$$

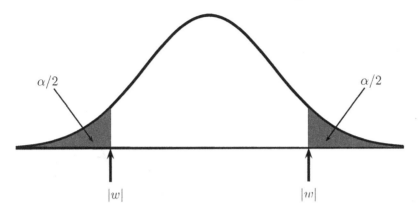

FIGURE 10.4. The p-value is the smallest α at which you would reject H_0. To find the p-value for the Wald test, we find α such that $|w|$ and $-|w|$ are just at the boundary of the rejection region. Here, w is the observed value of the Wald statistic: $w = (\hat{\theta} - \theta_0)/\hat{se}$. This implies that the p-value is the tail area $\mathbb{P}(|Z| > |w|)$ where $Z \sim N(0,1)$.

To compute the p-value, let $Z \sim N(0,1)$ denote a standard Normal random variable. Then,

$$\text{p-value} = \mathbb{P}(|Z| > 3.78) = 2\mathbb{P}(Z < -3.78) = .0002$$

which is very strong evidence against the null hypothesis. To test if the medians are different, let $\hat{\nu}_1$ and $\hat{\nu}_2$ denote the sample medians. Then,

$$W = \frac{\hat{\nu}_1 - \hat{\nu}_2}{\hat{se}} = \frac{212.5 - 194}{7.7} = 2.4$$

where the standard error 7.7 was found using the bootstrap. The p-value is

$$\text{p-value} = \mathbb{P}(|Z| > 2.4) = 2\mathbb{P}(Z < -2.4) = .02$$

which is strong evidence against the null hypothesis. ∎

10.3   The χ^2 Distribution

Before proceeding we need to discuss the χ^2 distribution. Let $Z_1, \ldots, Z_k$ be independent, standard Normals. Let $V = \sum_{i=1}^{k} Z_i^2$. Then we say that V has a χ^2 distribution with k degrees of freedom, written $V \sim \chi_k^2$. The probability density of V is

$$f(v) = \frac{v^{(k/2)-1}e^{-v/2}}{2^{k/2}\Gamma(k/2)}$$

for $v > 0$. It can be shown that $\mathbb{E}(V) = k$ and $\mathbb{V}(V) = 2k$. We define the upper α quantile $\chi_{k,\alpha}^2 = F^{-1}(1-\alpha)$ where F is the CDF. That is, $\mathbb{P}(\chi_k^2 > \chi_{k,\alpha}^2) = \alpha$.

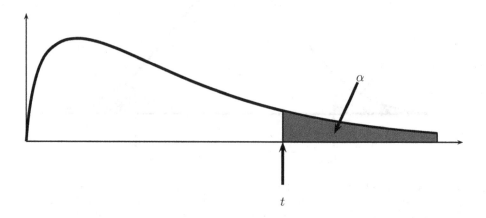

FIGURE 10.5. The p-value is the smallest α at which we would reject H_0. To find the p-value for the χ^2_{k-1} test, we find α such that the observed value t of the test statistic is just at the boundary of the rejection region. This implies that the p-value is the tail area $\mathbb{P}(\chi^2_{k-1} > t)$.

10.4   Pearson's χ^2 Test For Multinomial Data

Pearson's χ^2 test is used for multinomial data. Recall that if $X = (X_1, \ldots, X_k)$ has a multinomial (n, p) distribution, then the MLE of p is $\hat{p} = (\hat{p}_1, \ldots, \hat{p}_k) = (X_1/n, \ldots, X_k/n)$.

Let $p_0 = (p_{01}, \ldots, p_{0k})$ be some fixed vector and suppose we want to test

$$H_0 : p = p_0 \text{ versus } H_1 : p \neq p_0.$$

10.16 Definition. Pearson's χ^2 statistic *is*

$$T = \sum_{j=1}^{k} \frac{(X_j - np_{0j})^2}{np_{0j}} = \sum_{j=1}^{k} \frac{(X_j - E_j)^2}{E_j}$$

where $E_j = \mathbb{E}(X_j) = np_{0j}$ *is the expected value of* X_j *under* H_0.

10.17 Theorem. *Under* H_0, $T \rightsquigarrow \chi^2_{k-1}$. *Hence the test: reject* H_0 *if* $T > \chi^2_{k-1,\alpha}$ *has asymptotic level* α. *The p-value is* $\mathbb{P}(\chi^2_{k-1} > t)$ *where* t *is the observed value of the test statistic.*

Theorem 10.17 is illustrated in Figure 10.5.

10.18 Example (Mendel's peas). Mendel bred peas with round yellow seeds and wrinkled green seeds. There are four types of progeny: round yellow, wrinkled yellow, round green, and wrinkled green. The number of each type is multinomial with probability $p = (p_1, p_2, p_3, p_4)$. His theory of inheritance predicts that p is equal to

$$p_0 \equiv \left(\frac{9}{16}, \frac{3}{16}, \frac{3}{16}, \frac{1}{16} \right).$$

In $n = 556$ trials he observed $X = (315, 101, 108, 32)$. We will test $H_0 : p = p_0$ versus $H_1 : p \neq p_0$. Since, $np_{01} = 312.75, np_{02} = np_{03} = 104.25$, and $np_{04} = 34.75$, the test statistic is

$$
\begin{aligned}
\chi^2 &= \frac{(315 - 312.75)^2}{312.75} + \frac{(101 - 104.25)^2}{104.25} \\
&+ \frac{(108 - 104.25)^2}{104.25} + \frac{(32 - 34.75)^2}{34.75} = 0.47.
\end{aligned}
$$

The $\alpha = .05$ value for a χ_3^2 is 7.815. Since 0.47 is not larger than 7.815 we do not reject the null. The p-value is

$$\text{p-value} = \mathbb{P}(\chi_3^2 > .47) = .93$$

which is not evidence against H_0. Hence, the data do not contradict Mendel's theory.[3] ∎

In the previous example, one could argue that hypothesis testing is not the right tool. Hypothesis testing is useful to see if there is evidence to reject H_0. This is appropriate when H_0 corresponds to the status quo. It is not useful for proving that H_0 is true. Failure to reject H_0 might occur because H_0 is true, but it might occur just because the test has low power. Perhaps a confidence set for the distance between p and p_0 might be more useful in this example.

10.5   The Permutation Test

The permutation test is a nonparametric method for testing whether two distributions are the same. This test is "exact," meaning that it is not based on large sample theory approximations. Suppose that $X_1, \ldots, X_m \sim F_X$ and $Y_1, \ldots, Y_n \sim F_Y$ are two independent samples and H_0 is the hypothesis that

---

[3] There is some controversy about whether Mendel's results are "too good."

the two samples are identically distributed. This is the type of hypothesis we would consider when testing whether a treatment differs from a placebo. More precisely we are testing

$$H_0 : F_X = F_Y \quad \text{versus} \quad H_1 : F_X \neq F_Y.$$

Let $T(x_1, \ldots, x_m, y_1, \ldots, y_n)$ be some test statistic, for example,

$$T(X_1, \ldots, X_m, Y_1, \ldots, Y_n) = |\overline{X}_m - \overline{Y}_n|.$$

Let $N = m + n$ and consider forming all $N!$ permutations of the data $X_1, \ldots,$ $X_m, Y_1, \ldots, Y_n$. For each permutation, compute the test statistic T. Denote these values by $T_1, \ldots, T_{N!}$. Under the null hypothesis, each of these values is equally likely. [4] The distribution $\mathbb{P}_0$ that puts mass $1/N!$ on each T_j is called the **permutation distribution** of T. Let t_{obs} be the observed value of the test statistic. Assuming we reject when T is large, the p-value is

$$\text{p-value} = \mathbb{P}_0(T > t_{obs}) = \frac{1}{N!} \sum_{j=1}^{N!} I(T_j > t_{obs}).$$

10.19 Example. Here is a toy example to make the idea clear. Suppose the data are: $(X_1, X_2, Y_1) = (1, 9, 3)$. Let $T(X_1, X_2, Y_1) = |\overline{X} - \overline{Y}| = 2$. The permutations are:

permutation	value of T	probability
(1,9,3)	2	1/6
(9,1,3)	2	1/6
(1,3,9)	7	1/6
(3,1,9)	7	1/6
(3,9,1)	5	1/6
(9,3,1)	5	1/6

The p-value is $\mathbb{P}(T > 2) = 4/6$. ∎

Usually, it is not practical to evaluate all $N!$ permutations. We can approximate the p-value by sampling randomly from the set of permutations. The fraction of times $T_j > t_{obs}$ among these samples approximates the p-value.

---

[4] More precisely, under the null hypothesis, given the ordered data values, $X_1, \ldots, X_m, Y_1, \ldots, Y_n$ is uniformly distributed over the $N!$ permutations of the data.

Algorithm for Permutation Test

1. Compute the observed value of the test statistic
   $t_{\text{obs}} = T(X_1, \ldots, X_m, Y_1, \ldots, Y_n)$.

2. Randomly permute the data. Compute the statistic again using the permuted data.

3. Repeat the previous step B times and let $T_1, \ldots, T_B$ denote the resulting values.

4. The approximate p-value is

$$\frac{1}{B} \sum_{j=1}^{B} I(T_j > t_{\text{obs}}).$$

10.20 Example. DNA microarrays allow researchers to measure the expression levels of thousands of genes. The data are the levels of messenger RNA (mRNA) of each gene, which is thought to provide a measure of how much protein that gene produces. Roughly, the larger the number, the more active the gene. The table below, reproduced from Efron et al. (2001) shows the expression levels for genes from ten patients with two types of liver cancer cells. There are 2,638 genes in this experiment but here we show just the first two. The data are log-ratios of the intensity levels of two different color dyes used on the arrays.

	Type I					Type II				
Patient	1	2	3	4	5	6	7	8	9	10
Gene 1	230	-1,350	-1,580	-400	-760	970	110	-50	-190	-200
Gene 2	470	-850	-.8	-280	120	390	-1730	-1360	-1	-330
⋮	⋮	⋮	⋮	⋮	⋮	⋮	⋮	⋮	⋮	⋮

Let's test whether the median level of gene 1 is different between the two groups. Let ν_1 denote the median level of gene 1 of Type I and let ν_2 denote the median level of gene 1 of Type II. The absolute difference of sample medians is $T = |\widehat{\nu}_1 - \widehat{\nu}_2| = 710$. Now we estimate the permutation distribution by simulation and we find that the estimated p-value is .045. Thus, if we use a $\alpha = .05$ level of significance, we would say that there is evidence to reject the null hypothesis of no difference. ∎

In large samples, the permutation test usually gives similar results to a test that is based on large sample theory. The permutation test is thus most useful for small samples.

10.6   The Likelihood Ratio Test

The Wald test is useful for testing a scalar parameter. The likelihood ratio test is more general and can be used for testing a vector-valued parameter.

10.21 Definition. *Consider testing*

$$H_0 : \theta \in \Theta_0 \quad \text{versus} \quad H_1 : \theta \notin \Theta_0.$$

*The **likelihood ratio statistic** is*

$$\lambda = 2 \log \left(\frac{\sup_{\theta \in \Theta} \mathcal{L}(\theta)}{\sup_{\theta \in \Theta_0} \mathcal{L}(\theta)} \right) = 2 \log \left(\frac{\mathcal{L}(\widehat{\theta})}{\mathcal{L}(\widehat{\theta}_0)} \right)$$

where $\widehat{\theta}$ is the MLE and $\widehat{\theta}_0$ is the MLE when θ is restricted to lie in Θ_0.

You might have expected to see the maximum of the likelihood over Θ_0^c instead of Θ in the numerator. In practice, replacing Θ_0^c with Θ has little effect on the test statistic. Moreover, the theoretical properties of λ are much simpler if the test statistic is defined this way.

The likelihood ratio test is most useful when Θ_0 consists of all parameter values θ such that some coordinates of θ are fixed at particular values.

10.22 Theorem. *Suppose that $\theta = (\theta_1, \ldots, \theta_q, \theta_{q+1}, \ldots, \theta_r)$. Let*

$$\Theta_0 = \{\theta : (\theta_{q+1}, \ldots, \theta_r) = (\theta_{0,q+1}, \ldots, \theta_{0,r})\}.$$

Let λ be the likelihood ratio test statistic. Under $H_0 : \theta \in \Theta_0$,

$$\lambda(x^n) \rightsquigarrow \chi^2_{r-q,\alpha}$$

where $r - q$ is the dimension of Θ minus the dimension of Θ_0. The p-value for the test is $\mathbb{P}(\chi^2_{r-q} > \lambda)$.

For example, if $\theta = (\theta_1, \theta_2, \theta_3, \theta_4, \theta_5)$ and we want to test the null hypothesis that $\theta_4 = \theta_5 = 0$ then the limiting distribution has $5 - 3 = 2$ degrees of freedom.

10.23 Example (Mendel's Peas Revisited). Consider example 10.18 again. The likelihood ratio test statistic for $H_0 : p = p_0$ versus $H_1 : p \neq p_0$ is

$$
\begin{aligned}
\lambda &= 2 \log \left(\frac{L(\hat{p})}{L(p_0)} \right) \\
&= 2 \sum_{j=1}^{4} X_j \log \left(\frac{\hat{p}_j}{p_{0j}} \right) \\
&= 2 \left(315 \log \left(\frac{\frac{315}{556}}{\frac{9}{16}} \right) + 101 \log \left(\frac{\frac{101}{556}}{\frac{3}{16}} \right) \right. \\
&\qquad \left. + 108 \log \left(\frac{\frac{108}{556}}{\frac{3}{16}} \right) + 32 \log \left(\frac{\frac{32}{556}}{\frac{1}{16}} \right) \right) \\
&= 0.48.
\end{aligned}
$$

Under H_1 there are four parameters. However, the parameters must sum to one so the dimension of the parameter space is three. Under H_0 there are no free parameters so the dimension of the restricted parameter space is zero. The difference of these two dimensions is three. Therefore, the limiting distribution of λ under H_0 is χ_3^2 and the p-value is

$$
\text{p-value} = \mathbb{P}(\chi_3^2 > .48) = .92.
$$

The conclusion is the same as with the χ^2 test. ∎

When the likelihood ratio test and the χ^2 test are both applicable, as in the last example, they usually lead to similar results as long as the sample size is large.

10.7 Multiple Testing

In some situations we may conduct many hypothesis tests. In example 10.20, there were actually 2,638 genes. If we tested for a difference for each gene, we would be conducting 2,638 separate hypothesis tests. Suppose each test is conducted at level α. For any one test, the chance of a false rejection of the null is α. But the chance of at least one false rejection is much higher. This is the **multiple testing problem**. The problem comes up in many data mining situations where one may end up testing thousands or even millions of hypotheses. There are many ways to deal with this problem. Here we discuss two methods.

Consider m hypothesis tests:

$$H_{0i} \quad \text{versus} \quad H_{1i}, \quad i = 1, \ldots, m$$

and let $P_1, \ldots, P_m$ denote the m p-values for these tests.

The Bonferroni Method

Given p-values $P_1, \ldots, P_m$, reject null hypothesis H_{0i} if

$$P_i < \frac{\alpha}{m}.$$

10.24 Theorem. *Using the Bonferroni method, the probability of falsely rejecting any null hypotheses is less than or equal to α.*

PROOF. Let R be the event that at least one null hypothesis is falsely rejected. Let R_i be the event that the ith null hypothesis is falsely rejected. Recall that if $A_1, \ldots, A_k$ are events then $\mathbb{P}(\bigcup_{i=1}^{k} A_i) \leq \sum_{i=1}^{k} \mathbb{P}(A_i)$. Hence,

$$\mathbb{P}(R) = \mathbb{P}\left(\bigcup_{i=1}^{m} R_i\right) \leq \sum_{i=1}^{m} \mathbb{P}(R_i) = \sum_{i=1}^{m} \frac{\alpha}{m} = \alpha$$

from Theorem 10.14. ∎

10.25 Example. In the gene example, using $\alpha = .05$, we have that $.05/2,638 = .00001895375$. Hence, for any gene with p-value less than $.00001895375$, we declare that there is a significant difference. ∎

The Bonferroni method is very conservative because it is trying to make it unlikely that you would make even one false rejection. Sometimes, a more reasonable idea is to control the **false discovery rate** (FDR) which is defined as the mean of the number of false rejections divided by the number of rejections.

Suppose we reject all null hypotheses whose p-values fall below some threshold. Let m_0 be the number of null hypotheses that are true and let $m_1 = m - m_0$. The tests can be categorized in a 2×2 as in Table 10.2.

Define the **False Discovery Proportion** (FDP)

$$\text{FDP} = \begin{cases} V/R & \text{if } R > 0 \\ 0 & \text{if } R = 0. \end{cases}$$

The FDP is the proportion of rejections that are incorrect. Next define FDR = $\mathbb{E}(\text{FDP})$.

	H_0 Not Rejected	H_0 Rejected	Total
H_0 True	U	V	m_0
H_0 False	T	S	m_1
Total	$m - R$	R	m

TABLE 10.2. Types of outcomes in multiple testing.

The Benjamini-Hochberg (BH) Method

1. Let $P_{(1)} < \cdots < P_{(m)}$ denote the ordered p-values.

2. Define

$$\ell_i = \frac{i\alpha}{C_m m}, \quad \text{and} \quad R = \max\left\{ i : P_{(i)} < \ell_i \right\} \tag{10.8}$$

where C_m is defined to be 1 if the p-values are independent and $C_m = \sum_{i=1}^{m}(1/i)$ otherwise.

3. Let $T = P_{(R)}$; we call T the **BH rejection threshold**.

4. Reject all null hypotheses H_{0i} for which $P_i \leq T$.

10.26 Theorem (Benjamini and Hochberg). *If the procedure above is applied, then regardless of how many nulls are true and regardless of the distribution of the p-values when the null hypothesis is false,*

$$\text{FDR} = \mathbb{E}(\text{FDP}) \leq \frac{m_0}{m}\alpha \leq \alpha.$$

10.27 Example. Figure 10.6 shows six ordered p-values plotted as vertical lines. If we tested at level α without doing any correction for multiple testing, we would reject all hypotheses whose p-values are less than α. In this case, the four hypotheses corresponding to the four smallest p-values are rejected. The Bonferroni method rejects all hypotheses whose p-values are less than α/m. In this case, this leads to no rejections. The BH threshold corresponds to the last p-value that falls under the line with slope α. This leads to two hypotheses being rejected in this case. ∎

10.28 Example. Suppose that 10 independent hypothesis tests are carried leading to the following ordered p-values:

```
0.00017 0.00448 0.00671 0.00907 0.01220
0.33626 0.39341 0.53882 0.58125 0.98617
```

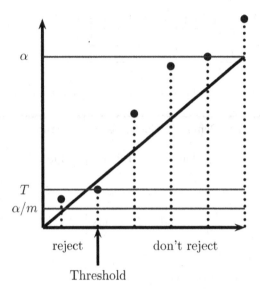

FIGURE 10.6. The Benjamini-Hochberg (BH) procedure. For uncorrected testing we reject when $P_i < \alpha$. For Bonferroni testing we reject when $P_i < \alpha/m$. The BH procedure rejects when $P_i \leq T$. The BH threshold T corresponds to the rightmost undercrossing of the upward sloping line.

With $\alpha = 0.05$, the Bonferroni test rejects any hypothesis whose p-value is less than $\alpha/10 = 0.005$. Thus, only the first two hypotheses are rejected. For the BH test, we find the largest i such that $P_{(i)} < i\alpha/m$, which in this case is $i = 5$. Thus we reject the first five hypotheses. ∎

10.8   Goodness-of-fit Tests

There is another situation where testing arises, namely, when we want to check whether the data come from an assumed parametric model. There are many such tests; here is one.

Let $\mathfrak{F} = \{f(x; \theta) : \theta \in \Theta\}$ be a parametric model. Suppose the data take values on the real line. Divide the line into k disjoint intervals $I_1, \ldots, I_k$. For $j = 1, \ldots, k$, let

$$p_j(\theta) = \int_{I_j} f(x; \theta) \, dx$$

be the probability that an observation falls into interval I_j under the assumed model. Here, $\theta = (\theta_1, \ldots, \theta_s)$ are the parameters in the assumed model. Let N_j be the number of observations that fall into I_j. The likelihood for θ based on the counts $N_1, \ldots, N_k$ is the multinomial likelihood

$$Q(\theta) = \prod_{j=1}^{k} p_i(\theta)^{N_j}.$$

Maximizing $Q(\theta)$ yields estimates $\widetilde{\theta} = (\widetilde{\theta}_1, \ldots, \widetilde{\theta}_s)$ of θ. Now define the test statistic

$$Q = \sum_{j=1}^{k} \frac{(N_j - np_j(\widetilde{\theta}))^2}{np_j(\widetilde{\theta})}. \tag{10.9}$$

10.29 Theorem. *Let H_0 be the null hypothesis that the data are* IID *draws from the model $\mathfrak{F} = \{f(x; \theta) : \theta \in \Theta\}$. Under $H - 0$, the statistic Q defined in equation (10.9) converges in distribution to a χ^2_{k-1-s} random variable. Thus, the (approximate) p-value for the test is $\mathbb{P}(\chi^2_{k-1-s} > q)$ where q denotes the observed value of Q.*

It is tempting to replace $\widetilde{\theta}$ in (10.9) with the MLE $\widehat{\theta}$. However, this will not result in a statistic whose limiting distribution is a χ^2_{k-1-s}. However, it can be shown — due to a theorem of Herman Chernoff and Erich Lehmann from 1954 — that the p-value is bounded approximately by the p-values obtained using a χ^2_{k-1-s} and a χ^2_{k-1}.

Goodness-of-fit testing has some serious limitations. If reject H_0 then we conclude we should not use the model. But if we do not reject H_0 we cannot conclude that the model is correct. We may have failed to reject simply because the test did not have enough power. This is why it is better to use nonparametric methods whenever possible rather than relying on parametric assumptions.

10.9   Bibliographic Remarks

The most complete book on testing is Lehmann (1986). See also Chapter 8 of Casella and Berger (2002) and Chapter 9 of Rice (1995). The FDR method is due to Benjamini and Hochberg (1995). Some of the exercises are from Rice (1995).

10.10   Appendix

10.10.1   The Neyman-Pearson Lemma

In the special case of a simple null $H_0 : \theta = \theta_0$ and a simple alternative $H_1 : \theta = \theta_1$ we can say precisely what the most powerful test is.

10.30 Theorem (Neyman-Pearson). *Suppose we test $H_0 : \theta = \theta_0$ versus $H_1 : \theta = \theta_1$. Let*

$$T = \frac{\mathcal{L}(\theta_1)}{\mathcal{L}(\theta_0)} = \frac{\prod_{i=1}^{n} f(x_i; \theta_1)}{\prod_{i=1}^{n} f(x_i; \theta_0)}.$$

Suppose we reject H_0 when $T > k$. If we choose k so that $\mathbb{P}_{\theta_0}(T > k) = \alpha$ then this test is the most powerful, size α test. That is, among all tests with size α, this test maximizes the power $\beta(\theta_1)$.

10.10.2   The t-test

To test $H_0 : \mu = \mu_0$ where $\mu = \mathbb{E}(X_i)$ is the mean, we can use the Wald test. When the data are assumed to be Normal and the sample size is small, it is common instead to use the **t-test**. A random variable T has a *t-distribution with k degrees of freedom* if it has density

$$f(t) = \frac{\Gamma\left(\frac{k+1}{2}\right)}{\sqrt{k\pi}\,\Gamma\left(\frac{k}{2}\right)\left(1 + \frac{t^2}{k}\right)^{(k+1)/2}}.$$

When the degrees of freedom $k \to \infty$, this tends to a Normal distribution. When $k = 1$ it reduces to a Cauchy.

Let $X_1, \ldots, X_n \sim N(\mu, \sigma^2)$ where $\theta = (\mu, \sigma^2)$ are both unknown. Suppose we want to test $\mu = \mu_0$ versus $\mu \neq \mu_0$. Let

$$T = \frac{\sqrt{n}(\overline{X}_n - \mu_0)}{S_n}$$

where S_n^2 is the sample variance. For large samples $T \approx N(0, 1)$ under H_0. The exact distribution of T under H_0 is t_{n-1}. Hence if we reject when $|T| > t_{n-1,\alpha/2}$ then we get a size α test. However, when n is moderately large, the t-test is essentially identical to the Wald test.

10.11   Exercises

1. Prove Theorem 10.6.

2. Prove Theorem 10.14.

3. Prove Theorem 10.10.

4. Prove Theorem 10.12.

5. Let $X_1, ..., X_n \sim \text{Uniform}(0, \theta)$ and let $Y = \max\{X_1, ..., X_n\}$. We want to test

$H_0 : \theta = 1/2$ versus $H_1 : \theta > 1/2$.

The Wald test is not appropriate since Y does not converge to a Normal. Suppose we decide to test this hypothesis by rejecting H_0 when $Y > c$.

(a) Find the power function.

(b) What choice of c will make the size of the test .05?

(c) In a sample of size $n = 20$ with Y=0.48 what is the p-value? What conclusion about H_0 would you make?

(d) In a sample of size $n = 20$ with Y=0.52 what is the p-value? What conclusion about H_0 would you make?

6. There is a theory that people can postpone their death until after an important event. To test the theory, Phillips and King (1988) collected data on deaths around the Jewish holiday Passover. Of 1919 deaths, 922 died the week before the holiday and 997 died the week after. Think of this as a binomial and test the null hypothesis that $\theta = 1/2$. Report and interpret the p-value. Also construct a confidence interval for θ.

7. In 1861, 10 essays appeared in the *New Orleans Daily Crescent*. They were signed "Quintus Curtius Snodgrass" and some people suspected they were actually written by Mark Twain. To investigate this, we will consider the proportion of three letter words found in an author's work. From eight Twain essays we have:

.225 .262 .217 .240 .230 .229 .235 .217

From 10 Snodgrass essays we have:

.209 .205 .196 .210 .202 .207 .224 .223 .220 .201

(a) Perform a Wald test for equality of the means. Use the nonparametric plug-in estimator. Report the p-value and a 95 per cent confidence interval for the difference of means. What do you conclude?

(b) Now use a permutation test to avoid the use of large sample methods. What is your conclusion? (Brinegar (1963)).

8. Let $X_1, \ldots, X_n \sim N(\theta, 1)$. Consider testing

$$H_0 : \theta = 0 \text{ versus } \theta = 1.$$

Let the rejection region be $R = \{x^n : T(x^n) > c\}$ where $T(x^n) = n^{-1} \sum_{i=1}^{n} X_i$.

(a) Find c so that the test has size α.

(b) Find the power under H_1, that is, find $\beta(1)$.

(c) Show that $\beta(1) \to 1$ as $n \to \infty$.

9. Let $\widehat{\theta}$ be the MLE of a parameter θ and let $\widehat{se} = \{nI(\widehat{\theta})\}^{-1/2}$ where $I(\theta)$ is the Fisher information. Consider testing

$$H_0 : \theta = \theta_0 \text{ versus } \theta \neq \theta_0.$$

Consider the Wald test with rejection region $R = \{x^n : |Z| > z_{\alpha/2}\}$ where $Z = (\widehat{\theta} - \theta_0)/\widehat{se}$. Let $\theta_1 > \theta_0$ be some alternative. Show that $\beta(\theta_1) \to 1$.

10. Here are the number of elderly Jewish and Chinese women who died just before and after the Chinese Harvest Moon Festival.

Week	Chinese	Jewish
-2	55	141
-1	33	145
1	70	139
2	49	161

Compare the two mortality patterns. (Phillips and Smith (1990)).

11. A randomized, double-blind experiment was conducted to assess the effectiveness of several drugs for reducing postoperative nausea. The data are as follows.

	Number of Patients	Incidence of Nausea
Placebo	80	45
Chlorpromazine	75	26
Dimenhydrinate	85	52
Pentobarbital (100 mg)	67	35
Pentobarbital (150 mg)	85	37

(a) Test each drug versus the placebo at the 5 per cent level. Also, report the estimated odds–ratios. Summarize your findings.

(b) Use the Bonferroni and the FDR method to adjust for multiple testing. (Beecher (1959)).

12. Let $X_1, ..., X_n \sim \text{Poisson}(\lambda)$.

(a) Let $\lambda_0 > 0$. Find the size α Wald test for

$$H_0 : \lambda = \lambda_0 \quad \text{versus} \quad H_1 : \lambda \neq \lambda_0.$$

(b) (Computer Experiment.) Let $\lambda_0 = 1$, $n = 20$ and $\alpha = .05$. Simulate $X_1, \ldots, X_n \sim \text{Poisson}(\lambda_0)$ and perform the Wald test. Repeat many times and count how often you reject the null. How close is the type I error rate to .05?

13. Let $X_1, \ldots, X_n \sim N(\mu, \sigma^2)$. Construct the likelihood ratio test for

$$H_0 : \mu = \mu_0 \quad \text{versus} \quad H_1 : \mu \neq \mu_0.$$

Compare to the Wald test.

14. Let $X_1, \ldots, X_n \sim N(\mu, \sigma^2)$. Construct the likelihood ratio test for

$$H_0 : \sigma = \sigma_0 \quad \text{versus} \quad H_1 : \sigma \neq \sigma_0.$$

Compare to the Wald test.

15. Let $X \sim \text{Binomial}(n, p)$. Construct the likelihood ratio test for

$$H_0 : p = p_0 \quad \text{versus} \quad H_1 : p \neq p_0.$$

Compare to the Wald test.

16. Let θ be a scalar parameter and suppose we test

$$H_0 : \theta = \theta_0 \quad \text{versus} \quad H_1 : \theta \neq \theta_0.$$

Let W be the Wald test statistic and let λ be the likelihood ratio test statistic. Show that these tests are equivalent in the sense that

$$\frac{W^2}{\lambda} \xrightarrow{\text{P}} 1$$

as $n \to \infty$. Hint: Use a Taylor expansion of the log-likelihood $\ell(\theta)$ to show that

$$\lambda \approx \left(\sqrt{n}(\widehat{\theta} - \theta_0) \right)^2 \left(-\frac{1}{n} \ell''(\widehat{\theta}) \right).$$

11
Bayesian Inference

11.1 The Bayesian Philosophy

The statistical methods that we have discussed so far are known as **frequentist (or classical)** methods. The frequentist point of view is based on the following postulates:

F1 Probability refers to limiting relative frequencies. Probabilities are objective properties of the real world.

F2 Parameters are fixed, unknown constants. Because they are not fluctuating, no useful probability statements can be made about parameters.

F3 Statistical procedures should be designed to have well-defined long run frequency properties. For example, a 95 percent confidence interval should trap the true value of the parameter with limiting frequency at least 95 percent.

There is another approach to inference called **Bayesian inference.** The Bayesian approach is based on the following postulates:

B1 Probability describes degree of belief, not limiting frequency. As such, we can make probability statements about lots of things, not just data which are subject to random variation. For example, I might say that "the probability that Albert Einstein drank a cup of tea on August 1, 1948" is .35. This does not refer to any limiting frequency. It reflects my strength of belief that the proposition is true.

B2 We can make probability statements about parameters, even though they are fixed constants.

B3 We make inferences about a parameter θ by producing a probability distribution for θ. Inferences, such as point estimates and interval estimates, may then be extracted from this distribution.

Bayesian inference is a controversial approach because it inherently embraces a subjective notion of probability. In general, Bayesian methods provide no guarantees on long run performance. The field of statistics puts more emphasis on frequentist methods although Bayesian methods certainly have a presence. Certain data mining and machine learning communities seem to embrace Bayesian methods very strongly. Let's put aside philosophical arguments for now and see how Bayesian inference is done. We'll conclude this chapter with some discussion on the strengths and weaknesses of the Bayesian approach.

11.2   The Bayesian Method

Bayesian inference is usually carried out in the following way.

1. We choose a probability density $f(\theta)$ — called the **prior distribution** — that expresses our beliefs about a parameter θ before we see any data.

2. We choose a statistical model $f(x|\theta)$ that reflects our beliefs about x given θ. Notice that we now write this as $f(x|\theta)$ instead of $f(x; \theta)$.

3. After observing data $X_1, \ldots, X_n$, we update our beliefs and calculate the **posterior** distribution $f(\theta|X_1, \ldots, X_n)$.

To see how the third step is carried out, first suppose that θ is discrete and that there is a single, discrete observation X. We should use a capital letter now to denote the parameter since we are treating it like a random variable, so let Θ denote the parameter. Now, in this discrete setting,

$$
\begin{aligned}
\mathbb{P}(\Theta = \theta | X = x) &= \frac{\mathbb{P}(X = x, \Theta = \theta)}{\mathbb{P}(X = x)} \\
&= \frac{\mathbb{P}(X = x | \Theta = \theta)\mathbb{P}(\Theta = \theta)}{\sum_\theta \mathbb{P}(X = x | \Theta = \theta)\mathbb{P}(\Theta = \theta)}
\end{aligned}
$$

which you may recognize from Chapter 1 as **Bayes' theorem.** The version for continuous variables is obtained by using density functions:

$$
f(\theta | x) = \frac{f(x | \theta)f(\theta)}{\int f(x | \theta)f(\theta)d\theta}. \tag{11.1}
$$

If we have n IID observations $X_1, \ldots, X_n$, we replace $f(x | \theta)$ with

$$
f(x_1, \ldots, x_n | \theta) = \prod_{i=1}^{n} f(x_i | \theta) = \mathcal{L}_n(\theta).
$$

NOTATION. We will write X^n to mean $(X_1, \ldots, X_n)$ and x^n to mean $(x_1, \ldots, x_n)$. Now,

$$
f(\theta | x^n) = \frac{f(x^n | \theta)f(\theta)}{\int f(x^n | \theta)f(\theta)d\theta} = \frac{\mathcal{L}_n(\theta)f(\theta)}{c_n} \propto \mathcal{L}_n(\theta)f(\theta) \tag{11.2}
$$

where

$$
c_n = \int \mathcal{L}_n(\theta)f(\theta)d\theta \tag{11.3}
$$

is called the **normalizing constant**. Note that c_n does not depend on θ. We can summarize by writing:

   Posterior is proportional to Likelihood times Prior

or, in symbols,

$$
f(\theta | x^n) \propto \mathcal{L}(\theta)f(\theta).
$$

You might wonder, doesn't it cause a problem to throw away the constant c_n? The answer is that we can always recover the constant later if we need to.

What do we do with the posterior distribution? First, we can get a point estimate by summarizing the center of the posterior. Typically, we use the mean or mode of the posterior. The posterior mean is

$$
\bar{\theta}_n = \int \theta f(\theta | x^n)d\theta = \frac{\int \theta \mathcal{L}_n(\theta)f(\theta)}{\int \mathcal{L}_n(\theta)f(\theta)d\theta}. \tag{11.4}
$$

We can also obtain a Bayesian interval estimate. We find a and b such that $\int_{-\infty}^{a} f(\theta|x^n)d\theta = \int_{b}^{\infty} f(\theta|x^n)d\theta = \alpha/2$. Let $C = (a,b)$. Then

$$\mathbb{P}(\theta \in C|x^n) = \int_{a}^{b} f(\theta|x^n)\, d\theta = 1 - \alpha$$

so C is a $1 - \alpha$ **posterior interval.**

11.1 Example. Let $X_1, \ldots, X_n \sim \text{Bernoulli}(p)$. Suppose we take the uniform distribution $f(p) = 1$ as a prior. By Bayes' theorem, the posterior has the form

$$f(p|x^n) \propto f(p)\mathcal{L}_n(p) = p^s(1-p)^{n-s} = p^{s+1-1}(1-p)^{n-s+1-1}$$

where $s = \sum_{i=1}^{n} x_i$ is the number of successes. Recall that a random variable has a Beta distribution with parameters α and β if its density is

$$f(p; \alpha, \beta) = \frac{\Gamma(\alpha + \beta)}{\Gamma(\alpha)\Gamma(\beta)} p^{\alpha-1}(1-p)^{\beta-1}.$$

We see that the posterior for p is a Beta distribution with parameters $s + 1$ and $n - s + 1$. That is,

$$f(p|x^n) = \frac{\Gamma(n + 2)}{\Gamma(s+1)\Gamma(n-s+1)} p^{(s+1)-1}(1-p)^{(n-s+1)-1}.$$

We write this as

$$p|x^n \sim \text{Beta}(s+1, n-s+1).$$

Notice that we have figured out the normalizing constant without actually doing the integral $\int \mathcal{L}_n(p)f(p)dp$. The mean of a $\text{Beta}(\alpha, \beta)$ distribution is $\alpha/(\alpha + \beta)$ so the Bayes estimator is

$$\overline{p} = \frac{s+1}{n+2}. \tag{11.5}$$

It is instructive to rewrite the estimator as

$$\overline{p} = \lambda_n \widehat{p} + (1 - \lambda_n)\widetilde{p} \tag{11.6}$$

where $\widehat{p} = s/n$ is the MLE, $\widetilde{p} = 1/2$ is the prior mean and $\lambda_n = n/(n+2) \approx 1$. A 95 percent posterior interval can be obtained by numerically finding a and b such that $\int_{a}^{b} f(p|x^n)\, dp = .95$.

Suppose that instead of a uniform prior, we use the prior $p \sim \text{Beta}(\alpha, \beta)$. If you repeat the calculations above, you will see that $p|x^n \sim \text{Beta}(\alpha + s, \beta +$

$n - s$). The flat prior is just the special case with $\alpha = \beta = 1$. The posterior
mean is

$$\overline{p} = \frac{\alpha + s}{\alpha + \beta + n} = \left(\frac{n}{\alpha + \beta + n}\right)\widehat{p} + \left(\frac{\alpha + \beta}{\alpha + \beta + n}\right)p_0$$

where $p_0 = \alpha/(\alpha + \beta)$ is the prior mean. ∎

In the previous example, the prior was a Beta distribution and the posterior
was a Beta distribution. When the prior and the posterior are in the same
family, we say that the prior is **conjugate** with respect to the model.

11.2 Example. Let $X_1, \ldots, X_n \sim N(\theta, \sigma^2)$. For simplicity, let us assume that
σ is known. Suppose we take as a prior $\theta \sim N(a, b^2)$. In problem 1 in the
exercises it is shown that the posterior for θ is

$$\theta | X^n \sim N(\overline{\theta}, \tau^2) \tag{11.7}$$

where

$$\overline{\theta} = w\overline{X} + (1 - w)a,$$

$$w = \frac{\frac{1}{\mathsf{se}^2}}{\frac{1}{\mathsf{se}^2} + \frac{1}{b^2}}, \quad \frac{1}{\tau^2} = \frac{1}{\mathsf{se}^2} + \frac{1}{b^2},$$

and $\mathsf{se} = \sigma/\sqrt{n}$ is the standard error of the MLE $\overline{X}$. This is another example
of a conjugate prior. Note that $w \to 1$ and $\tau/\mathsf{se} \to 1$ as $n \to \infty$. So, for large
n, the posterior is approximately $N(\widehat{\theta}, \mathsf{se}^2)$. The same is true if n is fixed but
$b \to \infty$, which corresponds to letting the prior become very flat.

Continuing with this example, let us find $C = (c, d)$ such that $\mathbb{P}(\theta \in C | X^n) = .95$. We can do this by choosing c and d such that $\mathbb{P}(\theta < c | X^n) = .025$ and $\mathbb{P}(\theta > d | X^n) = .025$. So, we want to find c such that

$$\mathbb{P}(\theta < c | X^n) = \mathbb{P}\left(\frac{\theta - \overline{\theta}}{\tau} < \frac{c - \overline{\theta}}{\tau} \,\Big|\, X^n\right)$$

$$= \mathbb{P}\left(Z < \frac{c - \overline{\theta}}{\tau}\right) = .025.$$

We know that $\mathbb{P}(Z < -1.96) = .025$. So,

$$\frac{c - \overline{\theta}}{\tau} = -1.96$$

implying that $c = \overline{\theta} - 1.96\tau$. By similar arguments, $d = \overline{\theta} + 1.96$. So a 95 percent
Bayesian interval is $\overline{\theta} \pm 1.96\,\tau$. Since $\overline{\theta} \approx \widehat{\theta}$ and $\tau \approx \mathsf{se}$, the 95 percent Bayesian
interval is approximated by $\widehat{\theta} \pm 1.96\,\mathsf{se}$ which is the frequentist confidence
interval. ∎

11.3   Functions of Parameters

How do we make inferences about a function $\tau = g(\theta)$? Remember in Chapter 3 we solved the following problem: given the density f_X for X, find the density for $Y = g(X)$. We now simply apply the same reasoning. The posterior CDF for τ is

$$H(\tau|x^n) = \mathbb{P}(g(\theta) \leq \tau|x^n) = \int_A f(\theta|x^n)d\theta$$

where $A = \{\theta : g(\theta) \leq \tau\}$. The posterior density is $h(\tau|x^n) = H'(\tau|x^n)$.

11.3 Example. Let $X_1, \ldots, X_n \sim$ Bernoulli(p) and $f(p) = 1$ so that $p|X^n \sim$ Beta($s + 1, n - s + 1$) with $s = \sum_{i=1}^n x_i$. Let $\psi = \log(p/(1-p))$. Then

$$
\begin{aligned}
H(\psi|x^n) &= \mathbb{P}(\Psi \leq \psi|x^n) = \mathbb{P}\left(\log\left(\frac{P}{1-P}\right) \leq \psi \,\middle|\, x^n\right) \\
&= \mathbb{P}\left(P \leq \frac{e^\psi}{1+e^\psi} \,\middle|\, x^n\right) \\
&= \int_0^{e^\psi/(1+e^\psi)} f(p|x^n)\, dp \\
&= \frac{\Gamma(n+2)}{\Gamma(s+1)\Gamma(n-s+1)} \int_0^{e^\psi/(1+e^\psi)} p^s (1-p)^{n-s}\, dp
\end{aligned}
$$

and

$$
\begin{aligned}
h(\psi|x^n) &= H'(\psi|x^n) \\
&= \frac{\Gamma(n+2)}{\Gamma(s+1)\Gamma(n-s+1)} \left(\frac{e^\psi}{1+e^\psi}\right)^s \left(\frac{1}{1+e^\psi}\right)^{n-s} \left(\frac{\partial\left(\frac{e^\psi}{1+e^\psi}\right)}{\partial\psi}\right) \\
&= \frac{\Gamma(n+2)}{\Gamma(s+1)\Gamma(n-s+1)} \left(\frac{e^\psi}{1+e^\psi}\right)^s \left(\frac{1}{1+e^\psi}\right)^{n-s} \left(\frac{1}{1+e^\psi}\right)^2 \\
&= \frac{\Gamma(n+2)}{\Gamma(s+1)\Gamma(n-s+1)} \left(\frac{e^\psi}{1+e^\psi}\right)^s \left(\frac{1}{1+e^\psi}\right)^{n-s+2}
\end{aligned}
$$

for $\psi \in \mathbb{R}$. ∎

11.4   Simulation

The posterior can often be approximated by simulation. Suppose we draw $\theta_1, \ldots, \theta_B \sim p(\theta|x^n)$. Then a histogram of $\theta_1, \ldots, \theta_B$ approximates the posterior density $p(\theta|x^n)$. An approximation to the posterior mean $\bar{\theta}_n = \mathbb{E}(\theta|x^n)$ is

$B^{-1} \sum_{j=1}^{B} \theta_j$. The posterior $1-\alpha$ interval can be approximated by $(\theta_{\alpha/2}, \theta_{1-\alpha/2})$ where $\theta_{\alpha/2}$ is the $\alpha/2$ sample quantile of $\theta_1, \ldots, \theta_B$.

Once we have a sample $\theta_1, \ldots, \theta_B$ from $f(\theta|x^n)$, let $\tau_i = g(\theta_i)$. Then $\tau_1, \ldots, \tau_B$ is a sample from $f(\tau|x^n)$. This avoids the need to do any analytical calculations. Simulation is discussed in more detail in Chapter 24.

11.4 Example. Consider again Example 11.3. We can approximate the posterior for ψ without doing any calculus. Here are the steps:

1. Draw $P_1, \ldots, P_B \sim \text{Beta}(s+1, n-s+1)$.

2. Let $\psi_i = \log(P_i/(1-P_i))$ for $i = 1, \ldots, B$.

Now $\psi_1, \ldots, \psi_B$ are IID draws from $h(\psi|x^n)$. A histogram of these values provides an estimate of $h(\psi|x^n)$. ∎

11.5 Large Sample Properties of Bayes' Procedures

In the Bernoulli and Normal examples we saw that the posterior mean was close to the MLE. This is true in greater generality.

11.5 Theorem. *Let $\widehat{\theta}_n$ be the MLE and let $\widehat{se} = 1/\sqrt{nI(\widehat{\theta}_n)}$. Under appropriate regularity conditions, the posterior is approximately Normal with mean $\widehat{\theta}_n$ and standard deviation $\widehat{se}$. Hence, $\overline{\theta}_n \approx \widehat{\theta}_n$. Also, if $C_n = (\widehat{\theta}_n - z_{\alpha/2}\widehat{se}, \widehat{\theta}_n + z_{\alpha/2}\widehat{se})$ is the asymptotic frequentist $1-\alpha$ confidence interval, then C_n is also an approximate $1-\alpha$ Bayesian posterior interval:*

$$\mathbb{P}(\theta \in C_n|X^n) \to 1-\alpha.$$

There is also a Bayesian delta method. Let $\tau = g(\theta)$. Then

$$\tau|X^n \approx N(\widehat{\tau}, \widetilde{se}^2)$$

where $\widehat{\tau} = g(\widehat{\theta})$ and $\widetilde{se} = \widehat{se}\,|g'(\widehat{\theta})|$.

11.6 Flat Priors, Improper Priors, and "Noninformative" Priors

An important question in Bayesian inference is: where does one get the prior $f(\theta)$? One school of thought, called **subjectivism** says that the prior should reflect our subjective opinion about θ (before the data are collected). This may be possible in some cases but is impractical in complicated problems especially if there are many parameters. Moreover, injecting subjective opinion into the analysis is contrary to the goal of making scientific inference as objective as possible. An alternative is to try to define some sort of "noninformative prior." An obvious candidate for a noninformative prior is to use a flat prior $f(\theta) \propto$ constant.

In the Bernoulli example, taking $f(p) = 1$ leads to $p|X^n \sim \text{Beta}(s+1, n-s+1)$ as we saw earlier, which seemed very reasonable. But unfettered use of flat priors raises some questions.

IMPROPER PRIORS. Let $X \sim N(\theta, \sigma^2)$ with σ known. Suppose we adopt a flat prior $f(\theta) \propto c$ where $c > 0$ is a constant. Note that $\int f(\theta)d\theta = \infty$ so this is not a probability density in the usual sense. We call such a prior an **improper prior.** Nonetheless, we can still formally carry out Bayes' theorem and compute the posterior density by multiplying the prior and the likelihood: $f(\theta) \propto \mathcal{L}_n(\theta)f(\theta) \propto \mathcal{L}_n(\theta)$. This gives $\theta|X^n \sim N(\overline{X}, \sigma^2/n)$ and the resulting point and interval estimators agree exactly with their frequentist counterparts. In general, improper priors are not a problem as long as the resulting posterior is a well-defined probability distribution.

FLAT PRIORS ARE NOT INVARIANT. Let $X \sim \text{Bernoulli}(p)$ and suppose we use the flat prior $f(p) = 1$. This flat prior presumably represents our lack of information about p before the experiment. Now let $\psi = \log(p/(1-p))$. This is a transformation of p and we can compute the resulting distribution for ψ, namely,

$$f_\Psi(\psi) = \frac{e^\psi}{(1+e^\psi)^2}$$

which is not flat. But if we are ignorant about p then we are also ignorant about ψ so we should use a flat prior for ψ. This is a contradiction. In short, the notion of a flat prior is not well defined because a flat prior on a parameter does not imply a flat prior on a transformed version of the parameter. Flat priors are not **transformation invariant.**

JEFFREYS' PRIOR. Jeffreys came up with a rule for creating priors. The rule is: take

$$f(\theta) \propto I(\theta)^{1/2}$$

where $I(\theta)$ is the Fisher information function. This rule turns out to be transformation invariant. There are various reasons for thinking that this prior might be a useful prior but we will not go into details here.

11.6 Example. Consider the Bernoulli (p) model. Recall that

$$I(p) = \frac{1}{p(1-p)}.$$

Jeffreys' rule says to use the prior

$$f(p) \propto \sqrt{I(p)} = p^{-1/2}(1-p)^{-1/2}.$$

This is a Beta $(1/2,1/2)$ density. This is very close to a uniform density. ∎

In a multiparameter problem, the Jeffreys' prior is defined to be $f(\theta) \propto \sqrt{|I(\theta)|}$ where $|A|$ denotes the determinant of a matrix A and $I(\theta)$ is the Fisher information matrix.

11.7 Multiparameter Problems

Suppose that $\theta = (\theta_1, \ldots, \theta_p)$. The posterior density is still given by

$$f(\theta|x^n) \propto \mathcal{L}_n(\theta)f(\theta). \tag{11.8}$$

The question now arises of how to extract inferences about one parameter. The key is to find the marginal posterior density for the parameter of interest. Suppose we want to make inferences about θ_1. The marginal posterior for θ_1 is

$$f(\theta_1|x^n) = \int \cdots \int f(\theta_1, \cdots, \theta_p|x^n)d\theta_2 \ldots d\theta_p. \tag{11.9}$$

In practice, it might not be feasible to do this integral. Simulation can help. Draw randomly from the posterior:

$$\theta^1, \ldots, \theta^B \sim f(\theta|x^n)$$

where the superscripts index the different draws. Each θ^j is a vector $\theta^j = (\theta_1^j, \ldots, \theta_p^j)$. Now collect together the first component of each draw:

$$\theta_1^1, \ldots, \theta_1^B.$$

These are a sample from $f(\theta_1|x^n)$ and we have avoided doing any integrals.

11.7 Example (Comparing Two Binomials). Suppose we have n_1 control patients and n_2 treatment patients and that X_1 control patients survive while X_2 treatment patients survive. We want to estimate $\tau = g(p_1, p_2) = p_2 - p_1$. Then,

$$X_1 \sim \text{Binomial}(n_1, p_1) \quad \text{and} \quad X_2 \sim \text{Binomial}(n_2, p_2).$$

If $f(p_1, p_2) = 1$, the posterior is

$$f(p_1, p_2 | x_1, x_2) \propto p_1^{x_1}(1 - p_1)^{n_1 - x_1} p_2^{x_2}(1 - p_2)^{n_2 - x_2}.$$

Notice that (p_1, p_2) live on a rectangle (a square, actually) and that

$$f(p_1, p_2 | x_1, x_2) = f(p_1 | x_1) f(p_2 | x_2)$$

where

$$f(p_1 | x_1) \propto p_1^{x_1}(1 - p_1)^{n_1 - x_1} \quad \text{and} \quad f(p_2 | x_2) \propto p_2^{x_2}(1 - p_2)^{n_2 - x_2}$$

which implies that p_1 and p_2 are independent under the posterior. Also, $p_1 | x_1 \sim \text{Beta}(x_1 + 1, n_1 - x_1 + 1)$ and $p_2 | x_2 \sim \text{Beta}(x_2 + 1, n_2 - x_2 + 1)$. If we simulate $P_{1,1}, \ldots, P_{1,B} \sim \text{Beta}(x_1 + 1, n_1 - x_1 + 1)$ and $P_{2,1}, \ldots, P_{2,B} \sim \text{Beta}(x_2 + 1, n_2 - x_2 + 1)$, then $\tau_b = P_{2,b} - P_{1,b}$, $b = 1, \ldots, B$, is a sample from $f(\tau | x_1, x_2)$. ∎

11.8   Bayesian Testing

Hypothesis testing from a Bayesian point of view is a complex topic. We will only give a brief sketch of the main idea here. The Bayesian approach to testing involves putting a prior on H_0 and on the parameter θ and then computing $\mathbb{P}(H_0 | X^n)$. Consider the case where θ is scalar and we are testing

$$H_0 : \theta = \theta_0 \quad \text{versus} \quad H_1 : \theta \neq \theta_0.$$

It is usually reasonable to use the prior $\mathbb{P}(H_0) = \mathbb{P}(H_1) = 1/2$ (although this is not essential in what follows). Under H_1 we need a prior for θ. Denote this prior density by $f(\theta)$. From Bayes' theorem

$$
\begin{aligned}
\mathbb{P}(H_0 | X^n = x^n) &= \frac{f(x^n | H_0)\mathbb{P}(H_0)}{f(x^n | H_0)\mathbb{P}(H_0) + f(x^n | H_1)\mathbb{P}(H_1)} \\
&= \frac{\frac{1}{2}f(x^n \mid \theta_0)}{\frac{1}{2}f(x^n \mid \theta_0) + \frac{1}{2}f(x^n \mid H_1)} \\
&= \frac{f(x^n \mid \theta_0)}{f(x^n \mid \theta_0) + \int f(x^n \mid \theta)f(\theta)d\theta} \\
&= \frac{\mathcal{L}(\theta_0)}{\mathcal{L}(\theta_0) + \int \mathcal{L}(\theta)f(\theta)d\theta}.
\end{aligned}
$$

We saw that, in estimation problems, the prior was not very influential and that the frequentist and Bayesian methods gave similar answers. This is not the case in hypothesis testing. Also, one can't use improper priors in testing because this leads to an undefined constant in the denominator of the expression above. Thus, if you use Bayesian testing you must choose the prior $f(\theta)$ very carefully. It is possible to get a prior-free bound on $\mathbb{P}(H_0|X^n = x^n)$. Notice that $0 \le \int \mathcal{L}(\theta)f(\theta)d\theta \le \mathcal{L}(\widehat{\theta})$. Hence,

$$\frac{\mathcal{L}(\theta_0)}{\mathcal{L}(\theta_0) + \mathcal{L}(\widehat{\theta})} \le \mathbb{P}(H_0|X^n = x^n) \le 1.$$

The upper bound is not very interesting, but the lower bound is non-trivial.

11.9 Strengths and Weaknesses of Bayesian Inference

Bayesian inference is appealing when prior information is available since Bayes' theorem is a natural way to combine prior information with data. Some people find Bayesian inference psychologically appealing because it allows us to make probability statements about parameters. In contrast, frequentist inference provides confidence sets C_n which trap the parameter 95 percent of the time, but we cannot say that $\mathbb{P}(\theta \in C_n|X^n)$ is .95. In the frequentist approach we can make probability statements about C_n, not θ. However, psychological appeal is not a compelling scientific argument for using one type of inference over another.

In parametric models, with large samples, Bayesian and frequentist methods give approximately the same inferences. In general, they need not agree.

Here are three examples that illustrate the strengths and weakness of Bayesian inference. The first example is Example 6.14 revisited. This example shows the psychological appeal of Bayesian inference. The second and third show that Bayesian methods can fail.

11.8 Example (Example 6.14 revisited). We begin by reviewing the example. Let θ be a fixed, known real number and let X_1, X_2 be independent random variables such that $\mathbb{P}(X_i = 1) = \mathbb{P}(X_i = -1) = 1/2$. Now define $Y_i = \theta + X_i$ and suppose that you only observe Y_1 and Y_2. Let

$$C = \begin{cases} \{Y_1 - 1\} & \text{if } Y_1 = Y_2 \\ \{(Y_1 + Y_2)/2\} & \text{if } Y_1 \ne Y_2. \end{cases}$$

This is a 75 percent confidence set since, no matter what θ is, $\mathbb{P}_\theta(\theta \in C) = 3/4$.

Suppose we observe $Y_1 = 15$ and $Y_2 = 17$. Then our 75 percent confidence interval is $\{16\}$. However, we are certain, in this case, that $\theta = 16$. So calling this a 75 percent confidence set, bothers many people. Nonetheless, C is a valid 75 percent confidence set. It will trap the true value 75 percent of the time.

The Bayesian solution is more satisfying to many. For simplicity, assume that θ is an integer. Let $f(\theta)$ be a prior mass function such that $f(\theta) > 0$ for every integer θ. When $Y = (Y_1, Y_2) = (15, 17)$, the likelihood function is

$$\mathcal{L}(\theta) = \begin{cases} 1/4 & \theta = 16 \\ 0 & \text{otherwise.} \end{cases}$$

Applying Bayes' theorem we see that

$$\mathbb{P}(\Theta = \theta | Y = (15, 17)) = \begin{cases} 1 & \theta = 16 \\ 0 & \text{otherwise.} \end{cases}$$

Hence, $\mathbb{P}(\theta \in C | Y = (15, 17)) = 1$. There is nothing wrong with saying that $\{16\}$ is a 75 percent confidence interval. But is it not a probability statement about θ. ∎

11.9 Example. This is a simplified version of the example in Robins and Ritov (1997). The data consist of n IID triples

$$(X_1, R_1, Y_1), \ldots, (X_n, Y_n, R_n).$$

Let B be a finite but very large number, like $B = 100^{100}$. Any realistic sample size n will be small compared to B. Let

$$\theta = (\theta_1, \ldots, \theta_B)$$

be a vector of unknown parameters such that $0 \le \theta_j \le 1$ for $1 \le j \le B$. Let

$$\xi = (\xi_1, \ldots, \xi_B)$$

be a vector of **known** numbers such that

$$0 < \delta \le \xi_j \le 1 - \delta < 1, \quad 1 \le j \le B,$$

where δ is some, small, positive number. Each data point (X_i, R_i, Y_i) is drawn in the following way:

1. Draw X_i uniformly from $\{1, \ldots, B\}$.

2. Draw $R_i \sim \text{Bernoulli}(\xi_{X_i})$.

3. If $R_i = 1$, then draw $Y_i \sim \text{Bernoulli}(\theta_{X_i})$. If $R_i = 0$, do not draw Y_i.

The model may seem a little artificial but, in fact, it is caricature of some real **missing data** problems in which some data points are not observed. In this example, $R_i = 0$ can be thought of as meaning "missing." Our goal is to estimate

$$\psi = \mathbb{P}(Y_i = 1).$$

Note that

$$\psi \ = \ \mathbb{P}(Y_i = 1) = \sum_{j=1}^{B} \mathbb{P}(Y_i = 1 | X = j)\mathbb{P}(X = j)$$

$$= \ \frac{1}{B} \sum_{j=1}^{B} \theta_j \equiv g(\theta)$$

so $\psi = g(\theta)$ is a function of θ.

Let us consider a Bayesian analysis first. The likelihood of a single observation is

$$f(X_i, R_i, Y_i) = f(X_i)f(R_i|X_i)f(Y_i|X_i)^{R_i}.$$

The last term is raised to the power R_i since, if $R_i = 0$, then Y_i is not observed and hence that term drops out of the likelihood. Since $f(X_i) = 1/B$ and that Y_i and R_i are Bernoulli,

$$f(X_i)f(R_i|X_i)f(Y_i|X_i)^{R_i} = \frac{1}{B} \, \xi_{X_i}^{R_i} \, (1 - \xi_{X_i})^{1-R_i} \, \theta_{X_i}^{Y_i R_i} \, (1 - \theta_{X_i})^{(1-Y_i)R_i}.$$

Thus, the likelihood function is

$$\mathcal{L}(\theta) \ = \ \prod_{i=1}^{n} f(X_i)f(R_i|X_i)f(Y_i|X_i)^{R_i}$$

$$= \ \prod_{i=1}^{n} \frac{1}{B} \, \xi_{X_i}^{R_i} \, (1 - \xi_{X_i})^{1-R_i} \, \theta_{X_i}^{Y_i R_i} \, (1 - \theta_{X_i})^{(1-Y_i)R_i}$$

$$\propto \ \theta_{X_i}^{Y_i R_i} (1 - \theta_{X_i})^{(1-Y_i)R_i}.$$

We have dropped all the terms involving B and the ξ_j's since these are known constants, not parameters. The log-likelihood is

$$\ell(\theta) \ = \ \sum_{i=1}^{n} Y_i \, R_i \log \theta_{X_i} + (1 - Y_i) \, R_i \, \log(1 - \theta_{X_i})$$

$$= \ \sum_{j=1}^{B} n_j \log \theta_j + \sum_{j=1}^{B} m_j \log(1 - \theta_j)$$

where

$$n_j = \#\{i : Y_i = 1, R_i = 1, X_i = j\}$$
$$m_j = \#\{i : Y_i = 0, R_i = 1, X_i = j\}.$$

Now, $n_j = m_j = 0$ for most j since B is so much larger than n. This has several implications. First, the MLE for most θ_j is not defined. Second, for most θ_j, the posterior distribution is equal to the prior distribution, since those θ_j do not appear in the likelihood. Hence, $f(\theta|\text{Data}) \approx f(\theta)$. It follows that $f(\psi|\text{Data}) \approx f(\psi)$. In other words, the data provide little information about ψ in a Bayesian analysis.

Now we consider a frequentist solution. Define

$$\widehat{\psi} = \frac{1}{n} \sum_{i=1}^{n} \frac{R_i Y_i}{\xi_{X_i}}. \tag{11.10}$$

We will now show that this estimator is unbiased and has small mean-squared error. It can be shown (see Exercise 7) that

$$\mathbb{E}(\widehat{\psi}) = \psi \quad \text{and} \quad \mathbb{V}(\widehat{\psi}) \leq \frac{1}{n\delta^2}. \tag{11.11}$$

Therefore, the MSE is of order $1/n$ which goes to 0 fairly quickly as we collect more data, no matter how large B is. The estimator defined in (11.10) is called the **Horwitz-Thompson** estimator. It cannot be derived from a Bayesian or likelihood point of view since it involves the terms ξ_{X_i}. These terms drop out of the log-likelihood and hence will not show up in any likelihood-based method including Bayesian estimators.

The moral of the story is this. Bayesian methods are tied to the likelihood function. But in high dimensional (and nonparametric) problems, the likelihood may not yield accurate inferences. ∎

11.10 Example. Suppose that f is a probability density function and that

$$f(x) = cg(x)$$

where $g(x) > 0$ is a known function and c is unknown. In principle we can compute c since $\int f(x)\,dx = 1$ implies that $c = 1/\int g(x)\,dx$. But in many cases we can't do the integral $\int g(x)\,dx$ since g might be a complicated function and x could be high dimensional. Despite the fact that c is not known, it is often possible to draw a sample $X_1, \ldots, X_n$ from f; see Chapter 24. Can we use the sample to estimate the normalizing constant c? Here is a frequentist solution:

Let $\widehat{f}_n(x)$ be a consistent estimate of the density f. Chapter 20 explains how to construct such an estimate. Choose any point x and note that $c = f(x)/g(x)$. Hence, $\widehat{c} = \widehat{f}(x)/g(x)$ is a consistent estimate of c. Now let us try to solve this problem from a Bayesian approach. Let $\pi(c)$ be a prior such that $\pi(c) > 0$ for all $c > 0$. The likelihood function is

$$\mathcal{L}_n(c) = \prod_{i=1}^{n} f(X_i) = \prod_{i=1}^{n} cg(X_i) = c^n \prod_{i=1}^{n} g(X_i) \propto c^n.$$

Hence the posterior is proportional to $c^n \pi(c)$. The posterior does not depend on $X_1, \ldots, X_n$, so we come to the startling conclusion that, from the Bayesian point of view, there is no information in the data about c. Moreover, the posterior mean is

$$\frac{\int_0^\infty c^{n+1} \pi(c)\, dc}{\int_0^\infty c^n \pi(c)\, dc}$$

which tends to infinity as n increases. ∎

These last two examples illustrate an important point. Bayesians are slaves to the likelihood function. When the likelihood goes awry, so will Bayesian inference.

What should we conclude from all this? The important thing is to understand that frequentist and Bayesian methods are answering different questions. To combine prior beliefs with data in a principled way, use Bayesian inference. To construct procedures with guaranteed long run performance, such as confidence intervals, use frequentist methods. Generally, Bayesian methods run into problems when the parameter space is high dimensional. In particular, 95 percent posterior intervals need not contain the true value 95 percent of the time (in the frequency sense).

11.10   Bibliographic Remarks

Some references on Bayesian inference include Carlin and Louis (1996), Gelman et al. (1995), Lee (1997), Robert (1994), and Schervish (1995). See Cox (1993), Diaconis and Freedman (1986), Freedman (1999), Barron et al. (1999), Ghosal et al. (2000), Shen and Wasserman (2001), and Zhao (2000) for discussions of some of the technicalities of nonparametric Bayesian inference. The Robins-Ritov example is discussed in detail in Robins and Ritov (1997) where it is cast more properly as a nonparametric problem. Example 11.10 is due to Edward George (personal communication). See Berger and Delampady (1987)

and Kass and Raftery (1995) for a discussion of Bayesian testing. See Kass
and Wasserman (1996) for a discussion of noninformative priors.

11.11 Appendix

Proof of Theorem 11.5.

It can be shown that the effect of the prior diminishes as n increases so
that $f(\theta|X^n) \propto \mathcal{L}_n(\theta)f(\theta) \approx \mathcal{L}_n(\theta)$. Hence, $\log f(\theta|X^n) \approx \ell(\theta)$. Now, $\ell(\theta) \approx$
$\ell(\widehat{\theta}) + (\theta - \widehat{\theta})\ell'(\widehat{\theta}) + [(\theta - \widehat{\theta})^2/2]\ell''(\widehat{\theta}) = \ell(\widehat{\theta}) + [(\theta - \widehat{\theta})^2/2]\ell''(\widehat{\theta})$ since $\ell'(\widehat{\theta}) = 0$.
Exponentiating, we get approximately that

$$f(\theta|X^n) \propto \exp\left\{-\frac{1}{2}\frac{(\theta - \widehat{\theta})^2}{\sigma_n^2}\right\}$$

where $\sigma_n^2 = -1/\ell''(\widehat{\theta}_n)$. So the posterior of θ is approximately Normal with
mean $\widehat{\theta}$ and variance σ_n^2. Let $\ell_i = \log f(X_i|\theta)$, then

$$\frac{1}{\sigma_n^2} = -\ell''(\widehat{\theta}_n) = \sum_i -\ell_i''(\widehat{\theta}_n)$$

$$= n\left(\frac{1}{n}\right)\sum_i -\ell_i''(\widehat{\theta}_n) \approx n\mathbb{E}_\theta\left[-\ell_i''(\widehat{\theta}_n)\right]$$

$$= nI(\widehat{\theta}_n)$$

and hence $\sigma_n \approx \text{se}(\widehat{\theta})$. ∎

11.12 Exercises

1. Verify (11.7).

2. Let $X_1, ..., X_n \sim \text{Normal}(\mu, 1)$.

    (a) Simulate a data set (using $\mu = 5$) consisting of n=100 observations.

    (b) Take $f(\mu) = 1$ and find the posterior density. Plot the density.

    (c) Simulate 1,000 draws from the posterior. Plot a histogram of the
    simulated values and compare the histogram to the answer in (b).

    (d) Let $\theta = e^\mu$. Find the posterior density for θ analytically and by
    simulation.

    (e) Find a 95 percent posterior interval for μ.

    (f) Find a 95 percent confidence interval for θ.

3. Let $X_1, ..., X_n \sim$ Uniform$(0, \theta)$. Let $f(\theta) \propto 1/\theta$. Find the posterior density.

4. Suppose that 50 people are given a placebo and 50 are given a new treatment. 30 placebo patients show improvement while 40 treated patients show improvement. Let $\tau = p_2 - p_1$ where p_2 is the probability of improving under treatment and p_1 is the probability of improving under placebo.

(a) Find the MLE of τ. Find the standard error and 90 percent confidence interval using the delta method.

(b) Find the standard error and 90 percent confidence interval using the parametric bootstrap.

(c) Use the prior $f(p_1, p_2) = 1$. Use simulation to find the posterior mean and posterior 90 percent interval for τ.

(d) Let

$$\psi = \log \left(\left(\frac{p_1}{1 - p_1} \right) \div \left(\frac{p_2}{1 - p_2} \right) \right)$$

be the log-odds ratio. Note that $\psi = 0$ if $p_1 = p_2$. Find the MLE of ψ. Use the delta method to find a 90 percent confidence interval for ψ.

(e) Use simulation to find the posterior mean and posterior 90 percent interval for ψ.

5. Consider the Bernoulli(p) observations

0 1 0 1 0 0 0 0 0 0

Plot the posterior for p using these priors: Beta$(1/2,1/2)$, Beta$(1,1)$, Beta$(10,10)$, Beta$(100,100)$.

6. Let $X_1, \ldots, X_n \sim$ Poisson(λ).

(a) Let $\lambda \sim$ Gamma(α, β) be the prior. Show that the posterior is also a Gamma. Find the posterior mean.

(b) Find the Jeffreys' prior. Find the posterior.

7. In Example 11.9, verify (11.11).

8. Let $X \sim N(\mu, 1)$. Consider testing

$$H_0 : \mu = 0 \quad \text{versus} \quad H_1 : \mu \neq 0.$$

Take $\mathbb{P}(H_0) = \mathbb{P}(H_1) = 1/2$. Let the prior for μ under H_1 be $\mu \sim N(0, b^2)$. Find an expression for $\mathbb{P}(H_0|X = x)$. Compare $\mathbb{P}(H_0|X = x)$ to the p-value of the Wald test. Do the comparison numerically for a variety of values of x and b. Now repeat the problem using a sample of size n. You will see that the posterior probability of H_0 can be large even when the p-value is small, especially when n is large. This disagreement between Bayesian and frequentist testing is called the Jeffreys-Lindley paradox.

12
Statistical Decision Theory

12.1 Preliminaries

We have considered several point estimators such as the maximum likelihood estimator, the method of moments estimator, and the posterior mean. In fact, there are many other ways to generate estimators. How do we choose among them? The answer is found in **decision theory** which is a formal theory for comparing statistical procedures.

Consider a parameter θ which lives in a parameter space Θ. Let $\widehat{\theta}$ be an estimator of θ. In the language of decision theory, an estimator is sometimes called a **decision rule** and the possible values of the decision rule are called **actions**.

We shall measure the discrepancy between θ and $\widehat{\theta}$ using a **loss function** $L(\theta, \widehat{\theta})$. Formally, L maps $\Theta \times \Theta$ into $\mathbb{R}$. Here are some examples of loss functions:

$$
\begin{aligned}
&L(\theta, \widehat{\theta}) = (\theta - \widehat{\theta})^2 && \text{squared error loss,} \\
&L(\theta, \widehat{\theta}) = |\theta - \widehat{\theta}| && \text{absolute error loss,} \\
&L(\theta, \widehat{\theta}) = |\theta - \widehat{\theta}|^p && L_p \text{ loss,} \\
&L(\theta, \widehat{\theta}) = 0 \text{ if } \theta = \widehat{\theta} \text{ or } 1 \text{ if } \theta \neq \widehat{\theta} && \text{zero–one loss,} \\
&L(\theta, \widehat{\theta}) = \int \log\left(\frac{f(x; \theta)}{f(x; \widehat{\theta})}\right) f(x; \theta)dx && \text{Kullback–Leibler loss.}
\end{aligned}
$$

Bear in mind in what follows that an estimator $\widehat{\theta}$ is a function of the data. To emphasize this point, sometimes we will write $\widehat{\theta}$ as $\widehat{\theta}(X)$. To assess an estimator, we evaluate the average loss or risk.

12.1 Definition. *The* **risk** *of an estimator* $\widehat{\theta}$ *is*

$$R(\theta, \widehat{\theta}) = \mathbb{E}_\theta\left(L(\theta, \widehat{\theta})\right) = \int L(\theta, \widehat{\theta}(x))f(x; \theta)dx.$$

When the loss function is squared error, the risk is just the MSE (mean squared error):

$$R(\theta, \widehat{\theta}) = \mathbb{E}_\theta(\widehat{\theta} - \theta)^2 = \text{MSE} = \mathbb{V}_\theta(\widehat{\theta}) + \text{bias}_\theta^2(\widehat{\theta}).$$

In the rest of the chapter, if we do not state what loss function we are using, assume the loss function is squared error.

12.2 Comparing Risk Functions

To compare two estimators we can compare their risk functions. However, this does not provide a clear answer as to which estimator is better. Consider the following examples.

12.2 Example. Let $X \sim N(\theta, 1)$ and assume we are using squared error loss. Consider two estimators: $\widehat{\theta}_1 = X$ and $\widehat{\theta}_2 = 3$. The risk functions are $R(\theta, \widehat{\theta}_1) = \mathbb{E}_\theta(X - \theta)^2 = 1$ and $R(\theta, \widehat{\theta}_2) = \mathbb{E}_\theta(3 - \theta)^2 = (3 - \theta)^2$. If $2 < \theta < 4$ then $R(\theta, \widehat{\theta}_2) < R(\theta, \widehat{\theta}_1)$, otherwise, $R(\theta, \widehat{\theta}_1) < R(\theta, \widehat{\theta}_2)$. Neither estimator uniformly dominates the other; see Figure 12.1. ∎

12.3 Example. Let $X_1, \ldots, X_n \sim$ Bernoulli(p). Consider squared error loss and let $\widehat{p}_1 = \overline{X}$. Since this has 0 bias, we have that

$$R(p, \widehat{p}_1) = \mathbb{V}(\overline{X}) = \frac{p(1-p)}{n}.$$

Another estimator is

$$\widehat{p}_2 = \frac{Y + \alpha}{\alpha + \beta + n}$$

where $Y = \sum_{i=1}^n X_i$ and α and β are positive constants. This is the posterior mean using a Beta (α, β) prior. Now,

$$R(p, \widehat{p}_2) = \mathbb{V}_p(\widehat{p}_2) + (\text{bias}_p(\widehat{p}_2))^2$$

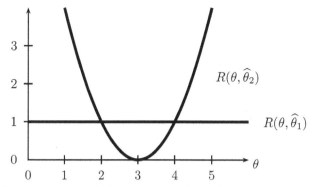

FIGURE 12.1. Comparing two risk functions. Neither risk function dominates the other at all values of θ.

$$
\begin{aligned}
&= \mathbb{V}_p \left(\frac{Y + \alpha}{\alpha + \beta + n} \right) + \left(\mathbb{E}_p \left(\frac{Y + \alpha}{\alpha + \beta + n} \right) - p \right)^2 \\
&= \frac{np(1 - p)}{(\alpha + \beta + n)^2} + \left(\frac{np + \alpha}{\alpha + \beta + n} - p \right)^2.
\end{aligned}
$$

Let $\alpha = \beta = \sqrt{n}/4$. (In Example 12.12 we will explain this choice.) The resulting estimator is

$$
\widehat{p}_2 = \frac{Y + \sqrt{n}/4}{n + \sqrt{n}}
$$

and the risk function is

$$
R(p, \widehat{p}_2) = \frac{n}{4(n + \sqrt{n})^2}.
$$

The risk functions are plotted in figure 12.2. As we can see, neither estimator uniformly dominates the other.

These examples highlight the need to be able to compare risk functions. To do so, we need a one-number summary of the risk function. Two such summaries are the maximum risk and the Bayes risk.

12.4 Definition. *The* **maximum risk** *is*

$$
\overline{R}(\widehat{\theta}) = \sup_{\theta} R(\theta, \widehat{\theta}) \tag{12.1}
$$

and the **Bayes risk** *is*

$$
r(f, \widehat{\theta}) = \int R(\theta, \widehat{\theta}) f(\theta) d\theta \tag{12.2}
$$

where $f(\theta)$ is a prior for θ.

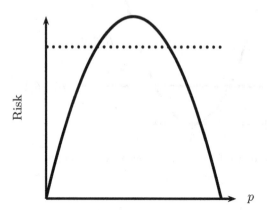

FIGURE 12.2. Risk functions for $\widehat{p}_1$ and $\widehat{p}_2$ in Example 12.3. The solid curve is $R(\widehat{p}_1)$. The dotted line is $R(\widehat{p}_2)$.

12.5 Example. Consider again the two estimators in Example 12.3. We have

$$\overline{R}(\widehat{p}_1) = \max_{0 \leq p \leq 1} \frac{p(1-p)}{n} = \frac{1}{4n}$$

and

$$\overline{R}(\widehat{p}_2) = \max_{p} \frac{n}{4(n + \sqrt{n})^2} = \frac{n}{4(n + \sqrt{n})^2}.$$

Based on maximum risk, $\widehat{p}_2$ is a better estimator since $\overline{R}(\widehat{p}_2) < \overline{R}(\widehat{p}_1)$. However, when n is large, $\overline{R}(\widehat{p}_1)$ has smaller risk except for a small region in the parameter space near $p = 1/2$. Thus, many people prefer $\widehat{p}_1$ to $\widehat{p}_2$. This illustrates that one-number summaries like maximum risk are imperfect. Now consider the Bayes risk. For illustration, let us take $f(p) = 1$. Then

$$r(f, \widehat{p}_1) = \int R(p, \widehat{p}_1) dp = \int \frac{p(1-p)}{n} dp = \frac{1}{6n}$$

and

$$r(f, \widehat{p}_2) = \int R(p, \widehat{p}_2) dp = \frac{n}{4(n + \sqrt{n})^2}.$$

For $n \geq 20$, $r(f, \widehat{p}_2) > r(f, \widehat{p}_1)$ which suggests that $\widehat{p}_1$ is a better estimator. This might seem intuitively reasonable but this answer depends on the choice of prior. The advantage of using maximum risk, despite its problems, is that it does not require one to choose a prior. ■

These two summaries of the risk function suggest two different methods for devising estimators: choosing $\widehat{\theta}$ to minimize the maximum risk leads to minimax estimators; choosing $\widehat{\theta}$ to minimize the Bayes risk leads to Bayes estimators.

12.6 Definition. *A decision rule that minimizes the Bayes risk is called a* **Bayes rule.** *Formally,* $\widehat{\theta}$ *is a Bayes rule with respect to the prior f if*

$$r(f,\widehat{\theta}) = \inf_{\widetilde{\theta}} r(f,\widetilde{\theta}) \tag{12.3}$$

where the infimum is over all estimators $\widetilde{\theta}$. An estimator that minimizes the maximum risk is called a **minimax rule.** *Formally,* $\widehat{\theta}$ *is minimax if*

$$\sup_{\theta} R(\theta,\widehat{\theta}) = \inf_{\widetilde{\theta}} \sup_{\theta} R(\theta,\widetilde{\theta}) \tag{12.4}$$

where the infimum is over all estimators $\widetilde{\theta}$.

12.3 Bayes Estimators

Let f be a prior. From Bayes' theorem, the posterior density is

$$f(\theta|x) = \frac{f(x|\theta)f(\theta)}{m(x)} = \frac{f(x|\theta)f(\theta)}{\int f(x|\theta)f(\theta)d\theta} \tag{12.5}$$

where $m(x) = \int f(x,\theta)d\theta = \int f(x|\theta)f(\theta)d\theta$ is the **marginal distribution** of X. Define the **posterior risk** of an estimator $\widehat{\theta}(x)$ by

$$r(\widehat{\theta}|x) = \int L(\theta,\widehat{\theta}(x))f(\theta|x)d\theta. \tag{12.6}$$

12.7 Theorem. *The Bayes risk $r(f,\widehat{\theta})$ satisfies*

$$r(f,\widehat{\theta}) = \int r(\widehat{\theta}|x)m(x)\,dx.$$

Let $\widehat{\theta}(x)$ be the value of θ that minimizes $r(\widehat{\theta}|x)$. Then $\widehat{\theta}$ is the Bayes estimator.

PROOF. We can rewrite the Bayes risk as follows:

$$
\begin{aligned}
r(f,\widehat{\theta}) &= \int R(\theta,\widehat{\theta})f(\theta)d\theta = \int\left(\int L(\theta,\widehat{\theta}(x))f(x|\theta)dx\right)f(\theta)d\theta \\
&= \int\int L(\theta,\widehat{\theta}(x))f(x,\theta)dxd\theta = \int\int L(\theta,\widehat{\theta}(x))f(\theta|x)m(x)dxd\theta \\
&= \int\left(\int L(\theta,\widehat{\theta}(x))f(\theta|x)d\theta\right)m(x)\,dx = \int r(\widehat{\theta}|x)m(x)\,dx.
\end{aligned}
$$

If we choose $\widehat{\theta}(x)$ to be the value of θ that minimizes $r(\widehat{\theta}|x)$ then we will minimize the integrand at every x and thus minimize the integral $\int r(\widehat{\theta}|x)m(x)dx$. ∎

Now we can find an explicit formula for the Bayes estimator for some specific loss functions.

12.8 Theorem. *If $L(\theta,\widehat{\theta}) = (\theta - \widehat{\theta})^2$ then the Bayes estimator is*

$$\widehat{\theta}(x) = \int \theta f(\theta|x)d\theta = \mathbb{E}(\theta|X = x). \tag{12.7}$$

If $L(\theta,\widehat{\theta}) = |\theta - \widehat{\theta}|$ then the Bayes estimator is the median of the posterior $f(\theta|x)$. If $L(\theta,\widehat{\theta})$ is zero–one loss, then the Bayes estimator is the mode of the posterior $f(\theta|x)$.

PROOF. We will prove the theorem for squared error loss. The Bayes rule $\widehat{\theta}(x)$ minimizes $r(\widehat{\theta}|x) = \int(\theta - \widehat{\theta}(x))^2 f(\theta|x)d\theta$. Taking the derivative of $r(\widehat{\theta}|x)$ with respect to $\widehat{\theta}(x)$ and setting it equal to 0 yields the equation $2\int(\theta - \widehat{\theta}(x))f(\theta|x)d\theta = 0$. Solving for $\widehat{\theta}(x)$ we get 12.7. ∎

12.9 Example. Let $X_1, \ldots, X_n \sim N(\mu, \sigma^2)$ where σ^2 is known. Suppose we use a $N(a, b^2)$ prior for μ. The Bayes estimator with respect to squared error loss is the posterior mean, which is

$$\widehat{\theta}(X_1, \ldots, X_n) = \frac{b^2}{b^2 + \frac{\sigma^2}{n}}\overline{X} + \frac{\frac{\sigma^2}{n}}{b^2 + \frac{\sigma^2}{n}}a. \quad ∎$$

12.4   Minimax Rules

Finding minimax rules is complicated and we cannot attempt a complete coverage of that theory here but we will mention a few key results. The main message to take away from this section is: Bayes estimators with a constant risk function are minimax.

12.10 Theorem. *Let $\widehat{\theta}^f$ be the Bayes rule for some prior f:*

$$r(f, \widehat{\theta}^f) = \inf_{\widehat{\theta}} r(f, \widehat{\theta}). \tag{12.8}$$

Suppose that

$$R(\theta, \widehat{\theta}^f) \le r(f, \widehat{\theta}^f) \quad \text{for all } \theta. \tag{12.9}$$

*Then $\widehat{\theta}^f$ is minimax and f is called a **least favorable prior**.*

PROOF. Suppose that $\widehat{\theta}^f$ is not minimax. Then there is another rule $\widehat{\theta}_0$ such that $\sup_\theta R(\theta, \widehat{\theta}_0) < \sup_\theta R(\theta, \widehat{\theta}^f)$. Since the average of a function is always less than or equal to its maximum, we have that $r(f, \widehat{\theta}_0) \leq \sup_\theta R(\theta, \widehat{\theta}_0)$. Hence,

$$r(f, \widehat{\theta}_0) \leq \sup_\theta R(\theta, \widehat{\theta}_0) < \sup_\theta R(\theta, \widehat{\theta}^f) \leq r(f, \widehat{\theta}^f)$$

which contradicts (12.8). ∎

12.11 Theorem. *Suppose that $\widehat{\theta}$ is the Bayes rule with respect to some prior f. Suppose further that $\widehat{\theta}$ has constant risk: $R(\theta, \widehat{\theta}) = c$ for some c. Then $\widehat{\theta}$ is minimax.*

PROOF. The Bayes risk is $r(f, \widehat{\theta}) = \int R(\theta, \widehat{\theta}) f(\theta) d\theta = c$ and hence $R(\theta, \widehat{\theta}) \leq r(f, \widehat{\theta})$ for all θ. Now apply the previous theorem. ∎

12.12 Example. Consider the Bernoulli model with squared error loss. In example 12.3 we showed that the estimator

$$\widehat{p}(X^n) = \frac{\sum_{i=1}^n X_i + \sqrt{n/4}}{n + \sqrt{n}}$$

has a constant risk function. This estimator is the posterior mean, and hence the Bayes rule, for the prior $\text{Beta}(\alpha, \beta)$ with $\alpha = \beta = \sqrt{n/4}$. Hence, by the previous theorem, this estimator is minimax. ∎

12.13 Example. Consider again the Bernoulli but with loss function

$$L(p, \widehat{p}) = \frac{(p - \widehat{p})^2}{p(1 - p)}.$$

Let

$$\widehat{p}(X^n) = \widehat{p} = \frac{\sum_{i=1}^n X_i}{n}.$$

The risk is

$$R(p, \widehat{p}) = E\left(\frac{(\widehat{p} - p)^2}{p(1 - p)}\right) = \frac{1}{p(1 - p)}\left(\frac{p(1 - p)}{n}\right) = \frac{1}{n}$$

which, as a function of p, is constant. It can be shown that, for this loss function, $\widehat{p}(X^n)$ is the Bayes estimator under the prior $f(p) = 1$. Hence, $\widehat{p}$ is minimax. ∎

A natural question to ask is: what is the minimax estimator for a Normal model?

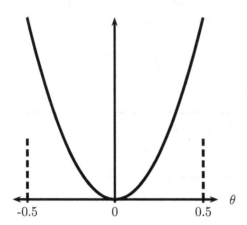

FIGURE 12.3. Risk function for constrained Normal with m=.5. The two short dashed lines show the least favorable prior which puts its mass at two points.

12.14 Theorem. *Let $X_1, \ldots, X_n \sim N(\theta, 1)$ and let $\widehat{\theta} = \overline{X}$. Then $\widehat{\theta}$ is minimax with respect to any well-behaved loss function.* [1] *It is the only estimator with this property.*

If the parameter space is restricted, then the theorem above does not apply as the next example shows.

12.15 Example. Suppose that $X \sim N(\theta, 1)$ and that θ is known to lie in the interval $[-m, m]$ where $0 < m < 1$. The unique, minimax estimator under squared error loss is

$$\widehat{\theta}(X) = m \tanh(mX)$$

where $\tanh(z) = (e^z - e^{-z})/(e^z + e^{-z})$. It can be shown that this is the Bayes rule with respect to the prior that puts mass $1/2$ at m and mass $1/2$ at $-m$. Moreover, it can be shown that the risk is not constant but it does satisfy $R(\theta, \widehat{\theta}) \leq r(f, \widehat{\theta})$ for all θ; see Figure 12.3. Hence, Theorem 12.10 implies that $\widehat{\theta}$ is minimax. ∎

---

[1] "Well-behaved" means that the level sets must be convex and symmetric about the origin. The result holds up to sets of measure 0.

12.5   Maximum Likelihood, Minimax, and Bayes

For parametric models that satisfy weak regularity conditions, the maximum likelihood estimator is approximately minimax. Consider squared error loss which is squared bias plus variance. In parametric models with large samples, it can be shown that the variance term dominates the bias so the risk of the MLE $\widehat{\theta}$ roughly equals the variance:[2]

$$R(\theta, \widehat{\theta}) = \mathbb{V}_\theta(\widehat{\theta}) + \text{bias}^2 \approx \mathbb{V}_\theta(\widehat{\theta}).$$

As we saw in Chapter 9, the variance of the MLE is approximately

$$\mathbb{V}(\widehat{\theta}) \approx \frac{1}{nI(\theta)}$$

where $I(\theta)$ is the Fisher information. Hence,

$$nR(\theta, \widehat{\theta}) \approx \frac{1}{I(\theta)}. \tag{12.10}$$

For any other estimator θ', it can be shown that for large n, $R(\theta, \theta') \geq R(\theta, \widehat{\theta})$. More precisely,

$$\lim_{\epsilon \to 0} \limsup_{n \to \infty} \sup_{|\theta - \theta'| < \epsilon} n\, R(\theta', \widehat{\theta}) \geq \frac{1}{I(\theta)}. \tag{12.11}$$

This says that, in a local, large sample sense, the MLE is minimax. It can also be shown that the MLE is approximately the Bayes rule.

In summary:

> In most parametric models, with large samples, the MLE is approximately minimax and Bayes.

There is a caveat: these results break down when the number of parameters is large as the next example shows.

12.16 Example (Many Normal means). Let $Y_i \sim N(\theta_i, \sigma^2/n)$, $i = 1, \ldots, n$. Let $Y = (Y_1, \ldots, Y_n)$ denote the data and let $\theta = (\theta_1, \ldots, \theta_n)$ denote the unknown parameters. Assume that

$$\theta \in \Theta_n \equiv \left\{ (\theta_1, \ldots, \theta_n) : \sum_{i=1}^n \theta_i^2 \leq c^2 \right\}$$

---

[2] Typically, the squared bias is order $O(n^{-2})$ while the variance is of order $O(n^{-1})$.

for some $c > 0$. In this model, there are as many parameters as observations. [3] The MLE is $\widehat{\theta} = Y = (Y_1, \ldots, Y_n)$. Under the loss function $L(\theta, \widehat{\theta}) = \sum_{i=1}^{n}(\widehat{\theta}_i - \theta_i)^2$, the risk of the MLE is $R(\theta, \widehat{\theta}) = \sigma^2$. It can be shown that the minimax risk is approximately $\sigma^2/(\sigma^2 + c^2)$ and one can find an estimator $\widetilde{\theta}$ that achieves this risk. Since $\sigma^2/(\sigma^2 + c^2) < \sigma^2$, we see that $\widetilde{\theta}$ has smaller risk than the MLE. In practice, the difference between the risks can be substantial. This shows that maximum likelihood is not an optimal estimator in high dimensional problems. ∎

12.6 Admissibility

Minimax estimators and Bayes estimators are "good estimators" in the sense that they have small risk. It is also useful to characterize bad estimators.

12.17 Definition. *An estimator $\widehat{\theta}$ is* **inadmissible** *if there exists another rule $\widehat{\theta}'$ such that*

$$R(\theta, \widehat{\theta}') \leq R(\theta, \widehat{\theta}) \text{ for all } \theta \text{ and}$$
$$R(\theta, \widehat{\theta}') < R(\theta, \widehat{\theta}) \text{ for at least one } \theta.$$

Otherwise, $\widehat{\theta}$ is **admissible**.

12.18 Example. Let $X \sim N(\theta, 1)$ and consider estimating θ with squared error loss. Let $\widehat{\theta}(X) = 3$. We will show that $\widehat{\theta}$ is admissible. Suppose not. Then there exists a different rule $\widehat{\theta}'$ with smaller risk. In particular, $R(3, \widehat{\theta}') \leq R(3, \widehat{\theta}) = 0$. Hence, $0 = R(3, \widehat{\theta}') = \int (\widehat{\theta}'(x) - 3)^2 f(x; 3) dx$. Thus, $\widehat{\theta}'(x) = 3$. So there is no rule that beats $\widehat{\theta}$. Even though $\widehat{\theta}$ is admissible it is clearly a bad decision rule. ∎

12.19 Theorem (Bayes Rules Are Admissible). *Suppose that $\Theta \subset \mathbb{R}$ and that $R(\theta, \widehat{\theta})$ is a continuous function of θ for every $\widehat{\theta}$. Let f be a prior density with full support, meaning that, for every θ and every $\epsilon > 0$, $\int_{\theta-\epsilon}^{\theta+\epsilon} f(\theta)d\theta > 0$. Let $\widehat{\theta}^f$ be the Bayes' rule. If the Bayes risk is finite then $\widehat{\theta}^f$ is admissible.*

Proof. Suppose $\widehat{\theta}^f$ is inadmissible. Then there exists a better rule $\widehat{\theta}$ such that $R(\theta, \widehat{\theta}) \leq R(\theta, \widehat{\theta}^f)$ for all θ and $R(\theta_0, \widehat{\theta}) < R(\theta_0, \widehat{\theta}^f)$ for some θ_0. Let

---

[3] The many Normal means problem is more general than it looks. Many nonparametric estimation problems are mathematically equivalent to this model.

$\nu = R(\theta_0, \widehat{\theta^f}) - R(\theta_0, \widehat{\theta}) > 0$. Since R is continuous, there is an $\epsilon > 0$ such that $R(\theta, \widehat{\theta^f}) - R(\theta, \widehat{\theta}) > \nu/2$ for all $\theta \in (\theta_0 - \epsilon, \theta_0 + \epsilon)$. Now,

$$
\begin{aligned}
r(f, \widehat{\theta^f}) - r(f, \widehat{\theta}) &= \int R(\theta, \widehat{\theta^f}) f(\theta) d\theta - \int R(\theta, \widehat{\theta}) f(\theta) d\theta \\
&= \int \left[R(\theta, \widehat{\theta^f}) - R(\theta, \widehat{\theta}) \right] f(\theta) d\theta \\
&\geq \int_{\theta_0 - \epsilon}^{\theta_0 + \epsilon} \left[R(\theta, \widehat{\theta^f}) - R(\theta, \widehat{\theta}) \right] f(\theta) d\theta \\
&\geq \frac{\nu}{2} \int_{\theta_0 - \epsilon}^{\theta_0 + \epsilon} f(\theta) d\theta \\
&> 0.
\end{aligned}
$$

Hence, $r(f, \widehat{\theta^f}) > r(f, \widehat{\theta})$. This implies that $\widehat{\theta^f}$ does not minimize $r(f, \widehat{\theta})$ which contradicts the fact that $\widehat{\theta^f}$ is the Bayes rule. ∎

12.20 Theorem. *Let $X_1, \ldots, X_n \sim N(\mu, \sigma^2)$. Under squared error loss, $\overline{X}$ is admissible.*

The proof of the last theorem is quite technical and is omitted but the idea is as follows: The posterior mean is admissible for any strictly positive prior. Take the prior to be $N(a, b^2)$. When b^2 is very large, the posterior mean is approximately equal to $\overline{X}$.

How are minimaxity and admissibility linked? In general, a rule may be one, both, or neither. But here are some facts linking admissibility and minimaxity.

12.21 Theorem. *Suppose that $\widehat{\theta}$ has constant risk and is admissible. Then it is minimax.*

PROOF. The risk is $R(\theta, \widehat{\theta}) = c$ for some c. If $\widehat{\theta}$ were not minimax then there exists a rule $\widehat{\theta}'$ such that

$$
R(\theta, \widehat{\theta}') \leq \sup_\theta R(\theta, \widehat{\theta}') < \sup_\theta R(\theta, \widehat{\theta}) = c.
$$

This would imply that $\widehat{\theta}$ is inadmissible. ∎

Now we can prove a restricted version of Theorem 12.14 for squared error loss.

12.22 Theorem. *Let $X_1, \ldots, X_n \sim N(\theta, 1)$. Then, under squared error loss, $\widehat{\theta} = \overline{X}$ is minimax.*

PROOF. According to Theorem 12.20, $\widehat{\theta}$ is admissible. The risk of $\widehat{\theta}$ is $1/n$ which is constant. The result follows from Theorem 12.21. ∎

Although minimax rules are not guaranteed to be admissible they are "close to admissible." Say that $\widehat{\theta}$ is **strongly inadmissible** if there exists a rule $\widehat{\theta}'$ and an $\epsilon > 0$ such that $R(\theta, \widehat{\theta}') < R(\theta, \widehat{\theta}) - \epsilon$ for all θ.

12.23 Theorem. *If $\widehat{\theta}$ is minimax, then it is not strongly inadmissible.*

12.7   Stein's Paradox

Suppose that $X \sim N(\theta, 1)$ and consider estimating θ with squared error loss. From the previous section we know that $\widehat{\theta}(X) = X$ is admissible. Now consider estimating two, unrelated quantities $\theta = (\theta_1, \theta_2)$ and suppose that $X_1 \sim N(\theta_1, 1)$ and $X_2 \sim N(\theta_2, 1)$ independently, with loss $L(\theta, \widehat{\theta}) = \sum_{j=1}^{2}(\theta_j - \widehat{\theta}_j)^2$. Not surprisingly, $\widehat{\theta}(X) = X$ is again admissible where $X = (X_1, X_2)$. Now consider the generalization to k normal means. Let $\theta = (\theta_1, \ldots, \theta_k)$, $X = (X_1, \ldots, X_k)$ with $X_i \sim N(\theta_i, 1)$ (independent) and loss $L(\theta, \widehat{\theta}) = \sum_{j=1}^{k}(\theta_j - \widehat{\theta}_j)^2$. Stein astounded everyone when he proved that, if $k \geq 3$, then $\widehat{\theta}(X) = X$ is inadmissible. It can be shown that the **James-Stein estimator** $\widehat{\theta}^S$ has smaller risk, where $\widehat{\theta}^S = (\widehat{\theta}_1^S, \ldots, \widehat{\theta}_k^S)$,

$$\widehat{\theta}_i^S(X) = \left(1 - \frac{k-2}{\sum_i X_i^2}\right)^+ X_i \tag{12.12}$$

and $(z)^+ = \max\{z, 0\}$. This estimator shrinks the X_i's towards 0. The message is that, when estimating many parameters, there is great value in shrinking the estimates. This observation plays an important role in modern nonparametric function estimation.

12.8   Bibliographic Remarks

Aspects of decision theory can be found in Casella and Berger (2002), Berger (1985), Ferguson (1967), and Lehmann and Casella (1998).

12.9   Exercises

1. In each of the following models, find the Bayes risk and the Bayes estimator, using squared error loss.

   (a) $X \sim \text{Binomial}(n, p)$, $p \sim \text{Beta}(\alpha, \beta)$.

(b) $X \sim \text{Poisson}(\lambda)$, $\lambda \sim \text{Gamma}(\alpha, \beta)$.

(c) $X \sim N(\theta, \sigma^2)$ where σ^2 is known and $\theta \sim N(a, b^2)$.

2. Let $X_1, \ldots, X_n \sim N(\theta, \sigma^2)$ and suppose we estimate θ with loss function $L(\theta, \widehat{\theta}) = (\theta - \widehat{\theta})^2 / \sigma^2$. Show that $\overline{X}$ is admissible and minimax.

3. Let $\Theta = \{\theta_1, \ldots, \theta_k\}$ be a finite parameter space. Prove that the posterior mode is the Bayes estimator under zero–one loss.

4. (Casella and Berger (2002).) Let $X_1, \ldots, X_n$ be a sample from a distribution with variance σ^2. Consider estimators of the form bS^2 where S^2 is the sample variance. Let the loss function for estimating σ^2 be

$$L(\sigma^2, \widehat{\sigma}^2) = \frac{\widehat{\sigma}^2}{\sigma^2} - 1 - \log\left(\frac{\widehat{\sigma}^2}{\sigma^2}\right).$$

Find the optimal value of b that minimizes the risk for all σ^2.

5. (Berliner (1983).) Let $X \sim \text{Binomial}(n, p)$ and suppose the loss function is

$$L(p, \widehat{p}) = \left(1 - \frac{\widehat{p}}{p}\right)^2$$

where $0 < p < 1$. Consider the estimator $\widehat{p}(X) = 0$. This estimator falls outside the parameter space $(0, 1)$ but we will allow this. Show that $\widehat{p}(X) = 0$ is the unique, minimax rule.

6. (Computer Experiment.) Compare the risk of the MLE and the James-Stein estimator (12.12) by simulation. Try various values of n and various vectors θ. Summarize your results.

Part III

Statistical Models and Methods

13

Linear and Logistic Regression

Regression is a method for studying the relationship between a **response variable** Y and a **covariate** X. The covariate is also called a **predictor variable** or a **feature**. [1] One way to summarize the relationship between X and Y is through the **regression function**

$$r(x) = \mathbb{E}(Y|X = x) = \int y\, f(y|x)dy. \tag{13.1}$$

Our goal is to estimate the regression function $r(x)$ from data of the form

$$(Y_1, X_1), \ldots, (Y_n, X_n) \sim F_{X,Y}.$$

In this Chapter, we take a parametric approach and assume that r is linear. In Chapters 20 and 21 we discuss nonparametric regression.

13.1 Simple Linear Regression

The simplest version of regression is when X_i is simple (one-dimensional) and $r(x)$ is assumed to be linear:

$$r(x) = \beta_0 + \beta_1 x.$$

---

[1] The term "regression" is due to Sir Francis Galton (1822-1911) who noticed that tall and short men tend to have sons with heights closer to the mean. He called this "regression towards the mean."

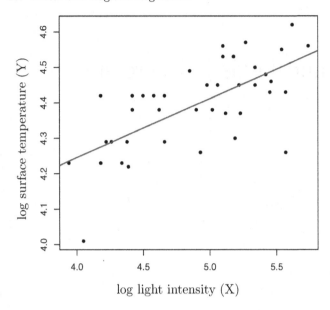

FIGURE 13.1. Data on nearby stars. The solid line is the least squares line.

This model is called the **the simple linear regression model**. We will make the further simplifying assumption that $\mathbb{V}(\epsilon_i|X = x) = \sigma^2$ does not depend on x. We can thus write the linear regression model as follows.

13.1 Definition. The Simple Linear Regression Model

$$Y_i = \beta_0 + \beta_1 X_i + \epsilon_i \tag{13.2}$$

where $\mathbb{E}(\epsilon_i|X_i) = 0$ and $\mathbb{V}(\epsilon_i|X_i) = \sigma^2$.

13.2 Example. Figure 13.1 shows a plot of log surface temperature (Y) versus log light intensity (X) for some nearby stars. Also on the plot is an estimated linear regression line which will be explained shortly. ∎

The unknown parameters in the model are the intercept β_0 and the slope β_1 and the variance σ^2. Let $\widehat{\beta}_0$ and $\widehat{\beta}_1$ denote estimates of β_0 and β_1. The **fitted line** is

$$\widehat{r}(x) = \widehat{\beta}_0 + \widehat{\beta}_1 x. \tag{13.3}$$

The **predicted values** or **fitted values** are $\widehat{Y}_i = \widehat{r}(X_i)$ and the **residuals** are defined to be

$$\widehat{\epsilon}_i = Y_i - \widehat{Y}_i = Y_i - \left(\widehat{\beta}_0 + \widehat{\beta}_1 X_i\right). \tag{13.4}$$

The **residual sums of squares** or RSS, which measures how well the line fits the data, is defined by RSS $= \sum_{i=1}^{n} \widehat{\epsilon}_i^2$.

13.3 Definition. *The* **least squares estimates** *are the values* $\widehat{\beta}_0$ *and* $\widehat{\beta}_1$ *that minimize* RSS $= \sum_{i=1}^{n} \widehat{\epsilon}_i^2$.

13.4 Theorem. *The least squares estimates are given by*

$$\widehat{\beta}_1 = \frac{\sum_{i=1}^{n}(X_i - \overline{X}_n)(Y_i - \overline{Y}_n)}{\sum_{i=1}^{n}(X_i - \overline{X}_n)^2}, \tag{13.5}$$

$$\widehat{\beta}_0 = \overline{Y}_n - \widehat{\beta}_1 \overline{X}_n. \tag{13.6}$$

An unbiased estimate of σ^2 *is*

$$\widehat{\sigma}^2 = \left(\frac{1}{n-2}\right) \sum_{i=1}^{n} \widehat{\epsilon}_i^2. \tag{13.7}$$

13.5 Example. Consider the star data from Example 13.2. The least squares estimates are $\widehat{\beta}_0 = 3.58$ and $\widehat{\beta}_1 = 0.166$. The fitted line $\widehat{r}(x) = 3.58 + 0.166\,x$ is shown in Figure 13.1. ∎

13.6 Example (The 2001 Presidential Election). Figure 13.2 shows the plot of votes for Buchanan (Y) versus votes for Bush (X) in Florida. The least squares estimates (omitting Palm Beach County) and the standard errors are

$$\begin{aligned}
\widehat{\beta}_0 &= 66.0991 & \widehat{\mathrm{se}}(\widehat{\beta}_0) &= 17.2926 \\
\widehat{\beta}_1 &= 0.0035 & \widehat{\mathrm{se}}(\widehat{\beta}_1) &= 0.0002.
\end{aligned}$$

The fitted line is

$$\mathrm{Buchanan} = 66.0991 + 0.0035\,\mathrm{Bush}.$$

(We will see later how the standard errors were computed.) Figure 13.2 also shows the residuals. The inferences from linear regression are most accurate when the residuals behave like random normal numbers. Based on the residual plot, this is not the case in this example. If we repeat the analysis replacing votes with log(votes) we get

$$\begin{aligned}
\widehat{\beta}_0 &= -2.3298 & \widehat{\mathrm{se}}(\widehat{\beta}_0) &= 0.3529 \\
\widehat{\beta}_1 &= 0.730300 & \widehat{\mathrm{se}}(\widehat{\beta}_1) &= 0.0358.
\end{aligned}$$

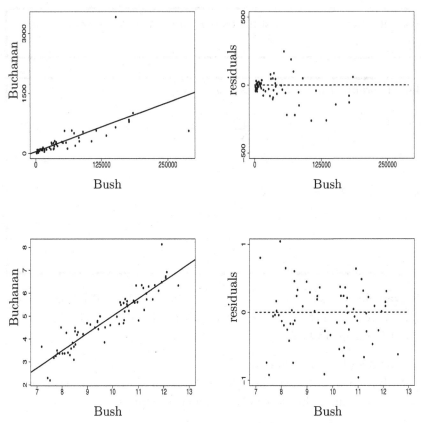

FIGURE 13.2. Voting Data for Election 2000. See example 13.6.

This gives the fit

$$\log(\text{Buchanan}) = -2.3298 + 0.7303 \ \log(\text{Bush}).$$

The residuals look much healthier. Later, we shall address the following question: how do we see if Palm Beach County has a statistically plausible outcome? ∎

13.2   Least Squares and Maximum Likelihood

Suppose we add the assumption that $\epsilon_i | X_i \sim N(0, \sigma^2)$, that is,

$$Y_i | X_i \sim N(\mu_i, \sigma^2)$$

where $\mu_i = \beta_0 + \beta_1 X_i$. The likelihood function is

$$
\prod_{i=1}^{n} f(X_i, Y_i) = \prod_{i=1}^{n} f_X(X_i) f_{Y|X}(Y_i|X_i)
$$

$$
= \prod_{i=1}^{n} f_X(X_i) \times \prod_{i=1}^{n} f_{Y|X}(Y_i|X_i)
$$

$$
= \mathcal{L}_1 \times \mathcal{L}_2
$$

where $\mathcal{L}_1 = \prod_{i=1}^{n} f_X(X_i)$ and

$$
\mathcal{L}_2 = \prod_{i=1}^{n} f_{Y|X}(Y_i|X_i). \tag{13.8}
$$

The term $\mathcal{L}_1$ does not involve the parameters β_0 and β_1. We shall focus on the second term $\mathcal{L}_2$ which is called the **conditional likelihood**, given by

$$
\mathcal{L}_2 \equiv \mathcal{L}(\beta_0, \beta_1, \sigma) = \prod_{i=1}^{n} f_{Y|X}(Y_i|X_i) \propto \sigma^{-n} \exp \left\{ -\frac{1}{2\sigma^2} \sum_i (Y_i - \mu_i)^2 \right\}.
$$

The conditional log-likelihood is

$$
\ell(\beta_0, \beta_1, \sigma) = -n \log \sigma - \frac{1}{2\sigma^2} \sum_{i=1}^{n} \left(Y_i - (\beta_0 + \beta_1 X_i) \right)^2. \tag{13.9}
$$

To find the MLE of (β_0, β_1) we maximize $\ell(\beta_0, \beta_1, \sigma)$. From (13.9) we see that maximizing the likelihood is the same as minimizing the RSS $\sum_{i=1}^{n} \left(Y_i - (\beta_0 + \beta_1 X_i) \right)^2$. Therefore, we have shown the following:

13.7 Theorem. *Under the assumption of Normality, the least squares estimator is also the maximum likelihood estimator.*

We can also maximize $\ell(\beta_0, \beta_1, \sigma)$ over σ, yielding the MLE

$$
\widehat{\sigma}^2 = \frac{1}{n} \sum_i \widehat{\epsilon}_i^2. \tag{13.10}
$$

This estimator is similar to, but not identical to, the unbiased estimator. Common practice is to use the unbiased estimator (13.7).

13.3   Properties of the Least Squares Estimators

We now record the standard errors and limiting distribution of the least squares estimator. In regression problems, we usually focus on the properties of the estimators conditional on $X^n = (X_1, \ldots, X_n)$. Thus, we state the means and variances as conditional means and variances.

13.8 Theorem. *Let $\widehat{\beta}^T = (\widehat{\beta}_0, \widehat{\beta}_1)^T$ denote the least squares estimators. Then,*

$$
\mathbb{E}(\widehat{\beta}|X^n) = \begin{pmatrix} \beta_0 \\ \beta_1 \end{pmatrix}
$$

$$
\mathbb{V}(\widehat{\beta}|X^n) = \frac{\sigma^2}{n\,s_X^2} \begin{pmatrix} \frac{1}{n}\sum_{i=1}^n X_i^2 & -\overline{X}_n \\ -\overline{X}_n & 1 \end{pmatrix}
\tag{13.11}
$$

where $s_X^2 = n^{-1}\sum_{i=1}^n (X_i - \overline{X}_n)^2$.

The estimated standard errors of $\widehat{\beta}_0$ and $\widehat{\beta}_1$ are obtained by taking the square roots of the corresponding diagonal terms of $\mathbb{V}(\widehat{\beta}|X^n)$ and inserting the estimate $\widehat{\sigma}$ for σ. Thus,

$$
\widehat{\mathsf{se}}(\widehat{\beta}_0) = \frac{\widehat{\sigma}}{s_X \sqrt{n}} \sqrt{\frac{\sum_{i=1}^n X_i^2}{n}}
\tag{13.12}
$$

$$
\widehat{\mathsf{se}}(\widehat{\beta}_1) = \frac{\widehat{\sigma}}{s_X \sqrt{n}}.
\tag{13.13}
$$

We should really write these as $\widehat{\mathsf{se}}(\widehat{\beta}_0|X^n)$ and $\widehat{\mathsf{se}}(\widehat{\beta}_1|X^n)$ but we will use the shorter notation $\widehat{\mathsf{se}}(\widehat{\beta}_0)$ and $\widehat{\mathsf{se}}(\widehat{\beta}_1)$.

13.9 Theorem. *Under appropriate conditions we have:*

1. *(Consistency):* $\widehat{\beta}_0 \overset{P}{\longrightarrow} \beta_0$ *and* $\widehat{\beta}_1 \overset{P}{\longrightarrow} \beta_1$.

2. *(Asymptotic Normality):*

$$
\frac{\widehat{\beta}_0 - \beta_0}{\widehat{\mathsf{se}}(\widehat{\beta}_0)} \rightsquigarrow N(0,1) \quad \text{and} \quad \frac{\widehat{\beta}_1 - \beta_1}{\widehat{\mathsf{se}}(\widehat{\beta}_1)} \rightsquigarrow N(0,1).
$$

3. *Approximate $1 - \alpha$ confidence intervals for β_0 and β_1 are*

$$
\widehat{\beta}_0 \pm z_{\alpha/2}\,\widehat{\mathsf{se}}(\widehat{\beta}_0) \quad \text{and} \quad \widehat{\beta}_1 \pm z_{\alpha/2}\,\widehat{\mathsf{se}}(\widehat{\beta}_1).
\tag{13.14}
$$

4. *The Wald test* [2] *for testing* $H_0 : \beta_1 = 0$ *versus* $H_1 : \beta_1 \neq 0$ *is: reject* H_0 *if* $|W| > z_{\alpha/2}$ *where* $W = \widehat{\beta}_1/\widehat{se}(\widehat{\beta}_1)$.

13.10 Example. For the election data, on the log scale, a 95 percent confidence interval is $.7303 \pm 2(.0358) = (.66, .80)$. The Wald statistics for testing $H_0 : \beta_1 = 0$ versus $H_1 : \beta_1 \neq 0$ is $|W| = |.7303 - 0|/.0358 = 20.40$ with a p-value of $\mathbb{P}(|Z| > 20.40) \approx 0$. This is strong evidence that that the true slope is not 0. ■

13.4   Prediction

Suppose we have estimated a regression model $\widehat{r}(x) = \widehat{\beta}_0 + \widehat{\beta}_1 x$ from data $(X_1, Y_1), \ldots, (X_n, Y_n)$. We observe the value $X = x_*$ of the covariate for a new subject and we want to predict their outcome Y_*. An estimate of Y_* is

$$\widehat{Y}_* = \widehat{\beta}_0 + \widehat{\beta}_1 x_*. \tag{13.15}$$

Using the formula for the variance of the sum of two random variables,

$$\mathbb{V}(\widehat{Y}_*) = \mathbb{V}(\widehat{\beta}_0 + \widehat{\beta}_1 x_*) = \mathbb{V}(\widehat{\beta}_0) + x_*^2 \mathbb{V}(\widehat{\beta}_1) + 2x_* \mathsf{Cov}(\widehat{\beta}_0, \widehat{\beta}_1).$$

Theorem 13.8 gives the formulas for all the terms in this equation. The estimated standard error $\widehat{se}(\widehat{Y}_*)$ is the square root of this variance, with $\widehat{\sigma}^2$ in place of σ^2. However, the confidence interval for Y_* is **not** of the usual form $\widehat{Y}_* \pm z_{\alpha/2}\widehat{se}$. The reason for this is explained in Exercise 10. The correct form of the confidence interval is given in the following theorem.

13.11 Theorem (Prediction Interval). *Let*

$$\widehat{\xi}_n^2 = \widehat{\sigma}^2 \left(\frac{\sum_{i=1}^n (X_i - X_*)^2}{n \sum_i (X_i - \overline{X})^2} + 1 \right). \tag{13.16}$$

An approximate $1 - \alpha$ *prediction interval for* Y_* *is*

$$\widehat{Y}_* \pm z_{\alpha/2}\, \widehat{\xi}_n. \tag{13.17}$$

---

[2]Recall from equation (10.5) that the Wald statistic for testing $H_0 : \beta = \beta_0$ versus $H_1 :$ $\beta \neq \beta_0$ is $W = (\widehat{\beta} - \beta_0)/\widehat{se}(\widehat{\beta})$.

13.12 Example (Election Data Revisited). On the log scale, our linear regression gives the following prediction equation:

$$\log(\text{Buchanan}) = -2.3298 + 0.7303 \log(\text{Bush}).$$

In Palm Beach, Bush had 152,954 votes and Buchanan had 3,467 votes. On the log scale this is 11.93789 and 8.151045. How likely is this outcome, assuming our regression model is appropriate? Our prediction for log Buchanan votes -2.3298 + .7303 (11.93789)=6.388441. Now, 8.151045 is bigger than 6.388441 but is it "significantly" bigger? Let us compute a confidence interval. We find that $\widehat{\xi}_n = .093775$ and the approximate 95 percent confidence interval is (6.200,6.578) which clearly excludes 8.151. Indeed, 8.151 is nearly 20 standard errors from $\widehat{Y}_*$. Going back to the vote scale by exponentiating, the confidence interval is (493,717) compared to the actual number of votes which is 3,467. ∎

13.5   Multiple Regression

Now suppose that the covariate is a vector of length k. The data are of the form

$$(Y_1, X_1), \ldots, (Y_i, X_i), \ldots, (Y_n, X_n)$$

where

$$X_i = (X_{i1}, \ldots, X_{ik}).$$

Here, X_i is the vector of k covariate values for the ith observation. The linear regression model is

$$Y_i = \sum_{j=1}^{k} \beta_j X_{ij} + \epsilon_i \tag{13.18}$$

for $i = 1, \ldots, n$, where $\mathbb{E}(\epsilon_i | X_{1i}, \ldots, X_{ki}) = 0$. Usually we want to include an intercept in the model which we can do by setting $X_{i1} = 1$ for $i = 1, \ldots, n$. At this point it will be more convenient to express the model in matrix notation. The outcomes will be denoted by

$$Y = \begin{pmatrix} Y_1 \\ Y_2 \\ \vdots \\ Y_n \end{pmatrix}$$

and the covariates will be denoted by

$$X = \begin{pmatrix} X_{11} & X_{12} & \cdots & X_{1k} \\ X_{21} & X_{22} & \cdots & X_{2k} \\ \vdots & \vdots & \vdots & \vdots \\ X_{n1} & X_{n2} & \cdots & X_{nk} \end{pmatrix}.$$

Each row is one observation; the columns correspond to the k covariates. Thus, X is a $(n \times k)$ matrix. Let

$$\beta = \begin{pmatrix} \beta_1 \\ \vdots \\ \beta_k \end{pmatrix} \quad \text{and} \quad \epsilon = \begin{pmatrix} \epsilon_1 \\ \vdots \\ \epsilon_n \end{pmatrix}.$$

Then we can write (13.18) as

$$Y = X\beta + \epsilon. \tag{13.19}$$

The form of the least squares estimate is given in the following theorem.

13.13 Theorem. *Assuming that the $(k \times k)$ matrix $X^T X$ is invertible,*

$$\begin{align} \widehat{\beta} &= (X^T X)^{-1} X^T Y \tag{13.20} \\ \mathbb{V}(\widehat{\beta}|X^n) &= \sigma^2 (X^T X)^{-1} \tag{13.21} \\ \widehat{\beta} &\approx N(\beta, \sigma^2 (X^T X)^{-1}). \tag{13.22} \end{align}$$

The estimate regression function is $\widehat{r}(x) = \sum_{j=1}^{k} \widehat{\beta}_j x_j$. An unbiased estimate of σ^2 is

$$\widehat{\sigma}^2 = \left(\frac{1}{n-k} \right) \sum_{i=1}^{n} \widehat{\epsilon}_i^2$$

where $\widehat{\epsilon} = X\widehat{\beta} - Y$ is the vector of residuals. An approximate $1 - \alpha$ confidence interval for β_j is

$$\widehat{\beta}_j \pm z_{\alpha/2} \widehat{se}(\widehat{\beta}_j) \tag{13.23}$$

where $\widehat{se}^2(\widehat{\beta}_j)$ is the jth diagonal element of the matrix $\widehat{\sigma}^2 (X^T X)^{-1}$.

13.14 Example. Crime data on 47 states in 1960 can be obtained from
http://lib.stat.cmu.edu/DASL/Stories/USCrime.html.
If we fit a linear regression of crime rate on 10 variables we get the following:

Covariate	$\widehat{\beta}_j$	$\widehat{se}(\widehat{\beta}_j)$	t value	p-value
(Intercept)	-589.39	167.59	-3.51	0.001 **
Age	1.04	0.45	2.33	0.025 *
Southern State	11.29	13.24	0.85	0.399
Education	1.18	0.68	1.7	0.093
Expenditures	0.96	0.25	3.86	0.000 ***
Labor	0.11	0.15	0.69	0.493
Number of Males	0.30	0.22	1.36	0.181
Population	0.09	0.14	0.65	0.518
Unemployment (14–24)	-0.68	0.48	-1.4	0.165
Unemployment (25–39)	2.15	0.95	2.26	0.030 *
Wealth	-0.08	0.09	-0.91	0.367

This table is typical of the output of a multiple regression program. The "t-value" is the Wald test statistic for testing $H_0 : \beta_j = 0$ versus $H_1 : \beta_j \neq 0$. The asterisks denote "degree of significance" and more asterisks denote smaller p-values. The example raises several important questions: (1) should we eliminate some variables from this model? (2) should we interpret these relationships as causal? For example, should we conclude that low crime prevention expenditures cause high crime rates? We will address question (1) in the next section. We will not address question (2) until Chapter 16. ∎

13.6   Model Selection

Example 13.14 illustrates a problem that often arises in multiple regression. We may have data on many covariates but we may not want to include all of them in the model. A smaller model with fewer covariates has two advantages: it might give better predictions than a big model and it is more parsimonious (simpler). Generally, as you add more variables to a regression, the bias of the predictions decreases and the variance increases. Too few covariates yields high bias; this called **underfitting**. Too many covariates yields high variance; this called **overfitting**. Good predictions result from achieving a good balance between bias and variance.

In model selection there are two problems: (i) assigning a "score" to each model which measures, in some sense, how good the model is, and (ii) searching through all the models to find the model with the best score.

Let us first discuss the problem of scoring models. Let $S \subset \{1, \ldots, k\}$ and let $\mathcal{X}_S = \{X_j : j \in S\}$ denote a subset of the covariates. Let β_S denote the coefficients of the corresponding set of covariates and let $\widehat{\beta}_S$ denote the least squares estimate of β_S. Also, let X_S denote the X matrix for this subset of covariates and define $\widehat{r}_S(x)$ to be the estimated regression function. The predicted values from model S are denoted by $\widehat{Y}_i(S) = \widehat{r}_S(X_i)$. The **prediction risk** is defined to be

$$R(S) = \sum_{i=1}^{n} \mathbb{E}(\widehat{Y}_i(S) - Y_i^*)^2 \tag{13.24}$$

where Y_i^* denotes the value of a future observation of Y_i at covariate value X_i. Our goal is to choose S to make $R(S)$ small.

The **training error** is defined to be

$$\widehat{R}_{tr}(S) = \sum_{i=1}^{n} (\widehat{Y}_i(S) - Y_i)^2.$$

This estimate is very biased as an estimate of $R(S)$.

13.15 Theorem. *The training error is a downward-biased estimate of the prediction risk:*

$$\mathbb{E}(\widehat{R}_{tr}(S)) < R(S).$$

In fact,

$$\mathsf{bias}(\widehat{R}_{tr}(S)) = \mathbb{E}(\widehat{R}_{tr}(S)) - R(S) = -2\sum_{i=1}^{n} \mathsf{Cov}(\widehat{Y}_i, Y_i). \tag{13.25}$$

The reason for the bias is that the data are being used twice: to estimate the parameters and to estimate the risk. When we fit a complex model with many parameters, the covariance $\mathsf{Cov}(\widehat{Y}_i, Y_i)$ will be large and the bias of the training error gets worse. Here are some better estimates of risk.

Mallow's C_p statistic is defined by

$$\widehat{R}(S) = \widehat{R}_{tr}(S) + 2|S|\widehat{\sigma}^2 \tag{13.26}$$

where $|S|$ denotes the number of terms in S and $\widehat{\sigma}^2$ is the estimate of σ^2 obtained from the full model (with all covariates in the model). This is simply the training error plus a bias correction. This estimate is named in honor of Colin Mallows who invented it. The first term in (13.26) measures the fit of the model while the second measure the complexity of the model. Think of the C_p statistic as:

lack of fit + complexity penalty.

Thus, **finding a good model involves trading off fit and complexity.**

A related method for estimating risk is **AIC (Akaike Information Criterion).** The idea is to choose S to maximize

$$\ell_S - |S| \tag{13.27}$$

where ℓ_S is the log-likelihood of the model evaluated at the MLE. [3] This can be thought of "goodness of fit" minus "complexity." In linear regression with Normal errors (and taking σ equal to its estimate from the largest model), maximizing AIC is equivalent to minimizing Mallow's C_p; see Exercise 8. The appendix contains more explanation about AIC.

Yet another method for estimating risk is **leave-one-out cross-validation.** In this case, the risk estimator is

$$\widehat{R}_{CV}(S) = \sum_{i=1}^{n}(Y_i - \widehat{Y}_{(i)})^2 \tag{13.28}$$

where $\widehat{Y}_{(i)}$ is the prediction for Y_i obtained by fitting the model with Y_i omitted. It can be shown that

$$\widehat{R}_{CV}(S) = \sum_{i=1}^{n}\left(\frac{Y_i - \widehat{Y}_i(S)}{1 - U_{ii}(S)}\right)^2 \tag{13.29}$$

where $U_{ii}(S)$ is the ith diagonal element of the matrix

$$U(S) = X_S(X_S^T X_S)^{-1}X_S^T. \tag{13.30}$$

Thus, one need not actually drop each observation and re-fit the model. A generalization is **k-fold cross-validation.** Here we divide the data into k groups; often people take $k = 10$. We omit one group of data and fit the models to the remaining data. We use the fitted model to predict the data in the group that was omitted. We then estimate the risk by $\sum_i(Y_i - \widehat{Y}_i)^2$ where the sum is over the the data points in the omitted group. This process is repeated for each of the k groups and the resulting risk estimates are averaged.

For linear regression, Mallows C_p and cross-validation often yield essentially the same results so one might as well use Mallows' method. In some of the more complex problems we will discuss later, cross-validation will be more useful.

Another scoring method is BIC (Bayesian information criterion). Here we choose a model to maximize

$$\text{BIC}(S) = \ell_S - \frac{|S|}{2}\log n. \tag{13.31}$$

---

[3] Some texts use a slightly different definition of AIC which involves multiplying the definition here by 2 or -2. This has no effect on which model is selected.

The BIC score has a Bayesian interpretation. Let $\mathcal{S} = \{S_1, \ldots, S_m\}$ denote a set of models. Suppose we assign the prior $\mathbb{P}(S_j) = 1/m$ over the models. Also, assume we put a smooth prior on the parameters within each model. It can be shown that the posterior probability for a model is approximately,

$$\mathbb{P}(S_j | \text{data}) \approx \frac{e^{BIC(S_j)}}{\sum_r e^{BIC(S_r)}}.$$

Hence, choosing the model with highest BIC is like choosing the model with highest posterior probability. The BIC score also has an information-theoretic interpretation in terms of something called minimum description length. The BIC score is identical to Mallows C_p except that it puts a more severe penalty for complexity. It thus leads one to choose a smaller model than the other methods.

Now let us turn to the problem of model search. If there are k covariates then there are 2^k possible models. We need to search through all these models, assign a score to each one, and choose the model with the best score. If k is not too large we can do a complete search over all the models. When k is large, this is infeasible. In that case we need to search over a subset of all the models. Two common methods are **forward and backward stepwise regression.** In forward stepwise regression, we start with no covariates in the model. We then add the one variable that leads to the best score. We continue adding variables one at a time until the score does not improve. Backwards stepwise regression is the same except that we start with the biggest model and drop one variable at a time. Both are greedy searches; nether is guaranteed to find the model with the best score. Another popular method is to do random searching through the set of all models. However, there is no reason to expect this to be superior to a deterministic search.

13.16 Example. We applied backwards stepwise regression to the crime data using AIC. The following was obtained from the program R. This program uses a slightly different definition of AIC. With their definition, we seek the smallest (not largest) possible AIC. This is the same is minimizing Mallows C_p.

The full model (which includes all covariates) has AIC= 310.37. In ascending order, the AIC scores for deleting one variable are as follows:

variable	Pop	Labor	South	Wealth	Males	U1	Educ.	U2	Age	Expend
AIC	308	309	309	309	310	310	312	314	315	324

For example, if we dropped Pop from the model and kept the other terms, then the AIC score would be 308. Based on this information we drop "population" from the model and the current AIC score is 308. Now we consider dropping a variable from the current model. The AIC scores are:

variable	South	Labor	Wealth	Males	U1	Education	U2	Age	Expend
AIC	308	308	308	309	309	310	313	313	329

We then drop "Southern" from the model. This process is continued until there is no gain in AIC by dropping any variables. In the end, we are left with the following model:

$$\text{Crime} \;=\; 1.2 \,\text{Age} + .75 \,\text{Education} + .87 \,\text{Expenditure}$$
$$+ .34 \,\text{Males} - .86 \,\text{U1} + 2.31 \,\text{U2}.$$

Warning! This does not yet address the question of which variables are **causes** of crime. ∎

There is another method for model selection that avoids having to search through all possible models. This method, which is due to Zheng and Loh (1995), does not seek to minimize prediction errors. Rather, it assumes some subset of the β_j's are exactly equal to 0 and tries to find the true model, that is, the smallest sub-model consisting of nonzero β_j terms. The method is carried out as follows.

Zheng-Loh Model Selection Method [4]

1. Fit the full model with all k covariates and let $W_j = \widehat{\beta}_j/\widehat{\text{se}}(\widehat{\beta}_j)$ denote the Wald test statistic for $H_0 : \beta_j = 0$ versus $H_1 : \beta_j \neq 0$.

2. Order the test statistics from largest to smallest in absolute value:

$$|W_{(1)}| \geq |W_{(2)}| \geq \cdots \geq |W_{(k)}|.$$

3. Let $\widehat{j}$ be the value of j that minimizes

$$\text{RSS}(j) + j\,\widehat{\sigma}^2 \log n$$

where $\text{RSS}(j)$ is the residual sums of squares from the model with the j largest Wald statistics.

4. Choose, as the final model, the regression with the $\widehat{j}$ terms with the largest absolute Wald statistics.

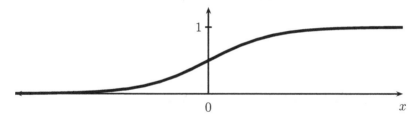

FIGURE 13.3. The logistic function $p = e^x/(1 + e^x)$.

Zheng and Loh showed that, under appropriate conditions, this method chooses the true model with probability tending to one as the sample size increases.

13.7 Logistic Regression

So far we have assumed that Y_i is real valued. **Logistic regression** is a parametric method for regression when $Y_i \in \{0, 1\}$ is binary. For a k-dimensional covariate X, the model is

$$p_i \equiv p_i(\beta) \equiv \mathbb{P}(Y_i = 1 | X = x) = \frac{e^{\beta_0 + \sum_{j=1}^{k} \beta_j x_{ij}}}{1 + e^{\beta_0 + \sum_{j=1}^{k} \beta_j x_{ij}}} \tag{13.32}$$

or, equivalently,

$$\text{logit}(p_i) = \beta_0 + \sum_{j=1}^{k} \beta_j x_{ij} \tag{13.33}$$

where

$$\text{logit}(p) = \log\left(\frac{p}{1-p}\right). \tag{13.34}$$

The name "logistic regression" comes from the fact that $e^x/(1 + e^x)$ is called the logistic function. A plot of the logistic for a one-dimensional covariate is shown in Figure 13.3.

Because the Y_i's are binary, the data are Bernoulli:

$$Y_i | X_i = x_i \sim \text{Bernoulli}(p_i).$$

Hence the (conditional) likelihood function is

$$\mathcal{L}(\beta) = \prod_{i=1}^{n} p_i(\beta)^{Y_i} (1 - p_i(\beta))^{1 - Y_i}. \tag{13.35}$$

---

[4] This is just one version of their method. In particular, the penalty $j \log n$ is only one choice from a set of possible penalty functions.

The MLE $\widehat{\beta}$ has to be obtained by maximizing $\mathcal{L}(\beta)$ numerically. There is a fast numerical algorithm called reweighted least squares. The steps are as follows:

Reweighted Least Squares Algorithm

Choose starting values $\widehat{\beta}^0 = (\widehat{\beta}_0^0, \ldots, \widehat{\beta}_k^0)$ and compute p_i^0 using equation (13.32), for $i = 1, \ldots, n$. Set $s = 0$ and iterate the following steps until convergence.

1. Set
$$Z_i = \text{logit}(p_i^s) + \frac{Y_i - p_i^s}{p_i^s(1 - p_i^s)}, \quad i = 1, \ldots, n.$$

2. Let W be a diagonal matrix with (i, i) element equal to $p_i^s(1 - p_i^s)$.

3. Set
$$\widehat{\beta}^s = (X^T W X)^{-1} X^T W Y.$$

   This corresponds to doing a (weighted) linear regression of Z on Y.

4. Set $s = s + 1$ and go back to the first step.

The Fisher information matrix I can also be obtained numerically. The estimate standard error of $\widehat{\beta}_j$ is the (j, j) element of $J = I^{-1}$. Model selection is usually done using the AIC score $\ell_S - |S|$.

13.17 Example. The Coronary Risk-Factor Study (CORIS) data involve 462 males between the ages of 15 and 64 from three rural areas in South Africa, (Rousseauw et al. (1983)). The outcome Y is the presence ($Y = 1$) or absence ($Y = 0$) of coronary heart disease. There are 9 covariates: systolic blood pressure, cumulative tobacco (kg), ldl (low density lipoprotein cholesterol), adiposity, famhist (family history of heart disease), typea (type-A behavior), obesity, alcohol (current alcohol consumption), and age. A logistic regression yields the following estimates and Wald statistics W_j for the coefficients:

Covariate	$\widehat{\beta}_j$	$\widehat{se}$	W_j	p-value
Intercept	-6.145	1.300	-4.738	0.000
sbp	0.007	0.006	1.138	0.255
tobacco	0.079	0.027	2.991	0.003
ldl	0.174	0.059	2.925	0.003
adiposity	0.019	0.029	0.637	0.524
famhist	0.925	0.227	4.078	0.000
typea	0.040	0.012	3.233	0.001
obesity	-0.063	0.044	-1.427	0.153
alcohol	0.000	0.004	0.027	0.979
age	0.045	0.012	3.754	0.000

Are you surprised by the fact that systolic blood pressure is not significant or by the minus sign for the obesity coefficient? If yes, then you are confusing association and causation. This issue is discussed in Chapter 16. The fact that blood pressure is not significant does not mean that blood pressure is not an important *cause* of heart disease. It means that it is not an important *predictor* of heart disease relative to the other variables in the model. ∎

13.8   Bibliographic Remarks

A succinct book on linear regression is Weisberg (1985). A data-mining view of regression is given in Hastie et al. (2001). The Akaike Information Criterion (AIC) is due to Akaike (1973). The Bayesian Information Criterion (BIC) is due to Schwarz (1978). References on logistic regression include Agresti (1990) and Dobson (2001).

13.9   Appendix

THE AKAIKE INFORMATION CRITERION (AIC). Consider a set of models $\{M_1, M_2, \ldots\}$. Let $\widehat{f}_j(x)$ denote the estimated probability function obtained by using the maximum likelihood estimator of model M_j. Thus, $\widehat{f}_j(x) = \widehat{f}(x; \widehat{\beta}_j)$ where $\widehat{\beta}_j$ is the MLE of the set of parameters β_j for model M_j. We will use the loss function $D(f, \widehat{f})$ where

$$D(f, g) = \sum_x f(x) \log\left(\frac{f(x)}{g(x)}\right)$$

is the Kullback-Leibler distance between two probability functions. The corresponding risk function is $R(f, \widehat{f}) = \mathbb{E}(D(f, \widehat{f}))$. Notice that $D(f, \widehat{f}) = c -$

$A(f, \widehat{f})$ where $c = \sum_x f(x) \log f(x)$ does not depend on $\widehat{f}$ and

$$A(f, \widehat{f}) = \sum_x f(x) \log \widehat{f}(x).$$

Thus, minimizing the risk is equivalent to maximizing $a(f, \widehat{f}) \equiv \mathbb{E}(A(f, \widehat{f}))$.

It is tempting to estimate $a(f, \widehat{f})$ by $\sum_x \widehat{f}(x) \log \widehat{f}(x)$ but, just as the training error in regression is a highly biased estimate of prediction risk, it is also the case that $\sum_x \widehat{f}(x) \log \widehat{f}(x)$ is a highly biased estimate of $a(f, \widehat{f})$. In fact, the bias is approximately equal to $|M_j|$. Thus:

13.18 Theorem. $AIC(M_j)$ *is an approximately unbiased estimate of* $a(f, \widehat{f})$.

13.10   Exercises

1. Prove Theorem 13.4.

2. Prove the formulas for the standard errors in Theorem 13.8. You should regard the X_i's as fixed constants.

3. Consider the **regression through the origin** model:

$$Y_i = \beta X_i + \epsilon.$$

   Find the least squares estimate for β. Find the standard error of the estimate. Find conditions that guarantee that the estimate is consistent.

4. Prove equation (13.25).

5. In the simple linear regression model, construct a Wald test for $H_0 : \beta_1 = 17\beta_0$ versus $H_1 : \beta_1 \neq 17\beta_0$.

6. Get the passenger car mileage data from

   http://lib.stat.cmu.edu/DASL/Datafiles/carmpgdat.html

   (a) Fit a simple linear regression model to predict MPG (miles per gallon) from HP (horsepower). Summarize your analysis including a plot of the data with the fitted line.

   (b) Repeat the analysis but use log(MPG) as the response. Compare the analyses.

7. Get the passenger car mileage data from

   http://lib.stat.cmu.edu/DASL/Datafiles/carmpgdat.html

   (a) Fit a multiple linear regression model to predict MPG (miles per gallon) from the other variables. Summarize your analysis.

   (b) Use Mallow C_p to select a best sub-model. To search through the models try (i) forward stepwise, (ii) backward stepwise. Summarize your findings.

   (c) Use the Zheng-Loh model selection method and compare to (b).

   (d) Perform all possible regressions. Compare C_p and BIC. Compare the results.

8. Assume a linear regression model with Normal errors. Take σ known. Show that the model with highest AIC (equation (13.27)) is the model with the lowest Mallows C_p statistic.

9. In this question we will take a closer look at the AIC method. Let $X_1, \ldots, X_n$ be IID observations. Consider two models $\mathcal{M}_0$ and $\mathcal{M}_1$. Under $\mathcal{M}_0$ the data are assumed to be $N(0,1)$ while under $\mathcal{M}_1$ the data are assumed to be $N(\theta, 1)$ for some unknown $\theta \in \mathbb{R}$:

$$\mathcal{M}_0 : X_1, \ldots, X_n \sim N(0,1)$$
$$\mathcal{M}_1 : X_1, \ldots, X_n \sim N(\theta, 1), \quad \theta \in \mathbb{R}.$$

This is just another way to view the hypothesis testing problem: $H_0 : \theta = 0$ versus $H_1 : \theta \neq 0$. Let $\ell_n(\theta)$ be the log-likelihood function. The AIC score for a model is the log-likelihood at the MLE minus the number of parameters. (Some people multiply this score by 2 but that is irrelevant.) Thus, the AIC score for $\mathcal{M}_0$ is $AIC_0 = \ell_n(0)$ and the AIC score for $\mathcal{M}_1$ is $AIC_1 = \ell_n(\widehat{\theta}) - 1$. Suppose we choose the model with the highest AIC score. Let J_n denote the selected model:

$$J_n = \begin{cases} 0 & \text{if } AIC_0 > AIC_1 \\ 1 & \text{if } AIC_1 > AIC_0. \end{cases}$$

(a) Suppose that $\mathcal{M}_0$ is the true model, i.e. $\theta = 0$. Find

$$\lim_{n \to \infty} \mathbb{P}\left(J_n = 0 \right).$$

Now compute $\lim_{n \to \infty} \mathbb{P}\left(J_n = 0 \right)$ when $\theta \neq 0$.

(b) The fact that $\lim_{n \to \infty} \mathbb{P}(J_n = 0) \neq 1$ when $\theta = 0$ is why some people say that AIC "overfits." But this is not quite true as we shall now see. Let $\phi_\theta(x)$ denote a Normal density function with mean θ and variance 1. Define

$$\widehat{f}_n(x) = \begin{cases} \phi_0(x) & \text{if } J_n = 0 \\ \phi_{\widehat{\theta}}(x) & \text{if } J_n = 1. \end{cases}$$

If $\theta = 0$, show that $D(\phi_0, \widehat{f}_n) \xrightarrow{P} 0$ as $n \to \infty$ where

$$D(f, g) = \int f(x) \log\left(\frac{f(x)}{g(x)}\right) dx$$

is the Kullback-Leibler distance. Show also that $D(\phi_\theta, \widehat{f}_n) \xrightarrow{P} 0$ if $\theta \neq 0$. Hence, AIC consistently estimates the true density even if it "overshoots" the correct model.

(c) Repeat this analysis for BIC which is the log-likelihood minus $(p/2) \log n$ where p is the number of parameters and n is sample size.

10. In this question we take a closer look at prediction intervals. Let $\theta = \beta_0 + \beta_1 X_*$ and let $\widehat{\theta} = \widehat{\beta}_0 + \widehat{\beta}_1 X_*$. Thus, $\widehat{Y}_* = \widehat{\theta}$ while $Y_* = \theta + \epsilon$. Now, $\widehat{\theta} \approx N(\theta, \mathsf{se}^2)$ where

$$\mathsf{se}^2 = \mathbb{V}(\widehat{\theta}) = \mathbb{V}(\widehat{\beta}_0 + \widehat{\beta}_1 x_*).$$

Note that $\mathbb{V}(\widehat{\theta})$ is the same as $\mathbb{V}(\widehat{Y}_*)$. Now, $\widehat{\theta} \pm 2\sqrt{\mathbb{V}(\widehat{\theta})}$ is an approximate 95 percent confidence interval for $\theta = \beta_0 + \beta_1 x_*$ using the usual argument for a confidence interval. But, as you shall now show, it is not a valid confidence interval for Y_*.

(a) Let $s = \sqrt{\mathbb{V}(\widehat{Y}_*)}$. Show that

$$\mathbb{P}(\widehat{Y}_* - 2s < Y_* < \widehat{Y}_* + 2s) \approx \mathbb{P}\left(-2 < N\left(0, 1 + \frac{\sigma^2}{s^2}\right) < 2\right)$$
$$\neq 0.95.$$

(b) The problem is that the quantity of interest Y_* is equal to a parameter θ plus a random variable. We can fix this by defining

$$\xi_n^2 = \mathbb{V}(\widehat{Y}_*) + \sigma^2 = \left[\frac{\sum_i (x_i - x_*)^2}{n \sum_i (x_i - \overline{x})^2} + 1\right]\sigma^2.$$

In practice, we substitute $\widehat{\sigma}$ for σ and we denote the resulting quantity by $\widehat{\xi}_n$. Now consider the interval $\widehat{Y}_* \pm 2\,\widehat{\xi}_n$. Show that

$$\mathbb{P}(\widehat{Y}_* - 2\widehat{\xi}_n < Y_* < \widehat{Y}_* + 2\widehat{\xi}_n) \approx \mathbb{P}(-2 < N(0,1) < 2) \approx 0.95.$$

11. Get the Coronary Risk-Factor Study (CORIS) data from the book web site. Use backward stepwise logistic regression based on AIC to select a model. Summarize your results.

14
Multivariate Models

In this chapter we revisit the Multinomial model and the multivariate Normal. Let us first review some notation from linear algebra. In what follows, x and y are vectors and A is a matrix.

Linear Algebra Notation

$x^T y$	inner product $\sum_j x_j y_j$
$\lvert A \rvert$	determinant
A^T	transpose of A
A^{-1}	inverse of A
I	the identity matrix
$\mathrm{tr}(A)$	trace of a square matrix; sum of its diagonal elements
$A^{1/2}$	square root matrix

The trace satisfies $\mathrm{tr}(AB) = \mathrm{tr}(BA)$ and $\mathrm{tr}(A) + \mathrm{tr}(B)$. Also, $\mathrm{tr}(a) = a$ if a is a scalar. A matrix is **positive definite** if $x^T \Sigma x > 0$ for all nonzero vectors x. If a matrix A is symmetric and positive definite, its square root $A^{1/2}$ exists and has the following properties: (1) $A^{1/2}$ is symmetric; (2) $A = A^{1/2} A^{1/2}$; (3) $A^{1/2} A^{-1/2} = A^{-1/2} A^{1/2} = I$ where $A^{-1/2} = (A^{1/2})^{-1}$.

14.1   Random Vectors

Multivariate models involve a random vector X of the form

$$X = \begin{pmatrix} X_1 \\ \vdots \\ X_k \end{pmatrix}.$$

The mean of a random vector X is defined by

$$\mu = \begin{pmatrix} \mu_1 \\ \vdots \\ \mu_k \end{pmatrix} = \begin{pmatrix} E(X_1) \\ \vdots \\ E(X_k) \end{pmatrix}. \tag{14.1}$$

The **covariance matrix** Σ, also written $\mathbb{V}(X)$, is defined to be

$$\Sigma = \begin{bmatrix} \mathbb{V}(X_1) & \mathsf{Cov}(X_1, X_2) & \cdots & \mathsf{Cov}(X_1, X_k) \\ \mathsf{Cov}(X_2, X_1) & \mathbb{V}(X_2) & \cdots & \mathsf{Cov}(X_2, X_k) \\ \vdots & \vdots & \vdots & \vdots \\ \mathsf{Cov}(X_k, X_1) & \mathsf{Cov}(X_k, X_2) & \cdots & \mathbb{V}(X_k) \end{bmatrix}. \tag{14.2}$$

This is also called the variance matrix or the variance–covariance matrix. The inverse Σ^{-1} is called the **precision matrix**.

14.1 Theorem. *Let a be a vector of length k and let X be a random vector of the same length with mean μ and variance Σ. Then $\mathbb{E}(a^T X) = a^T \mu$ and $\mathbb{V}(a^T X) = a^T \Sigma a$. If A is a matrix with k columns, then $\mathbb{E}(AX) = A\mu$ and $\mathbb{V}(AX) = A\Sigma A^T$.*

Now suppose we have a random sample of n vectors:

$$\begin{pmatrix} X_{11} \\ X_{21} \\ \vdots \\ X_{k1} \end{pmatrix}, \begin{pmatrix} X_{12} \\ X_{22} \\ \vdots \\ X_{k2} \end{pmatrix}, \ldots, \begin{pmatrix} X_{1n} \\ X_{2n} \\ \vdots \\ X_{kn} \end{pmatrix}. \tag{14.3}$$

The sample mean $\overline{X}$ is a vector defined by

$$\overline{X} = \begin{pmatrix} \overline{X}_1 \\ \vdots \\ \overline{X}_k \end{pmatrix}$$

where $\overline{X}_i = n^{-1}\sum_{j=1}^{n} X_{ij}$. The sample variance matrix, also called the covariance matrix or the variance–covariance matrix, is

$$
S = \begin{bmatrix}
s_{11} & s_{12} & \cdots & s_{1k} \\
s_{12} & s_{22} & \cdots & s_{2k} \\
\vdots & \vdots & \vdots & \vdots \\
s_{1k} & s_{2k} & \cdots & s_{kk}
\end{bmatrix}
\tag{14.4}
$$

where

$$
s_{ab} = \frac{1}{n-1}\sum_{j=1}^{n}(X_{aj} - \overline{X}_a)(X_{bj} - \overline{X}_b).
$$

It follows that $\mathbb{E}(\overline{X}) = \mu$. and $\mathbb{E}(S) = \Sigma$.

14.2 Estimating the Correlation

Consider n data points from a bivariate distribution:

$$
\begin{pmatrix} X_{11} \\ X_{21} \end{pmatrix}, \begin{pmatrix} X_{12} \\ X_{22} \end{pmatrix}, \cdots, \begin{pmatrix} X_{1n} \\ X_{2n} \end{pmatrix}.
$$

Recall that the correlation between X_1 and X_2 is

$$
\rho = \frac{\mathbb{E}((X_1 - \mu_1)(X_2 - \mu_2))}{\sigma_1 \sigma_2}
\tag{14.5}
$$

where $\sigma_j^2 = \mathbb{V}(X_{ji})$, $j = 1, 2$. The nonparametric plug-in estimator is the sample correlation [1]

$$
\widehat{\rho} = \frac{\sum_{i=1}^{n}(X_{1i} - \overline{X}_1)(X_{2i} - \overline{X}_2)}{s_1 s_2}
\tag{14.6}
$$

where

$$
s_j^2 = \frac{1}{n-1}\sum_{i=1}^{n}(X_{ji} - \overline{X}_j)^2.
$$

We can construct a confidence interval for ρ by applying the delta method. However, it turns out that we get a more accurate confidence interval by first constructing a confidence interval for a function $\theta = f(\rho)$ and then applying

---

[1] More precisely, the plug-in estimator has n rather than $n-1$ in the formula for s_j but this difference is small.

the inverse function f^{-1}. The method, due to Fisher, is as follows: Define f and its inverse by

$$f(r) = \frac{1}{2}\left(\log(1+r) - \log(1-r)\right)$$

$$f^{-1}(z) = \frac{e^{2z} - 1}{e^{2z} + 1}.$$

Approximate Confidence Interval for The Correlation

1. Compute

$$\widehat{\theta} = f(\widehat{\rho}) = \frac{1}{2}\left(\log(1+\widehat{\rho}) - \log(1-\widehat{\rho})\right).$$

2. Compute the approximate standard error of $\widehat{\theta}$ which can be shown to be

$$\widehat{\mathsf{se}}(\widehat{\theta}) = \frac{1}{\sqrt{n-3}}.$$

3. An approximate $1 - \alpha$ confidence interval for $\theta = f(\rho)$ is

$$(a, b) \equiv \left(\widehat{\theta} - \frac{z_{\alpha/2}}{\sqrt{n-3}}, \widehat{\theta} + \frac{z_{\alpha/2}}{\sqrt{n-3}}\right).$$

4. Apply the inverse transformation $f^{-1}(z)$ to get a confidence interval for ρ:

$$\left(\frac{e^{2a} - 1}{e^{2a} + 1}, \frac{e^{2b} - 1}{e^{2b} + 1}\right).$$

Yet another method for getting a confidence interval for ρ is to use the bootstrap.

14.3   Multivariate Normal

Recall that a vector X has a multivariate Normal distribution, denoted by $X \sim N(\mu, \Sigma)$, if its density is

$$f(x; \mu, \Sigma) = \frac{1}{(2\pi)^{k/2}|\Sigma|^{1/2}} \exp\left\{-\frac{1}{2}(x - \mu)^T \Sigma^{-1}(x - \mu)\right\} \qquad (14.7)$$

where μ is a vector of length k and Σ is a $k \times k$ symmetric, positive definite matrix. Then $\mathbb{E}(X) = \mu$ and $\mathbb{V}(X) = \Sigma$.

14.2 Theorem. *The following properties hold:*

1. *If* $Z \sim N(0,1)$ *and* $X = \mu + \Sigma^{1/2}Z$, *then* $X \sim N(\mu, \Sigma)$.

2. *If* $X \sim N(\mu, \Sigma)$, *then* $\Sigma^{-1/2}(X - \mu) \sim N(0,1)$.

3. *If* $X \sim N(\mu, \Sigma)$ *a is a vector of the same length as* X, *then* $a^T X \sim N(a^T \mu, a^T \Sigma a)$.

4. *Let*
$$V = (X - \mu)^T \Sigma^{-1} (X - \mu).$$
Then $V \sim \chi_k^2$.

14.3 Theorem. *Given a random sample of size n from a $N(\mu, \Sigma)$, the log-likelihood is (up to a constant not depending on μ or Σ) given by*

$$\ell(\mu, \Sigma) = -\frac{n}{2}(\overline{X} - \mu)^T \Sigma^{-1} (\overline{X} - \mu) - \frac{n}{2}\mathrm{tr}(\Sigma^{-1}S) - \frac{n}{2}\log|\Sigma|.$$

The MLE *is*

$$\widehat{\mu} = \overline{X} \quad \text{and} \quad \widehat{\Sigma} = \left(\frac{n-1}{n}\right)S. \tag{14.8}$$

14.4   Multinomial

Let us now review the Multinomial distribution. The data take the form $X = (X_1, \ldots, X_k)$ where each X_j is a count. Think of drawing n balls (with replacement) from an urn which has balls with k different colors. In this case, X_j is the number of balls of the kth color. Let $p = (p_1, \ldots, p_k)$ where $p_j \geq 0$ and $\sum_{j=1}^{k} p_j = 1$ and suppose that p_j is the probability of drawing a ball of color j.

14.4 Theorem. *Let* $X \sim$ Multinomial(n, p). *Then the marginal distribution of* X_j *is* $X_j \sim$ Binomial(n, p_j). *The mean and variance of* X *are*

$$\mathbb{E}(X) = \begin{pmatrix} np_1 \\ \vdots \\ np_k \end{pmatrix}$$

and

$$\mathbb{V}(X) = \begin{pmatrix} np_1(1 - p_1) & -np_1p_2 & \cdots & -np_1p_k \\ -np_1p_2 & np_2(1 - p_2) & \cdots & -np_2p_k \\ \vdots & \vdots & \vdots & \vdots \\ -np_1p_k & -np_2p_k & \cdots & np_k(1 - p_k) \end{pmatrix}.$$

PROOF. That $X_j \sim \text{Binomial}(n, p_j)$ follows easily. Hence, $\mathbb{E}(X_j) = np_j$ and $\mathbb{V}(X_j) = np_j(1 - p_j)$. To compute $\text{Cov}(X_i, X_j)$ we proceed as follows: Notice that $X_i + X_j \sim \text{Binomial}(n, p_i + p_j)$ and so $\mathbb{V}(X_i + X_j) = n(p_i + p_j)(1 - p_i - p_j)$. On the other hand,

$$
\begin{aligned}
\mathbb{V}(X_i + X_j) &= \mathbb{V}(X_i) + \mathbb{V}(X_j) + 2\text{Cov}(X_i, X_j) \\
&= np_i(1 - p_i) + np_j(1 - p_j) + 2\text{Cov}(X_i, X_j).
\end{aligned}
$$

Equating this last expression with $n(p_i + p_j)(1 - p_i - p_j)$ implies that $\text{Cov}(X_i, X_j) = -np_i p_j$. ∎

14.5 Theorem. *The maximum likelihood estimator of p is*

$$
\widehat{p} = \begin{pmatrix} \widehat{p}_1 \\ \vdots \\ \widehat{p}_k \end{pmatrix} = \begin{pmatrix} \frac{X_1}{n} \\ \vdots \\ \frac{X_k}{n} \end{pmatrix} = \frac{X}{n}.
$$

PROOF. The log-likelihood (ignoring a constant) is

$$
\ell(p) = \sum_{j=1}^{k} X_j \log p_j.
$$

When we maximize ℓ we have to be careful since we must enforce the constraint that $\sum_j p_j = 1$. We use the method of Lagrange multipliers and instead maximize

$$
A(p) = \sum_{j=1}^{k} X_j \log p_j + \lambda \left(\sum_j p_j - 1 \right).
$$

Now

$$
\frac{\partial A(p)}{\partial p_j} = \frac{X_j}{p_j} + \lambda.
$$

Setting $\frac{\partial A(p)}{\partial p_j} = 0$ yields $\widehat{p}_j = -X_j/\lambda$. Since $\sum_j \widehat{p}_j = 1$ we see that $\lambda = -n$ and hence $\widehat{p}_j = X_j/n$ as claimed. ∎

Next we would like to know the variability of the MLE. We can either compute the variance matrix of $\widehat{p}$ directly or we can approximate the variability of the MLE by computing the Fisher information matrix. These two approaches give the same answer in this case. The direct approach is easy: $\mathbb{V}(\widehat{p}) = \mathbb{V}(X/n) = n^{-2}\mathbb{V}(X)$, and so

$$
\mathbb{V}(\widehat{p}) = \frac{1}{n}\Sigma
$$

where

$$\Sigma = \begin{pmatrix} p_1(1-p_1) & -p_1p_2 & \cdots & -p_1p_k \\ -p_1p_2 & p_2(1-p_2) & \cdots & -p_2p_k \\ \vdots & \vdots & \vdots & \vdots \\ -p_1p_k & -p_2p_k & \cdots & p_k(1-p_k) \end{pmatrix}.$$

For large n, $\widehat{p}$ has approximately a multivariate Normal distribution.

14.6 Theorem. *As $n \to \infty$,*

$$\sqrt{n}(\widehat{p} - p) \rightsquigarrow N(0, \Sigma).$$

14.5 Bibliographic Remarks

Some references on multivariate analysis are Johnson and Wichern (1982) and Anderson (1984). The method for constructing the confidence interval for the correlation described in this chapter is due to Fisher (1921).

14.6 Appendix

PROOF of Theorem 14.3. Denote the ith random vector by X^i. The log-likelihood is

$$\begin{aligned} \ell(\mu, \Sigma) &= \sum_{i=1}^{n} f(X^i; \mu, \Sigma) \\ &= -\frac{kn}{2} \log(2\pi) - \frac{n}{2} \log |\Sigma| - \frac{1}{2} \sum_{i=1}^{n} (X^i - \mu)^T \Sigma^{-1} (X^i - \mu). \end{aligned}$$

Now,

$$\begin{aligned} \sum_{i=1}^{n} &(X^i - \mu)^T \Sigma^{-1} (X^i - \mu) \\ &= \sum_{i=1}^{n} [(X^i - \overline{X}) + (\overline{X} - \mu)]^T \Sigma^{-1} [(X^i - \overline{X}) + (\overline{X} - \mu)] \\ &= \sum_{i=1}^{n} \left[(X^i - \overline{X})^T \Sigma^{-1} (X^i - \overline{X}) \right] + n(\overline{X} - \mu)^T \Sigma^{-1} (\overline{X} - \mu) \end{aligned}$$

since $\sum_{i=1}^{n}(X^i - \overline{X})\Sigma^{-1}(\overline{X} - \mu) = 0$. Also, notice that $(X^i - \mu)^T \Sigma^{-1}(X^i - \mu)$ is a scalar, so

$$\sum_{i=1}^{n}(X^i - \mu)^T \Sigma^{-1}(X^i - \mu) = \sum_{i=1}^{n} \text{tr}\left[(X^i - \mu)^T \Sigma^{-1}(X^i - \mu) \right]$$

$$\begin{aligned}
&= \sum_{i=1}^{n} \operatorname{tr}\left[\Sigma^{-1}(X^i - \mu)(X^i - \mu)^T\right] \\
&= \operatorname{tr}\left[\Sigma^{-1}\sum_{i=1}^{n}(X^i - \mu)(X^i - \mu)^T\right] \\
&= n \operatorname{tr}\left[\Sigma^{-1}S\right]
\end{aligned}$$

and the conclusion follows. ■

14.7 Exercises

1. Prove Theorem 14.1.

2. Find the Fisher information matrix for the MLE of a Multinomial.

3. (Computer Experiment.) Write a function to generate nsim observations from a Multinomial(n, p) distribution.

4. (Computer Experiment.) Write a function to generate nsim observations from a Multivariate normal with given mean μ and covariance matrix Σ.

5. (Computer Experiment.) Generate 100 random vectors from a $N(\mu, \Sigma)$ distribution where

$$\mu = \begin{pmatrix} 3 \\ 8 \end{pmatrix}, \quad \Sigma = \begin{pmatrix} 1 & 1 \\ 1 & 2 \end{pmatrix}.$$

Plot the simulation as a scatterplot. Estimate the mean and covariance matrix Σ. Find the correlation ρ between X_1 and X_2. Compare this with the sample correlations from your simulation. Find a 95 percent confidence interval for ρ. Use two methods: the bootstrap and Fisher's method. Compare.

6. (Computer Experiment.) Repeat the previous exercise 1000 times. Compare the coverage of the two confidence intervals for ρ.

15
Inference About Independence

In this chapter we address the following questions:

(1) How do we test if two random variables are independent?

(2) How do we estimate the strength of dependence between two random variables?

When Y and Z are not independent, we say that they are **dependent** or **associated** or **related.** If Y and Z are associated, it does **not** imply that Y causes Z or that Z causes Y. Causation is discussed in Chapter 16.

Recall that we write $Y \amalg Z$ to mean that Y and Z are independent and we write $Y \not\!\amalg Z$ to mean that Y and Z are dependent.

15.1 Two Binary Variables

Suppose that Y and Z are both binary and consider data $(Y_1, Z_1), \ldots, (Y_n, Z_n)$. We can represent the data as a two-by-two table:

	$Y = 0$	$Y = 1$	
$Z = 0$	X_{00}	X_{01}	$X_{0\cdot}$
$Z = 1$	X_{10}	X_{11}	$X_{1\cdot}$
	$X_{\cdot 0}$	$X_{\cdot 1}$	$n = X_{\cdot\cdot}$

where

$$X_{ij} = \text{number of observations for which } Y = i \text{ and } Z = j.$$

The dotted subscripts denote sums. Thus,

$$X_{i.} = \sum_j X_{ij}, \quad X_{.j} = \sum_i X_{ij}, \quad n = X_{..} = \sum_{i,j} X_{ij}.$$

This is a convention we use throughout the remainder of the book. Denote the corresponding probabilities by:

	$Y = 0$	$Y = 1$	
$Z = 0$	p_{00}	p_{01}	$p_{0.}$
$Z = 1$	p_{10}	p_{11}	$p_{1.}$
	$p_{.0}$	$p_{.1}$	1

where $p_{ij} = \mathbb{P}(Z = i, Y = j)$. Let $X = (X_{00}, X_{01}, X_{10}, X_{11})$ denote the vector of counts. Then $X \sim \text{Multinomial}(n, p)$ where $p = (p_{00}, p_{01}, p_{10}, p_{11})$. It is now convenient to introduce two new parameters.

15.1 Definition. *The **odds ratio** is defined to be*

$$\psi = \frac{p_{00} p_{11}}{p_{01} p_{10}}. \tag{15.1}$$

*The **log odds ratio** is defined to be*

$$\gamma = \log(\psi). \tag{15.2}$$

15.2 Theorem. *The following statements are equivalent:*

1. $Y \amalg Z$.
2. $\psi = 1$.
3. $\gamma = 0$.
4. For $i, j \in \{0, 1\}$, $p_{ij} = p_{i.} p_{.j}$.

Now consider testing

$$H_0 : Y \amalg Z \quad \text{versus } H_1 : Y \ \text{\tiny $\infty\infty\infty$}\ Z. \tag{15.3}$$

First we consider the likelihood ratio test. Under H_1, $X \sim \text{Multinomial}(n, p)$ and the MLE is the vector $\hat{p} = X/n$. Under H_0, we again have that $X \sim \text{Multinomial}(n, p)$ but the restricted MLE is computed under the constraint $p_{ij} = p_{i.} p_{.j}$ This leads to the following test:

15.3 Theorem. *The likelihood ratio test statistic for (15.3) is*

$$T = 2 \sum_{i=0}^{1} \sum_{j=0}^{1} X_{ij} \log \left(\frac{X_{ij} X_{..}}{X_{i.} X_{.j}} \right).$$ (15.4)

Under H_0, $T \rightsquigarrow \chi_1^2$. Thus, an approximate level α test is obtained by rejecting H_0 when $T > \chi_{1,\alpha}^2$.

Another popular test for independence is Pearson's χ^2 test.

15.4 Theorem. *Pearson's χ^2 test statistic for independence is*

$$U = \sum_{i=0}^{1} \sum_{j=0}^{1} \frac{(X_{ij} - E_{ij})^2}{E_{ij}}$$ (15.5)

where

$$E_{ij} = \frac{X_{i.} X_{.j}}{n}.$$

Under H_0, $U \rightsquigarrow \chi_1^2$. Thus, an approximate level α test is obtained by rejecting H_0 when $U > \chi_{1,\alpha}^2$.

Here is the intuition for the Pearson test. Under H_0, $p_{ij} = p_{i.} p_{.j}$, so the maximum likelihood estimator of p_{ij} under H_0 is

$$\widehat{p}_{ij} = \widehat{p}_{i.} \widehat{p}_{.j} = \frac{X_{i.}}{n} \frac{X_{.j}}{n}.$$

Thus, the expected number of observations in the (i,j) cell is

$$E_{ij} = n\widehat{p}_{ij} = \frac{X_{i.} X_{.j}}{n}.$$

The statistic U compares the observed and expected counts.

15.5 Example. The following data from Johnson and Johnson (1972) relate tonsillectomy and Hodgkins disease. [1]

	Hodgkins Disease	No Disease	
Tonsillectomy	90	165	255
No Tonsillectomy	84	307	391
Total	174	472	646

---

[1] The data are actually from a case-control study; see the appendix for an explanation of case-control studies.

We would like to know if tonsillectomy is related to Hodgkins disease. The likelihood ratio statistic is $T = 14.75$ and the p-value is $\mathbb{P}(\chi_1^2 > 14.75) = .0001$. The χ^2 statistic is $U = 14.96$ and the p-value is $\mathbb{P}(\chi_1^2 > 14.96) = .0001$. We reject the null hypothesis of independence and conclude that tonsillectomy is associated with Hodgkins disease. This does not mean that tonsillectomies cause Hodgkins disease. Suppose, for example, that doctors gave tonsillectomies to the most seriously ill patients. Then the association between tonsillectomies and Hodgkins disease may be due to the fact that those with tonsillectomies were the most ill patients and hence more likely to have a serious disease. ∎

We can also estimate the strength of dependence by estimating the odds ratio ψ and the log-odds ratio γ.

15.6 Theorem. *The MLE's of ψ and γ are*

$$\widehat{\psi} = \frac{X_{00}X_{11}}{X_{01}X_{10}}, \quad \widehat{\gamma} = \log \widehat{\psi}. \tag{15.6}$$

The asymptotic standard errors (computed using the delta method) are

$$\widehat{\text{se}}(\widehat{\gamma}) = \sqrt{\frac{1}{X_{00}} + \frac{1}{X_{01}} + \frac{1}{X_{10}} + \frac{1}{X_{11}}} \tag{15.7}$$

$$\widehat{\text{se}}(\widehat{\psi}) = \widehat{\psi} \, \widehat{\text{se}}(\widehat{\gamma}). \tag{15.8}$$

15.7 Remark. For small sample sizes, $\widehat{\psi}$ and $\widehat{\gamma}$ can have a very large variance. In this case, we often use the modified estimator

$$\widehat{\psi} = \frac{\left(X_{00} + \frac{1}{2}\right)\left(X_{11} + \frac{1}{2}\right)}{\left(X_{01} + \frac{1}{2}\right)\left(X_{10} + \frac{1}{2}\right)}. \tag{15.9}$$

Another test for independence is the Wald test for $\gamma = 0$ given by $W = (\widehat{\gamma} - 0)/\widehat{\text{se}}(\widehat{\gamma})$. A $1 - \alpha$ confidence interval for γ is $\widehat{\gamma} \pm z_{\alpha/2}\widehat{\text{se}}(\widehat{\gamma})$.

A $1 - \alpha$ confidence interval for ψ can be obtained in two ways. First, we could use $\widehat{\psi} \pm z_{\alpha/2}\widehat{\text{se}}(\widehat{\psi})$. Second, since $\psi = e^\gamma$ we could use

$$\exp\left\{\widehat{\gamma} \pm z_{\alpha/2}\widehat{\text{se}}(\widehat{\gamma})\right\}. \tag{15.10}$$

This second method is usually more accurate.

15.8 Example. In the previous example,

$$\widehat{\psi} = \frac{90 \times 307}{165 \times 84} = 1.99$$

and

$$\widehat{\gamma} = \log(1.99) = .69.$$

So tonsillectomy patients were twice as likely to have Hodgkins disease. The standard error of $\hat{\gamma}$ is

$$\sqrt{\frac{1}{90} + \frac{1}{84} + \frac{1}{165} + \frac{1}{307}} = .18.$$

The Wald statistic is $W = .69/.18 = 3.84$ whose p-value is $\mathbb{P}(|Z| > 3.84) = .0001$, the same as the other tests. A 95 per cent confidence interval for γ is $\hat{\gamma} \pm 2(.18) = (.33, 1.05)$. A 95 per cent confidence interval for ψ is $(e^{.33}, e^{1.05}) = (1.39, 2.86)$. ∎

15.2   Two Discrete Variables

Now suppose that $Y \in \{1, \ldots, I\}$ and $Z \in \{1, \ldots, J\}$ are two discrete variables. The data can be represented as an $I \times J$ table of counts:

	$Y=1$	$Y=2$	$\cdots$	$Y=j$	$\cdots$	$Y=J$	
$Z=1$	X_{11}	X_{12}	$\cdots$	X_{1j}	$\cdots$	X_{1J}	$X_{1.}$
$\vdots$	$\vdots$	$\vdots$	$\vdots$	$\vdots$	$\vdots$	$\vdots$	$\vdots$
$Z=i$	X_{i1}	X_{i2}	$\cdots$	X_{ij}	$\cdots$	X_{iJ}	$X_{i.}$
$\vdots$	$\vdots$	$\vdots$	$\vdots$	$\vdots$	$\vdots$	$\vdots$	$\vdots$
$Z=I$	X_{I1}	X_{I2}	$\cdots$	X_{Ij}	$\cdots$	X_{IJ}	$X_{I.}$
	$X_{.1}$	$X_{.2}$	$\cdots$	$X_{.j}$	$\cdots$	$X_{.J}$	n

where

$$X_{ij} = \text{number of observations for which } Z = i \text{ and } Y = j.$$

Consider testing

$$H_0 : Y \amalg Z \quad \text{versus} \quad H_1 : Y \, \text{\textasciitilde} \, Z. \tag{15.11}$$

15.9 Theorem. *The likelihood ratio test statistic for (15.11) is*

$$T = 2 \sum_{i=1}^{I} \sum_{j=1}^{J} X_{ij} \log \left(\frac{X_{ij} X_{..}}{X_{i.} X_{.j}} \right). \tag{15.12}$$

The limiting distribution of T under the null hypothesis of independence is χ_ν^2 where $\nu = (I-1)(J-1)$. Pearson's χ^2 test statistic is

$$U = \sum_{i=1}^{I} \sum_{j=1}^{J} \frac{(X_{ij} - E_{ij})^2}{E_{ij}}. \tag{15.13}$$

> Asymptotically, under H_0, U has a χ^2_ν distribution where $\nu = (I-1)(J-1)$.

15.10 Example. These data are from Dunsmore et al. (1987). Patients with Hodgkins disease are classified by their response to treatment and by histological type.

Type	Positive Response	Partial Response	No Response	
LP	74	18	12	104
NS	68	16	12	96
MC	154	54	58	266
LD	18	10	44	72

The χ^2 test statistic is 75.89 with $2 \times 3 = 6$ degrees of freedom. The p-value is $\mathbb{P}(\chi^2_6 > 75.89) \approx 0$. The likelihood ratio test statistic is 68.30 with $2 \times 3 = 6$ degrees of freedom. The p-value is $\mathbb{P}(\chi^2_6 > 68.30) \approx 0$. Thus there is strong evidence that response to treatment and histological type are associated. ∎

15.3 Two Continuous Variables

Now suppose that Y and Z are both continuous. If we assume that the joint distribution of Y and Z is bivariate Normal, then we measure the dependence between Y and Z by means of the correlation coefficient ρ. Tests, estimates, and confidence intervals for ρ in the Normal case are given in the previous chapter in Section 14.2. If we do not assume Normality then we can still use the methods in Section 14.2 to draw inferences about the correlation ρ. However, if we conclude that ρ is 0, we cannot conclude that Y and Z are independent, only that they are uncorrelated. Fortunately, the reverse direction is valid: if we conclude that Y and Z are correlated than we can conclude they are dependent.

15.4 One Continuous Variable and One Discrete

Suppose that $Y \in \{1, \ldots, I\}$ is discrete and Z is continuous. Let $F_i(z) = \mathbb{P}(Z \leq z | Y = i)$ denote the CDF of Z conditional on $Y = i$.

15.11 Theorem. *When $Y \in \{1, \ldots, I\}$ is discrete and Z is continuous, then $Y \amalg Z$ if and only if $F_1 = \cdots = F_I$.*

It follows from the previous theorem that to test for independence, we need to test

$$H_0 : F_1 = \cdots = F_I \quad \text{versus} \quad H_1 : \text{not } H_0.$$

For simplicity, we consider the case where $I = 2$. To test the null hypothesis that $F_1 = F_2$ we will use the **two sample Kolmogorov-Smirnov test**. Let n_1 denote the number of observations for which $Y_i = 1$ and let n_2 denote the number of observations for which $Y_i = 2$. Let

$$\widehat{F}_1(z) = \frac{1}{n_1} \sum_{i=1}^{n} I(Z_i \leq z) I(Y_i = 1)$$

and

$$\widehat{F}_2(z) = \frac{1}{n_2} \sum_{i=1}^{n} I(Z_i \leq z) I(Y_i = 2)$$

denote the empirical distribution function of Z given $Y = 1$ and $Y = 2$ respectively. Define the test statistic

$$D = \sup_x |\widehat{F}_1(x) - \widehat{F}_2(x)|.$$

15.12 Theorem. *Let*

$$H(t) = 1 - 2 \sum_{j=1}^{\infty} (-1)^{j-1} e^{-2j^2 t^2}. \tag{15.14}$$

Under the null hypothesis that $F_1 = F_2$,

$$\lim_{n \to \infty} \mathbb{P}\left(\sqrt{\frac{n_1 n_2}{n_1 + n_2}} D \leq t \right) = H(t).$$

It follows from the theorem that an approximate level α test is obtained by rejecting H_0 when

$$\sqrt{\frac{n_1 n_2}{n_1 + n_2}} D > H^{-1}(1 - \alpha).$$

15.5 Appendix

INTERPRETING THE ODDS RATIOS. Suppose event A as probability $\mathbb{P}(A)$. The odds of A are defined as $\text{odds}(A) = \mathbb{P}(A)/(1 - \mathbb{P}(A))$. It follows that

$\mathbb{P}(A) = \text{odds}(A)/(1 + \text{odds}(A))$. Let E be the event that someone is exposed to something (smoking, radiation, etc) and let D be the event that they get a disease. The odds of getting the disease given that you are exposed are:

$$\text{odds}(D|E) = \frac{\mathbb{P}(D|E)}{1 - \mathbb{P}(D|E)}$$

and the odds of getting the disease given that you are not exposed are:

$$\text{odds}(D|E^c) = \frac{\mathbb{P}(D|E^c)}{1 - \mathbb{P}(D|E^c)}.$$

The *odds ratio* is defined to be

$$\psi = \frac{\text{odds}(D|E)}{\text{odds}(D|E^c)}.$$

If $\psi = 1$ then disease probability is the same for exposed and unexposed. This implies that these events are independent. Recall that the log-odds ratio is defined as $\gamma = \log(\psi)$. Independence corresponds to $\gamma = 0$.

Consider this table of probabilities and corresponding table of data:

	D^c	D	
E^c	p_{00}	p_{01}	$p_{0\cdot}$
E	p_{10}	p_{11}	$p_{1\cdot}$
	$p_{\cdot 0}$	$p_{\cdot 1}$	1

	D^c	D	
E^c	X_{00}	X_{01}	$X_{0\cdot}$
E	X_{10}	X_{11}	$X_{1\cdot}$
	$X_{\cdot 0}$	$X_{\cdot 1}$	$X_{\cdot\cdot}$

Now

$$\mathbb{P}(D|E) = \frac{p_{11}}{p_{10} + p_{11}} \quad \text{and} \quad \mathbb{P}(D|E^c) = \frac{p_{01}}{p_{00} + p_{01}},$$

and so

$$\text{odds}(D|E) = \frac{p_{11}}{p_{10}} \quad \text{and} \quad \text{odds}(D|E^c) = \frac{p_{01}}{p_{00}},$$

and therefore,

$$\psi = \frac{p_{11}p_{00}}{p_{01}p_{10}}.$$

To estimate the parameters, we have to first consider how the data were collected. There are three methods.

MULTINOMIAL SAMPLING. We draw a sample from the population and, for each person, record their exposure and disease status. In this case, $X = (X_{00}, X_{01}, X_{10}, X_{11}) \sim \text{Multinomial}(n, p)$. We then estimate the probabilities in the table by $\hat{p}_{ij} = X_{ij}/n$ and

$$\hat{\psi} = \frac{\hat{p}_{11}\hat{p}_{00}}{\hat{p}_{01}\hat{p}_{10}} = \frac{X_{11}X_{00}}{X_{01}X_{10}}.$$

PROSPECTIVE SAMPLING. (COHORT SAMPLING). We get some exposed and unexposed people and count the number with disease in each group. Thus,

$$X_{01} \sim \text{Binomial}(X_{0\cdot}, \mathbb{P}(D|E^c))$$
$$X_{11} \sim \text{Binomial}(X_{1\cdot}, \mathbb{P}(D|E)).$$

We should really write $x_{0\cdot}$ and $x_{1\cdot}$ instead of $X_{0\cdot}$ and $X_{1\cdot}$ since in this case, these are fixed not random, but for notational simplicity I'll keep using capital letters. We can estimate $\mathbb{P}(D|E)$ and $\mathbb{P}(D|E^c)$ but we cannot estimate all the probabilities in the table. Still, we can estimate ψ since ψ is a function of $\mathbb{P}(D|E)$ and $\mathbb{P}(D|E^c)$. Now

$$\widehat{\mathbb{P}}(D|E) = \frac{X_{11}}{X_{1\cdot}} \quad \text{and} \quad \widehat{\mathbb{P}}(D|E^c) = \frac{X_{01}}{X_{0\cdot}}.$$

Thus,

$$\widehat{\psi} = \frac{X_{11}X_{00}}{X_{01}X_{10}}$$

just as before.

CASE-CONTROL (RETROSPECTIVE) SAMPLING. Here we get some diseased and non-diseased people and we observe how many are exposed. This is much more efficient if the disease is rare. Hence,

$$X_{10} \sim \text{Binomial}(X_{\cdot 0}, \mathbb{P}(E|D^c))$$
$$X_{11} \sim \text{Binomial}(X_{\cdot 1}, \mathbb{P}(E|D)).$$

From these data we can estimate $\mathbb{P}(E|D)$ and $\mathbb{P}(E|D^c)$. Surprisingly, we can also still estimate ψ. To understand why, note that

$$\mathbb{P}(E|D) = \frac{p_{11}}{p_{01} + p_{11}}, \quad 1 - \mathbb{P}(E|D) = \frac{p_{01}}{p_{01} + p_{11}}, \quad \text{odds}(E|D) = \frac{p_{11}}{p_{01}}.$$

By a similar argument,

$$\text{odds}(E|D^c) = \frac{p_{10}}{p_{00}}.$$

Hence,

$$\frac{\text{odds}(E|D)}{\text{odds}(E|D^c)} = \frac{p_{11}p_{00}}{p_{01}p_{10}} = \psi.$$

From the data, we form the following estimates:

$$\widehat{P}(E|D) = \frac{X_{11}}{X_{\cdot 1}}, \quad 1 - \widehat{P}(E|D) = \frac{X_{01}}{X_{\cdot 1}}, \quad \widehat{\text{odds}}(E|D) = \frac{X_{11}}{X_{01}}, \quad \widehat{\text{odds}}(E|D^c) = \frac{X_{10}}{X_{00}}.$$

Therefore,

$$\widehat{\psi} = \frac{X_{00}X_{11}}{X_{01}X_{10}}.$$

So in all three data collection methods, the estimate of ψ turns out to be the same.

It is tempting to try to estimate $\mathbb{P}(D|E) - \mathbb{P}(D|E^c)$. In a case-control design, this quantity is not estimable. To see this, we apply Bayes' theorem to get

$$\mathbb{P}(D|E) - \mathbb{P}(D|E^c) = \frac{\mathbb{P}(E|D)\mathbb{P}(D)}{\mathbb{P}(E)} - \frac{\mathbb{P}(E^c|D)\mathbb{P}(D)}{\mathbb{P}(E^c)}.$$

Because of the way we obtained the data, $\mathbb{P}(D)$ is not estimable from the data. However, we can estimate $\xi = \mathbb{P}(D|E)/\mathbb{P}(D|E^c)$, which is called the **relative risk**, under the **rare disease assumption.**

15.13 Theorem. *Let* $\xi = \mathbb{P}(D|E)/\mathbb{P}(D|E^c)$. *Then*

$$\frac{\psi}{\xi} \to 1$$

as $\mathbb{P}(D) \to 0$.

Thus, under the rare disease assumption, the relative risk is approximately the same as the odds ratio and, as we have seen, we can estimate the odds ratio.

15.6   Exercises

1. Prove Theorem 15.2.

2. Prove Theorem 15.3.

3. Prove Theorem 15.6.

4. The *New York Times* (January 8, 2003, page A12) reported the following data on death sentencing and race, from a study in Maryland: [2]

	Death Sentence	No Death Sentence
Black Victim	14	641
White Victim	62	594

   Analyze the data using the tools from this chapter. Interpret the results. Explain why, based only on this information, you can't make causal conclusions. (The authors of the study did use much more information in their full report.)

---

[2] The data here are an approximate re-creation using the information in the article.

5. Analyze the data on the variables Age and Financial Status from:

   http://lib.stat.cmu.edu/DASL/Datafiles/montanadat.html

6. Estimate the correlation between temperature and latitude using the data from

   http://lib.stat.cmu.edu/DASL/Datafiles/USTemperatures.html

   Use the correlation coefficient. Provide estimates, tests, and confidence intervals.

7. Test whether calcium intake and drop in blood pressure are associated. Use the data in

   http://lib.stat.cmu.edu/DASL/Datafiles/Calcium.html

16
Causal Inference

Roughly speaking, the statement "X causes Y" means that changing the value of X will change the distribution of Y. When X causes Y, X and Y will be associated but the reverse is not, in general, true. Association does not necessarily imply causation. We will consider two frameworks for discussing causation. The first uses **counterfactual** random variables. The second, presented in the next chapter, uses **directed acyclic graphs.**

16.1 The Counterfactual Model

Suppose that X is a binary treatment variable where $X = 1$ means "treated" and $X = 0$ means "not treated." We are using the word "treatment" in a very broad sense. Treatment might refer to a medication or something like smoking. An alternative to "treated/not treated" is "exposed/not exposed" but we shall use the former.

Let Y be some outcome variable such as presence or absence of disease. To distinguish the statement "X is associated Y" from the statement "X causes Y" we need to enrich our probabilistic vocabulary. Specifically, we will decompose the response Y into a more fine-grained object.

We introduce two new random variables (C_0, C_1), called **potential outcomes** with the following interpretation: C_0 is the outcome if the subject is not treated ($X = 0$) and C_1 is the outcome if the subject is treated ($X = 1$). Hence,

$$Y = \begin{cases} C_0 & \text{if } X = 0 \\ C_1 & \text{if } X = 1. \end{cases}$$

We can express the relationship between Y and (C_0, C_1) more succinctly by

$$Y = C_X. \tag{16.1}$$

Equation (16.1) is called the **consistency relationship**.

Here is a toy dataset to make the idea clear:

X	Y	C_0	C_1
0	4	4	*
0	7	7	*
0	2	2	*
0	8	8	*
1	3	*	3
1	5	*	5
1	8	*	8
1	9	*	9

The asterisks denote unobserved values. When $X = 0$ we don't observe C_1, in which case we say that C_1 is a **counterfactual** since it is the outcome you would have had if, counter to the fact, you had been treated ($X = 1$). Similarly, when $X = 1$ we don't observe C_0, and we say that C_0 is **counterfactual**. There are four types of subjects:

Type	C_0	C_1
Survivors	1	1
Responders	0	1
Anti-responders	1	0
Doomed	0	0

Think of the potential outcomes (C_0, C_1) as hidden variables that contain all the relevant information about the subject.

Define the **average causal effect** or **average treatment effect** to be

$$\theta = \mathbb{E}(C_1) - \mathbb{E}(C_0). \tag{16.2}$$

The parameter θ has the following interpretation: θ is the mean if everyone were treated ($X = 1$) minus the mean if everyone were not treated ($X = 0$). There are other ways of measuring the causal effect. For example, if C_0 and C_1 are binary, we define the **causal odds ratio**

$$\frac{\mathbb{P}(C_1 = 1)}{\mathbb{P}(C_1 = 0)} \div \frac{\mathbb{P}(C_0 = 1)}{\mathbb{P}(C_0 = 0)}$$

and the **causal relative risk**

$$\frac{\mathbb{P}(C_1 = 1)}{\mathbb{P}(C_0 = 1)}.$$

The main ideas will be the same whatever causal effect we use. For simplicity, we shall work with the average causal effect θ.

Define the **association** to be

$$\alpha = \mathbb{E}(Y|X = 1) - \mathbb{E}(Y|X = 0). \tag{16.3}$$

Again, we could use odds ratios or other summaries if we wish.

16.1 Theorem (Association Is Not Causation). *In general, $\theta \neq \alpha$.*

16.2 Example. Suppose the whole population is as follows:

X	Y	C_0	C_1
0	0	0	0*
0	0	0	0*
0	0	0	0*
0	0	0	0*
1	1	1*	1
1	1	1*	1
1	1	1*	1
1	1	1*	1

Again, the asterisks denote unobserved values. Notice that $C_0 = C_1$ for every subject, thus, this treatment has no effect. Indeed,

$$\theta = \mathbb{E}(C_1) - \mathbb{E}(C_0) = \frac{1}{8}\sum_{i=1}^{8} C_{1i} - \frac{1}{8}\sum_{i=1}^{8} C_{0i}$$

$$= \frac{0+0+0+0+1+1+1+1}{8} - \frac{0+0+0+0+1+1+1+1}{8}$$

$$= 0.$$

Thus, the average causal effect is 0. The observed data are only the X's and Y's, from which we can estimate the association:

$$\alpha = \mathbb{E}(Y|X = 1) - \mathbb{E}(Y|X = 0)$$

$$= \frac{1+1+1+1}{4} - \frac{0+0+0+0}{4} = 1.$$

Hence, $\theta \neq \alpha$.

To add some intuition to this example, imagine that the outcome variable is 1 if "healthy" and 0 if "sick". Suppose that $X = 0$ means that the subject does not take vitamin C and that $X = 1$ means that the subject does take vitamin C. Vitamin C has no causal effect since $C_0 = C_1$ for each subject. In this example there are two types of people: healthy people $(C_0, C_1) = (1, 1)$ and unhealthy people $(C_0, C_1) = (0, 0)$. Healthy people tend to take vitamin C while unhealthy people don't. It is this association between (C_0, C_1) and X that creates an association between X and Y. If we only had data on X and Y we would conclude that X and Y are associated. Suppose we wrongly interpret this causally and conclude that vitamin C prevents illness. Next we might encourage everyone to take vitamin C. If most people comply with our advice, the population will look something like this:

X	Y	C_0	C_1
0	0	0	0*
1	0	0	0*
1	0	0	0*
1	0	0	0*
1	1	1*	1
1	1	1*	1
1	1	1*	1
1	1	1*	1

Now $\alpha = (4/7) - (0/1) = 4/7$. We see that α went down from 1 to 4/7. Of course, the causal effect never changed but the naive observer who does not distinguish association and causation will be confused because his advice seems to have made things worse instead of better. ∎

In the last example, $\theta = 0$ and $\alpha = 1$. It is not hard to create examples in which $\alpha > 0$ and yet $\theta < 0$. The fact that the association and causal effects can have different signs is very confusing to many people.

The example makes it clear that, in general, we cannot use the association to estimate the causal effect θ. The reason that $\theta \neq \alpha$ is that (C_0, C_1) was not independent of X. That is, treatment assignment was not independent of person type.

Can we ever estimate the causal effect? The answer is: sometimes. In particular, random assignment to treatment makes it possible to estimate θ.

16.3 Theorem. *Suppose we randomly assign subjects to treatment and that* $\mathbb{P}(X = 0) > 0$ *and* $\mathbb{P}(X = 1) > 0$. *Then* $\alpha = \theta$. *Hence, any consistent estimator of* α *is a consistent estimator of* θ. *In particular, a consistent estimator is*

$$\hat{\theta} = \hat{\mathbb{E}}(Y|X = 1) - \hat{\mathbb{E}}(Y|X = 0)$$

$$= \ \overline{Y}_1 - \overline{Y}_0$$

is a consistent estimator of θ, where

$$\overline{Y}_1 = \frac{1}{n_1} \sum_{i=1}^{n} Y_i X_i, \quad \overline{Y}_0 = \frac{1}{n_0} \sum_{i=1}^{n} Y_i (1 - X_i),$$

$n_1 = \sum_{i=1}^{n} X_i$, *and* $n_0 = \sum_{i=1}^{n} (1 - X_i)$.

PROOF. Since X is randomly assigned, X is independent of (C_0, C_1). Hence,

$$
\begin{aligned}
\theta &= \ \mathbb{E}(C_1) - \mathbb{E}(C_0) \\
&= \ \mathbb{E}(C_1|X = 1) - \mathbb{E}(C_0|X = 0) \quad \text{since } X \amalg (C_0, C_1) \\
&= \ \mathbb{E}(Y|X = 1) - \mathbb{E}(Y|X = 0) \quad \text{since } Y = C_X \\
&= \ \alpha.
\end{aligned}
$$

The consistency follows from the law of large numbers. ∎

If Z is a covariate, we define the **conditional causal effect** by

$$\theta_z = \mathbb{E}(C_1|Z = z) - \mathbb{E}(C_0|Z = z).$$

For example, if Z denotes gender with values $Z = 0$ (women) and $Z = 1$ (men), then θ_0 is the causal effect among women and θ_1 is the causal effect among men. In a randomized experiment, $\theta_z = \mathbb{E}(Y|X = 1, Z = z) - \mathbb{E}(Y|X = 0, Z = z)$ and we can estimate the conditional causal effect using appropriate sample averages.

Summary of the Counterfactual Model

Random variables: (C_0, C_1, X, Y).

Consistency relationship: $Y = C_X$.

Causal Effect: $\theta = \mathbb{E}(C_1) - \mathbb{E}(C_0)$.

Association: $\alpha = \mathbb{E}(Y|X = 1) - \mathbb{E}(Y|X = 0)$.

Random assignment $\implies (C_0, C_1) \amalg X \implies \theta = \alpha$.

16.2 Beyond Binary Treatments

Let us now generalize beyond the binary case. Suppose that $X \in \mathcal{X}$. For example, X could be the dose of a drug in which case $X \in \mathbb{R}$. The counterfactual vector (C_0, C_1) now becomes the **counterfactual function** $C(x)$ where

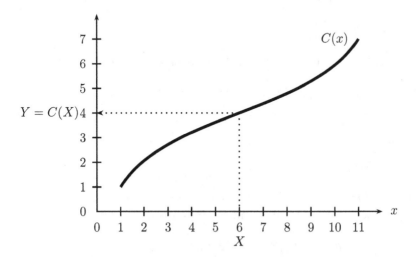

FIGURE 16.1. A counterfactual function $C(x)$. The outcome Y is the value of the curve $C(x)$ evaluated at the observed dose X.

$C(x)$ is the outcome a subject would have if he received dose x. The observed response is given by the consistency relation

$$Y \equiv C(X). \tag{16.4}$$

See Figure 16.1. The **causal regression function** is

$$\theta(x) = \mathbb{E}(C(x)). \tag{16.5}$$

The regression function, which measures association, is $r(x) = \mathbb{E}(Y|X = x)$.

16.4 Theorem. *In general, $\theta(x) \neq r(x)$. However, when X is randomly assigned, $\theta(x) = r(x)$.*

16.5 Example. An example in which $\theta(x)$ is constant but $r(x)$ is not constant is shown in Figure 16.2. The figure shows the counterfactual functions for four subjects. The dots represent their X values X_1, X_2, X_3, X_4. Since $C_i(x)$ is constant over x for all i, there is no causal effect and hence

$$\theta(x) = \frac{C_1(x) + C_2(x) + C_3(x) + C_4(x)}{4}$$

is constant. Changing the dose x will not change anyone's outcome. The four dots in the lower plot represent the observed data points $Y_1 = C_1(X_1), Y_2 = C_2(X_2), Y_3 = C_3(X_3), Y_4 = C_4(X_4)$. The dotted line represents the regression $r(x) = \mathbb{E}(Y|X = x)$. Although there is no causal effect, there is an association since the regression curve $r(x)$ is not constant. ■

16.3   Observational Studies and Confounding

A study in which treatment (or exposure) is not randomly assigned is called an **observational study.** In these studies, subjects select their own value of the exposure X. Many of the health studies you read about in the newspaper are like this. As we saw, association and causation could in general be quite different. This discrepancy occurs in non-randomized studies because the potential outcome C is not independent of treatment X. However, suppose we could find groupings of subjects such that, within groups, X and $\{C(x) : x \in \mathcal{X}\}$ are independent. This would happen if the subjects are very similar within groups. For example, suppose we find people who are very similar in age, gender, educational background, and ethnic background. Among these people we might feel it is reasonable to assume that the choice of X is essentially random. These other variables are called **confounding variables.**[1] If we denote these other variables collectively as Z, then we can express this idea by saying that

$$\{C(x) : x \in \mathcal{X}\} \amalg X|Z. \tag{16.6}$$

Equation (16.6) means that, within groups of Z, the choice of treatment X does not depend on type, as represented by $\{C(x) : x \in \mathcal{X}\}$. If (16.6) holds and we observe Z then we say that there is **no unmeasured confounding.**

16.6 Theorem. *Suppose that (16.6) holds. Then,*

$$\theta(x) = \int \mathbb{E}(Y|X = x, Z = z)dF_Z(z)dz. \tag{16.7}$$

If $\widehat{r}(x, z)$ is a consistent estimate of the regression function $\mathbb{E}(Y|X = x, Z = z)$, then a consistent estimate of $\theta(x)$ is

$$\widehat{\theta}(x) = \frac{1}{n}\sum_{i=1}^{n} \widehat{r}(x, Z_i).$$

---

[1]A more precise definition of confounding is given in the next chapter.

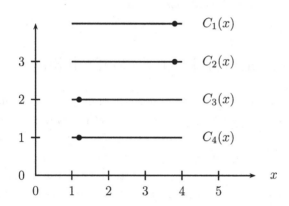

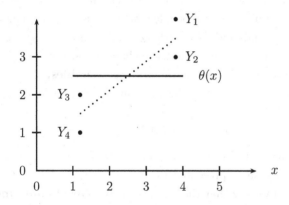

FIGURE 16.2. The top plot shows the counterfactual function $C(x)$ for four sub-jects. The dots represent their X values. Since $C_i(x)$ is constant over x for all i, there is no causal effect. Changing the dose will not change anyone's outcome. The lower plot shows the causal regression function $\theta(x) = (C_1(x)+C_2(x)+C_3(x)+C_4(x))/4$. The four dots represent the observed data points $Y_1 = C_1(X_1)$, $Y_2 = C_2(X_2)$, $Y_3 = C_3(X_3)$, $Y_4 = C_4(X_4)$. The dotted line represents the regression $r(x) = \mathbb{E}(Y|X = x)$. There is no causal effect since $C_i(x)$ is constant for all i. But there is an association since the regression curve $r(x)$ is not constant.

In particular, if $r(x, z) = \beta_0 + \beta_1 x + \beta_2 z$ is linear, then a consistent estimate of $\theta(x)$ is

$$\widehat{\theta}(x) = \widehat{\beta}_0 + \widehat{\beta}_1 x + \widehat{\beta}_2 \overline{Z}_n \tag{16.8}$$

where $(\widehat{\beta}_0, \widehat{\beta}_1, \widehat{\beta}_2)$ are the least squares estimators.

16.7 Remark. It is useful to compare equation (16.7) to $\mathbb{E}(Y|X = x)$ which can be written as $\mathbb{E}(Y|X = x) = \int \mathbb{E}(Y|X = x, Z = z) dF_{Z|X}(z|x)$.

Epidemiologists call (16.7) the **adjusted treatment effect**. The process of computing adjusted treatment effects is called **adjusting (or controlling) for confounding.** The selection of what confounders Z to measure and control for requires scientific insight. Even after adjusting for confounders, we cannot be sure that there are not other confounding variables that we missed. This is why observational studies must be treated with healthy skepticism. Results from observational studies start to become believable when: (i) the results are replicated in many studies, (ii) each of the studies controlled for plausible confounding variables, (iii) there is a plausible scientific explanation for the existence of a causal relationship.

A good example is smoking and cancer. Numerous studies have shown a relationship between smoking and cancer even after adjusting for many confounding variables. Moreover, in laboratory studies, smoking has been shown to damage lung cells. Finally, a causal link between smoking and cancer has been found in randomized animal studies. It is this collection of evidence over many years that makes this a convincing case. One single observational study is not, by itself, strong evidence. Remember that when you read the newspaper.

16.4 Simpson's Paradox

Simpson's paradox is a puzzling phenomenon that is discussed in most statistics texts. Unfortunately, most explanations are confusing (and in some cases incorrect). The reason is that it is nearly impossible to explain the paradox without using counterfactuals (or directed acyclic graphs).

Let X be a binary treatment variable, Y a binary outcome, and Z a third binary variable such as gender. Suppose the joint distribution of X, Y, Z is

	$Y = 1$	$Y = 0$	$Y = 1$	$Y = 0$
$X = 1$	.1500	.2250	.1000	.0250
$X = 0$	.0375	.0875	.2625	.1125
	$Z = 1$ (men)		$Z = 0$ (women)	

The marginal distribution for (X, Y) is

	$Y = 1$	$Y = 0$	
$X = 1$	.25	.25	.50
$X = 0$	.30	.20	.50
	.55	.45	1

From these tables we find that,

$$\mathbb{P}(Y = 1 | X = 1) - \mathbb{P}(Y = 1 | X = 0) = -0.1$$
$$\mathbb{P}(Y = 1 | X = 1, Z = 1) - \mathbb{P}(Y = 1 | X = 0, Z = 1) = 0.1$$
$$\mathbb{P}(Y = 1 | X = 1, Z = 0) - \mathbb{P}(Y = 1 | X = 0, Z = 0) = 0.1.$$

To summarize, we *seem* to have the following information:

Mathematical Statement	English Statement?
$\mathbb{P}(Y = 1 | X = 1) < \mathbb{P}(Y = 1 | X = 0)$	treatment is harmful
$\mathbb{P}(Y = 1 | X = 1, Z = 1) > \mathbb{P}(Y = 1 | X = 0, Z = 1)$	treatment is beneficial to men
$\mathbb{P}(Y = 1 | X = 1, Z = 0) > \mathbb{P}(Y = 1 | X = 0, Z = 0)$	treatment is beneficial to women

Clearly, something is amiss. There can't be a treatment which is good for men, good for women, but bad overall. This is nonsense. The problem is with the set of English statements in the table. Our translation from math into English is specious.

The inequality $\mathbb{P}(Y = 1 | X = 1) < \mathbb{P}(Y = 1 | X = 0)$ does not mean that treatment is harmful.

The phrase "treatment is harmful" should be written mathematically as $\mathbb{P}(C_1 = 1) < \mathbb{P}(C_0 = 1)$. The phrase "treatment is harmful for men" should be written $\mathbb{P}(C_1 = 1 | Z = 1) < \mathbb{P}(C_0 = 1 | Z = 1)$. The three mathematical statements in the table are not at all contradictory. It is only the translation into English that is wrong.

Let us now show that a real Simpson's paradox cannot happen, that is, there cannot be a treatment that is beneficial for men and women but harmful overall. Suppose that treatment is beneficial for both sexes. Then

$$\mathbb{P}(C_1 = 1 | Z = z) > \mathbb{P}(C_0 = 1 | Z = z)$$

for all z. It then follows that

$$
\begin{aligned}
\mathbb{P}(C_1 = 1) &= \sum_z \mathbb{P}(C_1 = 1 | Z = z) \mathbb{P}(Z = z) \\
&> \sum_z \mathbb{P}(C_0 = 1 | Z = z) \mathbb{P}(Z = z) \\
&= \mathbb{P}(C_0 = 1).
\end{aligned}
$$

Hence, $\mathbb{P}(C_1 = 1) > \mathbb{P}(C_0 = 1)$, so treatment is beneficial overall. No paradox.

16.5 Bibliographic Remarks

The use of potential outcomes to clarify causation is due mainly to Jerzy Neyman and Donald Rubin. Later developments are due to Jamie Robins, Paul Rosenbaum, and others. A parallel development took place in econometrics by various people including James Heckman and Charles Manski. Texts on causation include Pearl (2000), Rosenbaum (2002), Spirtes et al. (2000), and van der Laan and Robins (2003).

16.6 Exercises

1. Create an example like Example 16.2 in which $\alpha > 0$ and $\theta < 0$.

2. Prove Theorem 16.4.

3. Suppose you are given data $(X_1, Y_1), \ldots, (X_n, Y_n)$ from an observational study, where $X_i \in \{0, 1\}$ and $Y_i \in \{0, 1\}$. Although it is not possible to estimate the causal effect θ, it is possible to put bounds on θ. Find upper and lower bounds on θ that can be consistently estimated from the data. Show that the bounds have width 1.

   Hint: Note that $\mathbb{E}(C_1) = \mathbb{E}(C_1 | X = 1) \mathbb{P}(X = 1) + \mathbb{E}(C_1 | X = 0) \mathbb{P}(X = 0)$.

4. Suppose that $X \in \mathbb{R}$ and that, for each subject i, $C_i(x) = \beta_{1i} x$. Each subject has their own slope β_{1i}. Construct a joint distribution on (β_1, X) such that $\mathbb{P}(\beta_1 > 0) = 1$ but $\mathbb{E}(Y | X = x)$ is a decreasing function of x, where $Y = C(X)$. Interpret.

5. Let $X \in \{0, 1\}$ be a binary treatment variable and let (C_0, C_1) denote the corresponding potential outcomes. Let $Y = C_X$ denote the observed

response. Let F_0 and F_1 be the cumulative distribution functions for C_0 and C_1. Assume that F_0 and F_1 are both continuous and strictly increasing. Let $\theta = m_1 - m_0$ where $m_0 = F_0^{-1}(1/2)$ is the median of C_0 and $m_1 = F_1^{-1}(1/2)$ is the median of C_1. Suppose that the treatment X is assigned randomly. Find an expression for θ involving only the joint distribution of X and Y.

17
Directed Graphs and Conditional Independence

17.1 Introduction

A directed graph consists of a set of nodes with arrows between some nodes. An example is shown in Figure 17.1.

Graphs are useful for representing independence relations between variables. They can also be used as an alternative to counterfactuals to represent causal relationships. Some people use the phrase **Bayesian network** to refer to a directed graph endowed with a probability distribution. This is a poor choice of terminology. Statistical inference for directed graphs can be performed using

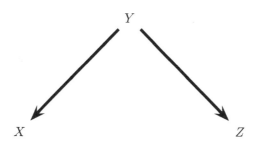

FIGURE 17.1. A directed graph with vertices $V = \{X, Y, Z\}$ and edges $E = \{(Y, X), (Y, Z)\}$.

frequentist or Bayesian methods, so it is misleading to call them Bayesian networks.

Before getting into details about directed acyclic graphs (DAGs), we need to discuss conditional independence.

17.2   Conditional Independence

17.1 Definition. *Let X, Y and Z be random variables. X and Y are* **conditionally independent given** Z*, written $X \amalg Y \mid Z$, if*

$$f_{X,Y|Z}(x,y|z) = f_{X|Z}(x|z)f_{Y|Z}(y|z). \tag{17.1}$$

for all x, y and z.

Intuitively, this means that, once you know Z, Y provides no extra information about X. An equivalent definition is that

$$f(x|y,z) = f(x|z). \tag{17.2}$$

The conditional independence relation satisfies some basic properties.

17.2 Theorem. *The following implications hold:* [1]

$$X \amalg Y \mid Z \implies Y \amalg X \mid Z$$
$$X \amalg Y \mid Z \text{ and } U = h(X) \implies U \amalg Y \mid Z$$
$$X \amalg Y \mid Z \text{ and } U = h(X) \implies X \amalg Y \mid (Z,U)$$
$$X \amalg Y \mid Z \text{ and } X \amalg W|(Y,Z) \implies X \amalg (W,Y) \mid Z$$
$$X \amalg Y \mid Z \text{ and } X \amalg Z \mid Y \implies X \amalg (Y,Z).$$

17.3   DAGs

A **directed graph** $\mathcal{G}$ consists of a set of vertices V and an edge set E of ordered pairs of vertices. For our purposes, each vertex will correspond to a random variable. If $(X,Y) \in E$ then there is an arrow pointing from X to Y. See Figure 17.1.

---

[1] The last property requires the assumption that all events have positive probability; the first four do not.

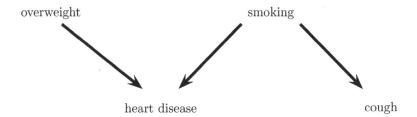

FIGURE 17.2. DAG for Example 17.4.

If an arrow connects two variables X and Y (in either direction) we say that X and Y are **adjacent.** If there is an arrow from X to Y then X is a **parent** of Y and Y is a **child** of X. The set of all parents of X is denoted by π_X or $\pi(X)$. A **directed path** between two variables is a set of arrows all pointing in the same direction linking one variable to the other such as:

$$X \longrightarrow \bullet \; \bullet \; \bullet \longrightarrow Y$$

A sequence of adjacent vertices staring with X and ending with Y but ignoring the direction of the arrows is called an **undirected path.** The sequence $\{X, Y, Z\}$ in Figure 17.1 is an undirected path. X is an **ancestor** of Y if there is a directed path from X to Y (or $X = Y$). We also say that Y is a **descendant** of X.

A configuration of the form:

$$X \longrightarrow Y \longleftarrow Z$$

is called a **collider** at Y. A configuration not of that form is called a **noncollider,** for example,

$$X \longrightarrow Y \longrightarrow Z$$

or

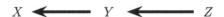

The collider property is path dependent. In Figure 17.7, Y is a collider on the path $\{X, Y, Z\}$ but it is a non-collider on the path $\{X, Y, W\}$. When the variables pointing into the collider are not adjacent, we say that the collider is **unshielded**. A directed path that starts and ends at the same variable is called a **cycle.** A directed graph is **acyclic** if it has no cycles. In this case we say that the graph is a **directed acyclic graph** or **DAG.** From now on, we only deal with acyclic graphs.

17.4   Probability and DAGs

Let $\mathcal{G}$ be a DAG with vertices $V = (X_1, \ldots, X_k)$.

17.3 Definition. *If* $\mathbb{P}$ *is a distribution for* V *with probability function* f, *we say that* $\mathbb{P}$ *is* **Markov to** $\mathcal{G}$, *or that* $\mathcal{G}$ **represents** $\mathbb{P}$, *if*

$$f(v) = \prod_{i=1}^{k} f(x_i \mid \pi_i) \tag{17.3}$$

where π_i *are the parents of* X_i. *The set of distributions represented by* $\mathcal{G}$ *is denoted by* $M(\mathcal{G})$.

17.4 Example. Figure 17.2 shows a DAG with four variables. The probability function for this example factors as

$$
\begin{aligned}
f(&\text{overweight}, \text{smoking}, \text{heart disease}, \text{cough}) \\
&= \quad f(\text{overweight}) \times f(\text{smoking}) \\
&\quad \times \quad f(\text{heart disease} \mid \text{overweight}, \text{smoking}) \\
&\quad \times \quad f(\text{cough} \mid \text{smoking}). \quad \blacksquare
\end{aligned}
$$

17.5 Example. For the DAG in Figure 17.3, $\mathbb{P} \in M(\mathcal{G})$ if and only if its probability function f has the form

$$f(x, y, z, w) = f(x)f(y)f(z \mid x, y)f(w \mid z). \quad \blacksquare$$

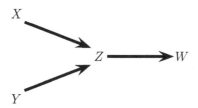

FIGURE 17.3. Another DAG.

The following theorem says that $\mathbb{P} \in M(\mathcal{G})$ if and only if the **Markov Condition** holds. Roughly speaking, the Markov Condition means that every variable W is independent of the "past" given its parents.

17.6 Theorem. *A distribution* $\mathbb{P} \in M(\mathcal{G})$ *if and only if the following* **Markov Condition** *holds: for every variable* W,

$$W \amalg \widetilde{W} \mid \pi_W \tag{17.4}$$

where $\widetilde{W}$ *denotes all the other variables except the parents and descendants of* W.

17.7 Example. In Figure 17.3, the Markov Condition implies that

$$X \amalg Y \quad \text{and} \quad W \amalg \{X, Y\} \mid Z. \quad \blacksquare$$

17.8 Example. Consider the DAG in Figure 17.4. In this case probability function must factor like

$$f(a, b, c, d, e) = f(a) f(b|a) f(c|a) f(d|b, c) f(e|d).$$

The Markov Condition implies the following independence relations:

$$D \amalg A \mid \{B, C\}, \quad E \amalg \{A, B, C\} \mid D \quad \text{and} \quad B \amalg C \mid A \quad \blacksquare$$

17.5 More Independence Relations

The Markov Condition allows us to list some independence relations implied by a DAG. These relations might imply other independence relations. Con-

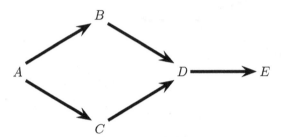

FIGURE 17.4. Yet another DAG.

sider the DAG in Figure 17.5. The Markov Condition implies:

$$X_1 \amalg X_2, \qquad X_2 \amalg \{X_1, X_4\}, \qquad X_3 \amalg X_4 \mid \{X_1, X_2\},$$

$$X_4 \amalg \{X_2, X_3\} \mid X_1, \qquad X_5 \amalg \{X_1, X_2\} \mid \{X_3, X_4\}$$

It turns out (but it is not obvious) that these conditions imply that

$$\{X_4, X_5\} \amalg X_2 \mid \{X_1, X_3\}.$$

How do we find these extra independence relations? The answer is "d-separation" which means "directed separation." d-separation can be summarized by three rules. Consider the four DAG's in Figure 17.6 and the DAG in Figure 17.7. The first 3 DAG's in Figure 17.6 have no colliders. The DAG in the lower right of Figure 17.6 has a collider. The DAG in Figure 17.7 has a collider with a descendant.

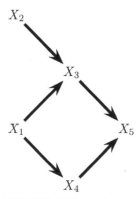

FIGURE 17.5. And yet another DAG.

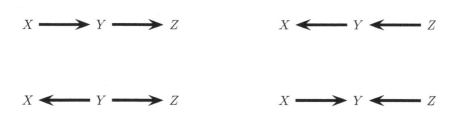

FIGURE 17.6. The first three DAG's have no colliders. The fourth DAG in the lower right corner has a collider at Y.

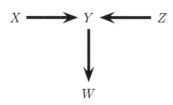

FIGURE 17.7. A collider with a descendant.

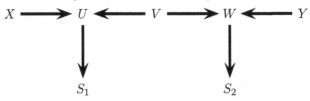

FIGURE 17.8. d-separation explained.

The Rules of d-Separation

Consider the DAGs in Figures 17.6 and 17.7.

1. When Y is not a collider, X and Z are **d-connected**, but they are **d-separated** given Y.

2. If X and Z collide at Y, then X and Z are **d-separated**, but they are **d-connected** given Y.

3. Conditioning on the descendant of a collider has the same effect as conditioning on the collider. Thus in Figure 17.7, X and Z are **d-separated** but they are **d-connected** given W.

Here is a more formal definition of d-separation. Let X and Y be distinct vertices and let W be a set of vertices not containing X or Y. Then X and Y are **d-separated given** W if there exists no undirected path U between X and Y such that (i) every collider on U has a descendant in W, and (ii) no other vertex on U is in W. If A, B, and W are distinct sets of vertices and A and B are not empty, then A and B are d-separated given W if for every $X \in A$ and $Y \in B$, X and Y are d-separated given W. Sets of vertices that are not d-separated are said to be d-connected.

17.9 Example. Consider the DAG in Figure 17.8. From the d-separation rules we conclude that:

X and Y are d-separated (given the empty set);
X and Y are d-connected given $\{S_1, S_2\}$;
X and Y are d-separated given $\{S_1, S_2, V\}$.

17.10 Theorem. [2] *Let A, B, and C be disjoint sets of vertices. Then $A \amalg B \mid C$ if and only if A and B are d-separated by C.*

---

[2] We implicitly assume that $\mathbb{P}$ is **faithful** to $\mathcal{G}$ which means that $\mathbb{P}$ has no extra independence relations other than those logically implied by the Markov Condition.

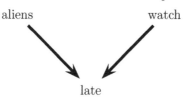

FIGURE 17.9. Jordan's alien example (Example 17.11). Was your friend kidnapped by aliens or did you forget to set your watch?

17.11 Example. The fact that conditioning on a collider creates dependence might not seem intuitive. Here is a whimsical example from Jordan (2004) that makes this idea more palatable. Your friend appears to be late for a meeting with you. There are two explanations: she was abducted by aliens or you forgot to set your watch ahead one hour for daylight savings time. (See Figure 17.9.) Aliens and Watch are blocked by a collider which implies they are marginally independent. This seems reasonable since — before we know anything about your friend being late — we would expect these variables to be independent. We would also expect that $\mathbb{P}(\mathsf{Aliens} = \mathsf{yes}|\mathsf{Late} = \mathsf{yes}) > \mathbb{P}(\mathsf{Aliens} = \mathsf{yes})$; learning that your friend is late certainly increases the probability that she was abducted. But when we learn that you forgot to set your watch properly, we would lower the chance that your friend was abducted. Hence, $\mathbb{P}(\mathsf{Aliens} = \mathsf{yes}|\mathsf{Late} = \mathsf{yes}) \neq \mathbb{P}(\mathsf{Aliens} = \mathsf{yes}|\mathsf{Late} = \mathsf{yes}, \mathsf{Watch} = \mathsf{no})$. Thus, Aliens and Watch are dependent given Late. ∎

17.12 Example. Consider the DAG in Figure 17.2. In this example, overweight and smoking are marginally independent but they are dependent given heart disease. ∎

Graphs that look different may actually imply the same independence relations. If $\mathcal{G}$ is a DAG, we let $\mathcal{I}(\mathcal{G})$ denote all the independence statements implied by $\mathcal{G}$. Two DAGs $\mathcal{G}_1$ and $\mathcal{G}_2$ for the same variables V are **Markov equivalent** if $\mathcal{I}(\mathcal{G}_1) = \mathcal{I}(\mathcal{G}_2)$. Given a DAG $\mathcal{G}$, let skeleton$(\mathcal{G})$ denote the undirected graph obtained by replacing the arrows with undirected edges.

17.13 Theorem. *Two DAGs $\mathcal{G}_1$ and $\mathcal{G}_2$ are Markov equivalent if and only if (i) skeleton$(\mathcal{G}_1)$ = skeleton$(\mathcal{G}_2)$ and (ii) $\mathcal{G}_1$ and $\mathcal{G}_2$ have the same unshielded colliders.*

17.14 Example. The first three DAGs in Figure 17.6 are Markov equivalent. The DAG in the lower right of the Figure is not Markov equivalent to the others. ∎

17.6   Estimation for DAGs

Two estimation questions arise in the context of DAGs. First, given a DAG $\mathcal{G}$ and data $V_1, \ldots, V_n$ from a distribution f consistent with $\mathcal{G}$, how do we estimate f? Second, given data $V_1, \ldots, V_n$ how do we estimate $\mathcal{G}$? The first question is pure estimation while the second involves model selection. These are very involved topics and are beyond the scope of this book. We will just briefly mention the main ideas.

Typically, one uses some parametric model $f(x|\pi_x; \theta_x)$ for each conditional density. The likelihood function is then

$$\mathcal{L}(\theta) = \prod_{i=1}^{n} f(V_i; \theta) = \prod_{i=1}^{n} \prod_{j=1}^{m} f(X_{ij}|\pi_j; \theta_j),$$

where X_{ij} is the value of X_j for the ith data point and θ_j are the parameters for the jth conditional density. We can then estimate the parameters by maximum likelihood.

To estimate the structure of the DAG itself, we could fit every possible DAG using maximum likelihood and use AIC (or some other method) to choose a DAG. However, there are many possible DAGs so you would need much data for such a method to be reliable. Also, searching through all possible DAGs is a serious computational challenge. Producing a valid, accurate confidence set for the DAG structure would require astronomical sample sizes. If prior information is available about part of the DAG structure, the computational and statistical problems are at least partly ameliorated.

17.7   Bibliographic Remarks

There are a number of texts on DAGs including Edwards (1995) and Jordan (2004). The first use of DAGs for representing causal relationships was by Wright (1934). Modern treatments are contained in Spirtes et al. (2000) and Pearl (2000). Robins et al. (2003) discuss the problems with estimating causal structure from data.

17.8   Appendix

CAUSATION REVISITED. We discussed causation in Chapter 16 using the idea of counterfactual random variables. A different approach to causation uses

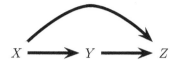

$$X \longrightarrow Y \longrightarrow Z$$

FIGURE 17.10. Conditioning versus intervening.

DAGs. The two approaches are mathematically equivalent though they appear to be quite different. In the DAG approach, the extra element is the idea of **intervention**. Consider the DAG in Figure 17.10.

The probability function for a distribution consistent with this DAG has the form $f(x, y, z) = f(x)f(y|x)f(z|x, y)$. The following is pseudocode for generating from this distribution.

$$\text{For } i = 1, \ldots, n :$$
$$x_i \leftarrow p_X(x_i)$$
$$y_i \leftarrow p_{Y|X}(y_i|x_i)$$
$$z_i \leftarrow p_{Z|X,Y}(z_i|x_i, y_i)$$

Suppose we repeat this code many times, yielding data $(x_1, y_1, z_1), \ldots, (x_n, y_n, z_n)$. Among all the times that we observe $Y = y$, how often is $Z = z$? The answer to this question is given by the conditional distribution of $Z|Y$. Specifically,

$$
\begin{aligned}
\mathbb{P}(Z = z|Y = y) &= \frac{\mathbb{P}(Y = y, Z = z)}{\mathbb{P}(Y = y)} = \frac{f(y, z)}{f(y)} \\
&= \frac{\sum_x f(x, y, z)}{f(y)} = \frac{\sum_x f(x) f(y|x) f(z|x, y)}{f(y)} \\
&= \sum_x f(z|x, y) \frac{f(y|x) f(x)}{f(y)} = \sum_x f(z|x, y) \frac{f(x, y)}{f(y)} \\
&= \sum_x f(z|x, y) f(x|y).
\end{aligned}
$$

Now suppose we **intervene** by changing the computer code. Specifically, suppose we fix Y at the value y. The code now looks like this:

$$\text{set } Y = y$$
$$\text{for } i = 1, \ldots, n$$
$$x_i \leftarrow p_X(x_i)$$
$$z_i \leftarrow p_{Z|X,Y}(z_i|x_i, y)$$

Having **set** $Y = y$, how often was $Z = z$? To answer, note that the intervention has changed the joint probability to be

$$f^*(x, z) = f(x)f(z|x, y).$$

The answer to our question is given by the marginal distribution

$$f^*(z) = \sum_x f^*(x, z) = \sum_x f(x)f(z|x, y).$$

We shall denote this as $\mathbb{P}(Z = z|Y := y)$ or $f(z|Y := y)$. We call $\mathbb{P}(Z = z|Y = y)$ **conditioning by observation** or **passive conditioning**. We call $\mathbb{P}(Z = z|Y := y)$ **conditioning by intervention** or **active conditioning**.

Passive conditioning is used to answer a predictive question like:

"Given that Joe smokes, what is the probability he will get lung cancer?"

Active conditioning is used to answer a causal question like:

"If Joe quits smoking, what is the probability he will get lung cancer?"

Consider a pair $(\mathcal{G}, \mathbb{P})$ where $\mathcal{G}$ is a DAG and $\mathbb{P}$ is a distribution for the variables V of the DAG. Let p denote the probability function for $\mathbb{P}$. Consider intervening and fixing a variable X to be equal to x. We represent the intervention by doing two things:

(1) Create a new DAG $\mathcal{G}^*$ by removing all arrows pointing into X;

(2) Create a new distribution $f^*(v) = \mathbb{P}(V = v|X := x)$ by removing the term $f(x|\pi_X)$ from $f(v)$.

The new pair $(\mathcal{G}^*, f^*)$ represents the intervention "set $X = x$."

17.15 Example. You may have noticed a correlation between rain and having a wet lawn, that is, the variable "Rain" is not independent of the variable "Wet Lawn" and hence $p_{R,W}(r, w) \neq p_R(r)p_W(w)$ where R denotes Rain and W denotes Wet Lawn. Consider the following two DAGs:

$$\text{Rain} \longrightarrow \text{Wet Lawn} \qquad \text{Rain} \longleftarrow \text{Wet Lawn}.$$

The first DAG implies that $f(w, r) = f(r)f(w|r)$ while the second implies that $f(w, r) = f(w)f(r|w)$ No matter what the joint distribution $f(w, r)$ is, both graphs are correct. Both imply that R and W are not independent. But, intuitively, if we want a graph to indicate causation, the first graph is right and the second is wrong. Throwing water on your lawn doesn't cause rain. The reason we feel the first is correct while the second is wrong is because the interventions implied by the first graph are correct.

Look at the first graph and form the intervention $W = 1$ where 1 denotes "wet lawn." Following the rules of intervention, we break the arrows into W to get the modified graph:

$$\boxed{\text{Rain} \quad \fbox{\textbf{set} \;\; \text{Wet Lawn} =1}}$$

with distribution $f^*(r) = f(r)$. Thus $\mathbb{P}(R = r \mid W := w) = \mathbb{P}(R = r)$ tells us that "wet lawn" does not cause rain.

Suppose we (wrongly) assume that the second graph is the correct causal graph and form the intervention $W = 1$ on the second graph. There are no arrows into W that need to be broken so the intervention graph is the same as the original graph. Thus $f^*(r) = f(r|w)$ which would imply that changing "wet" changes "rain." Clearly, this is nonsense.

Both are correct probability graphs but only the first is correct causally. We know the correct causal graph by using background knowledge.

17.16 Remark. We could try to learn the correct causal graph from data but this is dangerous. In fact it is impossible with two variables. With more than two variables there are methods that can find the causal graph under certain assumptions but they are large sample methods and, furthermore, there is no way to ever know if the sample size you have is large enough to make the methods reliable.

We can use DAGs to represent confounding variables. If X is a treatment and Y is an outcome, a confounding variable Z is a variable with arrows into both X and Y; see Figure 17.11. It is easy to check, using the formalism of interventions, that the following facts are true:

In a randomized study, the arrow between Z and X is broken. In this case, even with Z unobserved (represented by enclosing Z in a circle), the causal relationship between X and Y is estimable because it can be shown that $\mathbb{E}(Y|X := x) = \mathbb{E}(Y|X = x)$ which does not involve the unobserved Z. In an observational study, with all confounders observed, we get $\mathbb{E}(Y|X := x) = \int \mathbb{E}(Y|X = x, Z = z) dF_Z(z)$ as in formula (16.7). If Z is unobserved then we cannot estimate the causal effect because $\mathbb{E}(Y|X := x) = \int \mathbb{E}(Y|X = x, Z = z) dF_Z(z)$ involves the unobserved Z. We can't just use X and Y since in this case. $\mathbb{P}(Y = y|X = x) \neq \mathbb{P}(Y = y|X := x)$ which is just another way of saying that causation is not association.

In fact, we can make a precise connection between DAGs and counterfactuals as follows. Suppose that X and Y are binary. Define the confounding

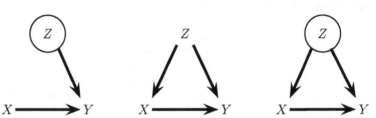

FIGURE 17.11. Randomized study; Observational study with measured confounders; Observational study with unmeasured confounders. The circled variables are unobserved.

variable Z by

$$Z = \begin{cases} 1 & \text{if } (C_0, C_1) = (0,0) \\ 2 & \text{if } (C_0, C_1) = (0,1) \\ 3 & \text{if } (C_0, C_1) = (1,0) \\ 4 & \text{if } (C_0, C_1) = (1,1). \end{cases}$$

From this, you can make the correspondence between the DAG approach and the counterfactual approach explicit. I leave this for the interested reader.

17.9   Exercises

1. Show that (17.1) and (17.2) are equivalent.

2. Prove Theorem 17.2.

3. Let X, Y and Z have the following joint distribution:

	$Y = 0$	$Y = 1$
$X = 0$	.405	.045
$X = 1$	.045	.005
$Z = 0$		

	$Y = 0$	$Y = 1$
$X = 0$	.125	.125
$X = 1$	.125	.125
$Z = 1$		

(a) Find the conditional distribution of X and Y given $Z = 0$ and the conditional distribution of X and Y given $Z = 1$.

(b) Show that $X \amalg Y | Z$.

(c) Find the marginal distribution of X and Y.

(d) Show that X and Y are not marginally independent.

4. Consider the three DAGs in Figure 17.6 without a collider. Prove that $X \amalg Z | Y$.

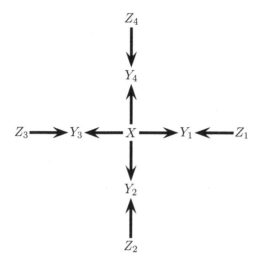

FIGURE 17.12. DAG for exercise 7.

5. Consider the DAG in Figure 17.6 with a collider. Prove that $X \amalg Z$ and that X and Z are dependent given Y.

6. Let $X \in \{0,1\}$, $Y \in \{0,1\}$, $Z \in \{0,1,2\}$. Suppose the distribution of (X,Y,Z) is Markov to:

$$X \longrightarrow Y \longrightarrow Z$$

Create a joint distribution $f(x,y,z)$ that is Markov to this DAG. Generate 1000 random vectors from this distribution. Estimate the distribution from the data using maximum likelihood. Compare the estimated distribution to the true distribution. Let $\theta = (\theta_{000}, \theta_{001}, \ldots, \theta_{112})$ where $\theta_{rst} = \mathbb{P}(X = r, Y = s, Z = t)$. Use the bootstrap to get standard errors and 95 percent confidence intervals for these 12 parameters.

7. Consider the DAG in Figure 17.12.

   (a) Write down the factorization of the joint density.

   (b) Prove that $X \amalg Z_j$.

8. Let $V = (X, Y, Z)$ have the following joint distribution

$$X \sim \text{Bernoulli} \left(\frac{1}{2} \right)$$

$$Y \mid X = x \quad \sim \quad \text{Bernoulli} \left(\frac{e^{4x-2}}{1 + e^{4x-2}} \right)$$

$$Z \mid X = x, Y = y \quad \sim \quad \text{Bernoulli} \left(\frac{e^{2(x+y)-2}}{1 + e^{2(x+y)-2}} \right).$$

(a) Find an expression for $\mathbb{P}(Z = z \mid Y = y)$. In particular, find $\mathbb{P}(Z = 1 \mid Y = 1)$.

(b) Write a program to simulate the model. Conduct a simulation and compute $\mathbb{P}(Z = 1 \mid Y = 1)$ empirically. Plot this as a function of the simulation size N. It should converge to the theoretical value you computed in (a).

(c) (Refers to material in the appendix.) Write down an expression for $\mathbb{P}(Z = 1 \mid Y := y)$. In particular, find $\mathbb{P}(Z = 1 \mid Y := 1)$.

(d) (Refers to material in the appendix.) Modify your program to simulate the intervention "set $Y = 1$." Conduct a simulation and compute $\mathbb{P}(Z = 1 \mid Y := 1)$ empirically. Plot this as a function of the simulation size N. It should converge to the theoretical value you computed in (c).

9. This is a continuous, Gaussian version of the last question. Let $V = (X, Y, Z)$ have the following joint distribution

$$X \quad \sim \quad \text{Normal} (0, 1)$$
$$Y \mid X = x \quad \sim \quad \text{Normal} (\alpha x, 1)$$
$$Z \mid X = x, Y = y \quad \sim \quad \text{Normal} (\beta y + \gamma x, 1).$$

Here, α, β and γ are fixed parameters. economists refer to models like this as **structural equation models.**

(a) Find an explicit expression for $f(z \mid y)$ and $\mathbb{E}(Z \mid Y = y) = \int z f(z \mid y) dz$.

(b) (Refers to material in the appendix.) Find an explicit expression for $f(z \mid Y := y)$ and then find $\mathbb{E}(Z \mid Y := y) \equiv \int z f(z \mid Y := y) dy$. Compare to (b).

(c) Find the joint distribution of (Y, Z). Find the correlation ρ between Y and Z.

(d) (Refers to material in the appendix.) Suppose that X is not observed and we try to make causal conclusions from the marginal distribution of (Y, Z). (Think of X as unobserved confounding variables.) In particular, suppose we declare that Y causes Z if $\rho \neq 0$ and we declare that Y does not cause Z if $\rho = 0$. Show that this will lead to erroneous conclusions.

(e) (Refers to material in the appendix.) Suppose we conduct a randomized experiment in which Y is randomly assigned. To be concrete, suppose that

$$
\begin{aligned}
X \quad &\sim \quad \text{Normal}(0,1) \\
Y \quad &\sim \quad \text{Normal}(\alpha,1) \\
Z \mid X = x, Y = y \quad &\sim \quad \text{Normal}(\beta y + \gamma x, 1).
\end{aligned}
$$

Show that the method in (d) now yields correct conclusions (i.e., $\rho = 0$ if and only if $f(z \mid Y := y)$ does not depend on y).

18
Undirected Graphs

Undirected graphs are an alternative to directed graphs for representing independence relations. Since both directed and undirected graphs are used in practice, it is a good idea to be facile with both. The main difference between the two is that the rules for reading independence relations from the graph are different.

18.1 Undirected Graphs

An **undirected graph** $\mathcal{G} = (V, E)$ has a finite set V of **vertices (or nodes)** and a set E of **edges (or arcs)** consisting of pairs of vertices. The vertices correspond to random variables $X, Y, Z, \ldots$ and edges are written as unordered pairs. For example, $(X, Y) \in E$ means that X and Y are joined by an edge. An example of a graph is in Figure 18.1.

Two vertices are **adjacent**, written $X \sim Y$, if there is an edge between them. In Figure 18.1, X and Y are adjacent but X and Z are not adjacent. A sequence $X_0, \ldots, X_n$ is called a **path** if $X_{i-1} \sim X_i$ for each i. In Figure 18.1, X, Y, Z is a path. A graph is **complete** if there is an edge between every pair of vertices. A subset $U \subset V$ of vertices together with their edges is called a **subgraph.**

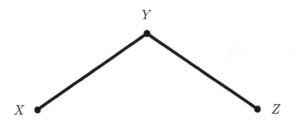

FIGURE 18.1. A graph with vertices $V = \{X, Y, Z\}$. The edge set is
$E = \{(X, Y), (Y, Z)\}$.

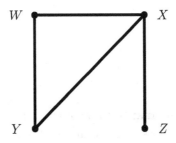

FIGURE 18.2. $\{Y, W\}$ and $\{Z\}$ are separated by $\{X\}$. Also, W and Z are separated
by $\{X, Y\}$.

If A, B and C are three distinct subsets of V, we say that C **separates**
A **and** B if every path from a variable in A to a variable in B intersects a
variable in C. In Figure 18.2 $\{Y, W\}$ and $\{Z\}$ are separated by $\{X\}$. Also, W
and Z are separated by $\{X, Y\}$.

18.2   Probability and Graphs

Let V be a set of random variables with distribution $\mathbb{P}$. Construct a graph
with one vertex for each random variable in V. Omit the edge between a pair
of variables if they are independent given the rest of the variables:

$$\text{no edge between } X \text{ and } Y \iff X \amalg Y | \text{rest}$$

FIGURE 18.3. $X \amalg Z | Y$.

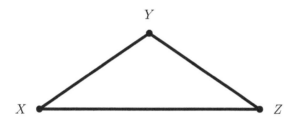

FIGURE 18.4. No implied independence relations.

where "rest" refers to all the other variables besides X and Y. The resulting graph is called a **pairwise Markov graph**. Some examples are shown in Figures 18.3, 18.4, 18.5, and 18.6.

The graph encodes a set of pairwise conditional independence relations. These relations imply other conditional independence relations. How can we figure out what they are? Fortunately, we can read these other conditional independence relations directly from the graph as well, as is explained in the next theorem.

18.1 Theorem. *Let* $\mathcal{G} = (V, E)$ *be a pairwise Markov graph for a distribution* $\mathbb{P}$. *Let* A, B *and* C *be distinct subsets of* V *such that* C *separates* A *and* B. *Then* $A \amalg B | C$.

18.2 Remark. If A and B are not connected (i.e., there is no path from A to B) then we may regard A and B as being separated by the empty set. Then Theorem 18.1 implies that $A \amalg B$.

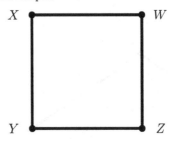

FIGURE 18.5. $X \amalg Z | \{Y, W\}$ and $Y \amalg W | \{X, Z\}$.

FIGURE 18.6. Pairwise independence implies that $X \amalg Z | \{Y, W\}$. But is $X \amalg Z | Y$?

The independence condition in Theorem 18.1 is called the **global Markov property**. We thus see that the pairwise and global Markov properties are equivalent. Let us state this more precisely. Given a graph $\mathcal{G}$, let $M_{\text{pair}}(\mathcal{G})$ be the set of distributions which satisfy the pairwise Markov property: thus $\mathbb{P} \in M_{\text{pair}}(\mathcal{G})$ if, under $\mathbb{P}$, $X \amalg Y | \text{rest}$ if and only if there is no edge between X and Y. Let $M_{\text{global}}(\mathcal{G})$ be the set of distributions which satisfy the global Markov property: thus $\mathbb{P} \in M_{\text{pair}}(\mathcal{G})$ if, under $\mathbb{P}$, $A \amalg B | C$ if and only if C separates A and B.

18.3 Theorem. *Let $\mathcal{G}$ be a graph. Then, $M_{\text{pair}}(\mathcal{G}) = M_{\text{global}}(\mathcal{G})$.*

Theorem 18.3 allows us to construct graphs using the simpler pairwise property and then we can deduce other independence relations using the global Markov property. Think how hard this would be to do algebraically. Returning to 18.6, we now see that $X \amalg Z | Y$ and $Y \amalg W | Z$.

18.4 Example. Figure 18.7 implies that $X \amalg Y$, $X \amalg Z$ and $X \amalg (Y, Z)$. ∎

18.5 Example. Figure 18.8 implies that $X \amalg W | (Y, Z)$ and $X \amalg Z | Y$. ∎

FIGURE 18.7. $X \amalg Y$, $X \amalg Z$ and $X \amalg (Y, Z)$.

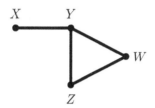

FIGURE 18.8. $X \amalg W | (Y, Z)$ and $X \amalg Z | Y$.

18.3   Cliques and Potentials

A **clique** is a set of variables in a graph that are all adjacent to each other. A set of variables is a **maximal clique** if it is a clique and if it is not possible to include another variable and still be a clique. A **potential** is any positive function. Under certain conditions, it can be shown that $\mathbb{P}$ is Markov $\mathcal{G}$ if and only if its probability function f can be written as

$$f(x) = \frac{\prod_{C \in \mathcal{C}} \psi_C(x_C)}{Z} \tag{18.1}$$

where $\mathcal{C}$ is the set of maximal cliques and

$$Z = \sum_x \prod_{C \in \mathcal{C}} \psi_C(x_C).$$

18.6 Example. The maximal cliques for the graph in Figure 18.1 are $C_1 = \{X, Y\}$ and $C_2 = \{Y, Z\}$. Hence, if $\mathbb{P}$ is Markov to the graph, then its probability function can be written

$$f(x, y, z) \propto \psi_1(x, y) \psi_2(y, z)$$

for some positive functions ψ_1 and ψ_2. ∎

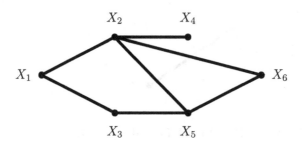

FIGURE   18.9.   The   maximumly   cliques   of   this   graph   are
$\{X_1, X_2\}, \{X_1, X_3\}, \{X_2, X_4\}, \{X_3, X_5\}, \{X_2, X_5, X_6\}$.

18.7 Example. The maximal cliques for the graph in Figure 18.9 are

$$\{X_1, X_2\}, \{X_1, X_3\}, \{X_2, X_4\}, \{X_3, X_5\}, \{X_2, X_5, X_6\}.$$

Thus we can write the probability function as

$$f(x_1, x_2, x_3, x_4, x_5, x_6) \quad \propto \quad \psi_{12}(x_1, x_2)\psi_{13}(x_1, x_3)\psi_{24}(x_2, x_4)$$
$$\times \psi_{35}(x_3, x_5)\psi_{256}(x_2, x_5, x_6). \quad \blacksquare$$

18.4   Fitting Graphs to Data

Given a data set, how do we find a graphical model that fits the data? As
with directed graphs, this is a big topic that we will not treat here. However,
in the discrete case, one way to fit a graph to data is to use a **log-linear
model**, which is the subject of the next chapter.

18.5   Bibliographic Remarks

Thorough treatments of undirected graphs can be found in Whittaker (1990)
and Lauritzen (1996). Some of the exercises below are from Whittaker (1990).

18.6   Exercises

1. Consider random variables (X_1, X_2, X_3). In each of the following cases,
   draw a graph that has the given independence relations.

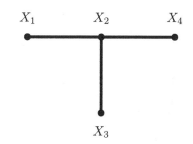

FIGURE 18.10.

$$X_1 \quad\quad X_2 \quad\quad X_3 \quad\quad X_4$$

FIGURE 18.11.

(a) $X_1 \amalg X_3 \mid X_2$.

(b) $X_1 \amalg X_2 \mid X_3$ and $X_1 \amalg X_3 \mid X_2$.

(c) $X_1 \amalg X_2 \mid X_3$ and $X_1 \amalg X_3 \mid X_2$ and $X_2 \amalg X_3 \mid X_1$.

2. Consider random variables (X_1, X_2, X_3, X_4). In each of the following cases, draw a graph that has the given independence relations.

(a) $X_1 \amalg X_3 \mid X_2, X_4$ and $X_1 \amalg X_4 \mid X_2, X_3$ and $X_2 \amalg X_4 \mid X_1, X_3$.

(b) $X_1 \amalg X_2 \mid X_3, X_4$ and $X_1 \amalg X_3 \mid X_2, X_4$ and $X_2 \amalg X_3 \mid X_1, X_4$.

(c) $X_1 \amalg X_3 \mid X_2, X_4$ and $X_2 \amalg X_4 \mid X_1, X_3$.

3. A conditional independence between a pair of variables is **minimal** if it is not possible to use the Separation Theorem to eliminate any variable from the conditioning set, i.e. from the right hand side of the bar Whittaker (1990). Write down the minimal conditional independencies from: (a) Figure 18.10; (b) Figure 18.11; (c) Figure 18.12; (d) Figure 18.13.

4. Let X_1, X_2, X_3 be binary random variables. Construct the likelihood ratio test for

$$H_0 : X_1 \amalg X_2 \mid X_3 \quad \text{versus} \quad H_1 : X_1 \text{ is not independent of } X_2 \mid X_3.$$

5. Here are breast cancer data from Morrison et al. (1973) on diagnostic center (X_1), nuclear grade (X_2), and survival (X_3):

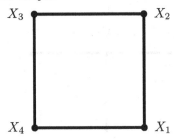

FIGURE 18.12.

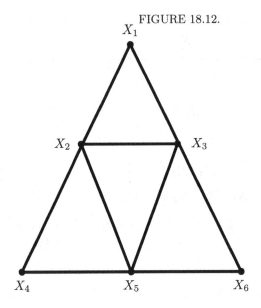

FIGURE 18.13.

	X_2 X_3	malignant died	malignant survived	benign died	benign survived
X_1	Boston	35	59	47	112
	Glamorgan	42	77	26	76

(a) Treat this as a multinomial and find the maximum likelihood estimator.

(b) If someone has a tumor classified as benign at the Glamorgan clinic, what is the estimated probability that they will die? Find the standard error for this estimate.

(c) Test the following hypotheses:

$$X_1 \amalg X_2 | X_3 \quad \text{versus} \quad X_1 \sim X_2 | X_3$$
$$X_1 \amalg X_3 | X_2 \quad \text{versus} \quad X_1 \sim X_3 | X_2$$
$$X_2 \amalg X_3 | X_1 \quad \text{versus} \quad X_2 \sim X_3 | X_1$$

Use the test from question 4. Based on the results of your tests, draw and interpret the resulting graph.

19
Log-Linear Models

In this chapter we study **log-linear models** which are useful for modeling multivariate discrete data. There is a strong connection between log-linear models and undirected graphs.

19.1 The Log-Linear Model

Let $X = (X_1, \ldots, X_m)$ be a discrete random vector with probability function

$$f(x) = \mathbb{P}(X = x) = \mathbb{P}(X_1 = x_1, \ldots, X_m = x_m)$$

where $x = (x_1, \ldots, x_m)$. Let r_j be the number of values that X_j takes. Without loss of generality, we can assume that $X_j \in \{0, 1, \ldots, r_j - 1\}$. Suppose now that we have n such random vectors. We can think of the data as a sample from a Multinomial with $N = r_1 \times r_2 \times \cdots \times r_m$ categories. The data can be represented as counts in a $r_1 \times r_2 \times \cdots \times r_m$ table. Let $p = (p_1, \ldots, p_N)$ denote the multinomial parameter.

Let $S = \{1, \ldots, m\}$. Given a vector $x = (x_1, \ldots, x_m)$ and a subset $A \subset S$, let $x_A = (x_j : j \in A)$. For example, if $A = \{1, 3\}$ then $x_A = (x_1, x_3)$.

19.1 Theorem. *The joint probability function $f(x)$ of a single random vector $X = (X_1, \ldots, X_m)$ can be written as*

$$\log f(x) = \sum_{A \subset S} \psi_A(x) \tag{19.1}$$

where the sum is over all subsets A of $S = \{1, \ldots, m\}$ and the ψ's satisfy the following conditions:

1. *$\psi_\emptyset(x)$ is a constant;*

2. *For every $A \subset S$, $\psi_A(x)$ is only a function of x_A and not the rest of the $x_j's$.*

3. *If $i \in A$ and $x_i = 0$, then $\psi_A(x) = 0$.*

The formula in equation (19.1) is called the **log-linear expansion** of f. Each $\psi_A(x)$ may depend on some unknown parameters β_A. Let $\beta = (\beta_A : A \subset S)$ be the set of all these parameters. We will write $f(x) = f(x; \beta)$ when we want to emphasize the dependence on the unknown parameters β.

In terms of the multinomial, the parameter space is

$$\mathcal{P} = \left\{ p = (p_1, \ldots, p_N) : \ p_j \geq 0, \ \sum_{j=1}^{N} p_j = 1 \right\}.$$

This is an $N - 1$ dimensional space. In the log-linear representation, the parameter space is

$$\Theta = \left\{ \beta = (\beta_1, \ldots, \beta_N) : \ \beta = \beta(p), p \in \mathcal{P} \right\}$$

where $\beta(p)$ is the set of β values associated with p. The set Θ is a $N - 1$ dimensional surface in $\mathbb{R}^N$. We can always go back and forth between the two parameterizations we can write $\beta = \beta(p)$ and $p = p(\beta)$.

19.2 Example. Let $X \sim \text{Bernoulli}(p)$ where $0 < p < 1$. We can write the probability mass function for X as

$$f(x) = p^x (1-p)^{1-x} = p_1^x p_2^{1-x}$$

for $x = 0, 1$, where $p_1 = p$ and $p_2 = 1 - p$. Hence,

$$\log f(x) = \psi_\emptyset(x) + \psi_1(x)$$

where

$$
\begin{aligned}
\psi_0(x) &= \log(p_2) \\
\psi_1(x) &= x \log\left(\frac{p_1}{p_2}\right).
\end{aligned}
$$

Notice that $\psi_0(x)$ is a constant (as a function of x) and $\psi_1(x) = 0$ when $x = 0$. Thus the three conditions of Theorem 19.1 hold. The log-linear parameters are

$$
\beta_0 = \log(p_2), \quad \beta_1 = \log\left(\frac{p_1}{p_2}\right).
$$

The original, multinomial parameter space is $\mathcal{P} = \{(p_1, p_2) : p_j \geq 0, p_1 + p_2 = 1\}$. The log-linear parameter space is

$$
\Theta = \left\{(\beta_0, \beta_1) \in \mathbb{R}^2 : e^{\beta_0 + \beta_1} + e^{\beta_0} = 1.\right\}
$$

Given (p_1, p_2) we can solve for (β_0, β_1). Conversely, given (β_0, β_1) we can solve for (p_1, p_2). ∎

19.3 Example. Let $X = (X_1, X_2)$ where $X_1 \in \{0, 1\}$ and $X_2 \in \{0, 1, 2\}$. The joint distribution of n such random vectors is a multinomial with 6 categories. The multinomial parameters can be written as a 2-by-3 table as follows:

multinomial	x_2	0	1	2
x_1	0	p_{00}	p_{01}	p_{02}
	1	p_{10}	p_{11}	p_{12}

The n data vectors can be summarized as counts:

data	x_2	0	1	2
x_1	0	C_{00}	C_{01}	C_{02}
	1	C_{10}	C_{11}	C_{12}

For $x = (x_1, x_2)$, the log-linear expansion takes the form

$$
\log f(x) = \psi_0(x) + \psi_1(x) + \psi_2(x) + \psi_{12}(x)
$$

where

$$
\begin{aligned}
\psi_0(x) &= \log p_{00} \\
\psi_1(x) &= x_1 \log\left(\frac{p_{10}}{p_{00}}\right) \\
\psi_2(x) &= I(x_2 = 1) \log\left(\frac{p_{01}}{p_{00}}\right) + I(x_2 = 2) \log\left(\frac{p_{02}}{p_{00}}\right) \\
\psi_{12}(x) &= I(x_1 = 1, x_2 = 1) \log\left(\frac{p_{11}p_{00}}{p_{01}p_{10}}\right) + I(x_1 = 1, x_2 = 2) \log\left(\frac{p_{12}p_{00}}{p_{02}p_{10}}\right).
\end{aligned}
$$

Convince yourself that the three conditions on the ψ's of the theorem are satisfied. The six parameters of this model are:

$$\beta_1 = \log p_{00} \qquad \beta_2 = \log\left(\frac{p_{10}}{p_{00}}\right) \qquad \beta_3 = \log\left(\frac{p_{01}}{p_{00}}\right)$$

$$\beta_4 = \log\left(\frac{p_{02}}{p_{00}}\right) \quad \beta_5 = \log\left(\frac{p_{11}p_{00}}{p_{01}p_{10}}\right) \quad \beta_6 = \log\left(\frac{p_{12}p_{00}}{p_{02}p_{10}}\right).$$

∎

The next theorem gives an easy way to check for conditional independence in a log-linear model.

19.4 Theorem. *Let (X_a, X_b, X_c) be a partition of a vectors $(X_1, \ldots, X_m)$. Then $X_b \amalg X_c | X_a$ if and only if all the ψ-terms in the log-linear expansion that have at least one coordinate in b and one coordinate in c are 0.*

To prove this theorem, we will use the following lemma whose proof follows easily from the definition of conditional independence.

19.5 Lemma. *A partition (X_a, X_b, X_c) satisfies $X_b \amalg X_c | X_a$ if and only if $f(x_a, x_b, x_c) = g(x_a, x_b)h(x_a, x_c)$ for some functions g and h*

PROOF. (Theorem 19.4.) Suppose that ψ_t is 0 whenever t has coordinates in b and c. Hence, ψ_t is 0 if $t \not\subset a \bigcup b$ or $t \not\subset a \bigcup c$. Therefore

$$\log f(x) = \sum_{t \subset a \bigcup b} \psi_t(x) + \sum_{t \subset a \bigcup c} \psi_t(x) - \sum_{t \subset a} \psi_t(x).$$

Exponentiating, we see that the joint density is of the form $g(x_a, x_b)h(x_a, x_c)$. By Lemma 19.5, $X_b \amalg X_c | X_a$. The converse follows by reversing the argument. ∎

19.2 Graphical Log-Linear Models

A log-linear model is **graphical** if missing terms correspond only to conditional independence constraints.

19.6 Definition. *Let $\log f(x) = \sum_{A \subset S} \psi_A(x)$ be a log-linear model. Then f is **graphical** if all ψ-terms are nonzero except for any pair of coordinates not in the edge set for some graph $\mathcal{G}$. In other words, $\psi_A(x) = 0$ if and only if $\{i, j\} \subset A$ and (i, j) is not an edge.*

Here is a way to think about the definition above:

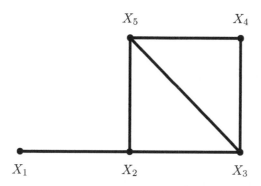

FIGURE 19.1. Graph for Example 19.7.

If you can add a term to the model and the graph does not change, then the model is not graphical.

19.7 Example. Consider the graph in Figure 19.1.

The graphical log-linear model that corresponds to this graph is

$$\log f(x) = \psi_\emptyset + \psi_1(x) + \psi_2(x) + \psi_3(x) + \psi_4(x) + \psi_5(x)$$
$$+ \psi_{12}(x) + \psi_{23}(x) + \psi_{25}(x) + \psi_{34}(x) + \psi_{35}(x) + \psi_{45}(x) + \psi_{235}(x) + \psi_{345}(x)$$

Let's see why this model is graphical. The edge $(1, 5)$ is missing in the graph. Hence any term containing that pair of indices is omitted from the model. For example,

$$\psi_{15}, \ \psi_{125}, \ \psi_{135}, \ \psi_{145}, \ \psi_{1235}, \ \psi_{1245}, \ \psi_{1345}, \ \psi_{12345}$$

are all omitted. Similarly, the edge $(2, 4)$ is missing and hence

$$\psi_{24}, \ \psi_{124}, \ \psi_{234}, \ \psi_{245}, \ \psi_{1234}, \ \psi_{1245}, \ \psi_{2345}, \ \psi_{12345}$$

are all omitted. There are other missing edges as well. You can check that the model omits all the corresponding ψ terms. Now consider the model

$$\log f(x) = \psi_\emptyset(x) + \psi_1(x) + \psi_2(x) + \psi_3(x) + \psi_4(x) + \psi_5(x)$$
$$+ \psi_{12}(x) + \psi_{23}(x) + \psi_{25}(x) + \psi_{34}(x) + \psi_{35}(x) + \psi_{45}(x).$$

This is the same model except that the three way interactions were removed. If we draw a graph for this model, we will get the same graph. For example, no ψ terms contain $(1, 5)$ so we omit the edge between X_1 and X_5. But this is not graphical since it has extra terms omitted. The independencies and graphs

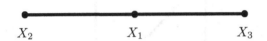

$$X_2 \qquad\qquad X_1 \qquad\qquad X_3$$

FIGURE 19.2. Graph for Example 19.10.

for the two models are the same but the latter model has other constraints besides conditional independence constraints. This is not a bad thing. It just means that if we are only concerned about presence or absence of conditional independences, then we need not consider such a model. The presence of the three-way interaction ψ_{235} means that the strength of association between X_2 and X_3 varies as a function of X_5. Its absence indicates that this is not so. ∎

19.3   Hierarchical Log-Linear Models

There is a set of log-linear models that is larger than the set of graphical models and that are used quite a bit. These are the hierarchical log-linear models.

19.8 Definition. *A log-linear model is* **hierarchical** *if* $\psi_A = 0$ *and* $A \subset B$ *implies that* $\psi_B = 0$.

19.9 Lemma. *A graphical model is hierarchical but the reverse need not be true.*

19.10 Example. Let

$$\log f(x) = \psi_0(x) + \psi_1(x) + \psi_2(x) + \psi_3(x) + \psi_{12}(x) + \psi_{13}(x).$$

The model is hierarchical; its graph is given in Figure 19.2. The model is graphical because all terms involving (2,3) are omitted. It is also hierarchical. ∎

19.11 Example. Let

$$\log f(x) = \psi_0(x) + \psi_1(x) + \psi_2(x) + \psi_3(x) + \psi_{12}(x) + \psi_{13}(x) + \psi_{23}(x).$$

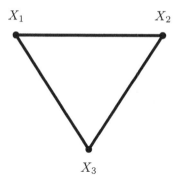

FIGURE 19.3. The graph is complete. The model is hierarchical but not graphical.

FIGURE 19.4. The model for this graph is not hierarchical.

The model is hierarchical. It is not graphical. The graph corresponding to this model is complete; see Figure 19.3. It is not graphical because $\psi_{123}(x) = 0$ which does not correspond to any pairwise conditional independence. ∎

19.12 Example. Let

$$\log f(x) = \psi_\emptyset(x) + \psi_3(x) + \psi_{12}(x).$$

The graph corresponding is in Figure 19.4. This model is not hierarchical since $\psi_2 = 0$ but ψ_{12} is not. Since it is not hierarchical, it is not graphical either. ∎

19.4   Model Generators

Hierarchical models can be written succinctly using **generators**. This is most easily explained by example. Suppose that $X = (X_1, X_2, X_3)$. Then, $M = 1.2 + 1.3$ stands for

$$\log f = \psi_\emptyset + \psi_1 + \psi_2 + \psi_3 + \psi_{12} + \psi_{13}.$$

The formula $M = 1.2 + 1.3$ says: "include ψ_{12} and ψ_{13}." We have to also include the lower order terms or it won't be hierarchical. The generator $M = 1.2.3$ is the **saturated** model

$$\log f = \psi_{\emptyset} + \psi_1 + \psi_2 + \psi_3 + \psi_{12} + \psi_{13} + \psi_{23} + \psi_{123}.$$

The saturated models corresponds to fitting an unconstrained multinomial. Consider $M = 1 + 2 + 3$ which means

$$\log f = \psi_{\emptyset} + \psi_1 + \psi_2 + \psi_3.$$

This is the mutual independence model. Finally, consider $M = 1.2$ which has log-linear expansion

$$\log f = \psi_{\emptyset} + \psi_1 + \psi_2 + \psi_{12}.$$

This model makes $X_3 | X_2 = x_2, X_1 = x_1$ a uniform distribution.

19.5   Fitting Log-Linear Models to Data

Let β denote all the parameters in a log-linear model M. The loglikelihood for β is

$$\ell(\beta) = \sum_{i=1}^{n} \log f(X_i; \beta)$$

where $f(X_i; \beta)$ is the probability function for the ith random vector $X_i = (X_{i1}, \ldots, X_{im})$ as give by equation (19.1). The MLE $\widehat{\beta}$ generally has to be found numerically. The Fisher information matrix is also found numerically and we can then get the estimated standard errors from the inverse Fisher information matrix.

When fitting log-linear models, one has to address the following model selection problem: which ψ terms should we include in the model? This is essentially the same as the model selection problem in linear regression.

One approach is is to use AIC. Let M denote some log-linear model. Different models correspond to setting different ψ terms to 0. Now we choose the model M which maximizes

$$\text{AIC}(M) = \widehat{\ell}(M) - |M| \tag{19.2}$$

where $|M|$ is the number of parameters in model M and $\widehat{\ell}(M)$ is the value of the log-likelihood evaluated at the MLE for that model. Usually the model search is restricted to hierarchical models. This reduces the search space. Some also claim that we should only search through the hierarchical models because other models are less interpretable.

A different approach is based on hypothesis testing. The model that includes all possible ψ-terms is called the **saturated model** and we denote it by M_{sat}. Now for each M we test the hypothesis

$$H_0 : \text{the true model is } M \quad \text{versus} \quad H_1 : \text{the true model is } M_{sat}.$$

The likelihood ratio test for this hypothesis is called the deviance.

19.13 Definition. *For any submodel M, define the **deviance** dev(M) by*

$$\mathrm{dev}(M) = 2(\widehat{\ell}_{sat} - \widehat{\ell}_M)$$

where $\widehat{\ell}_{sat}$ is the log-likelihood of the saturated model evaluated at the MLE and $\widehat{\ell}_M$ is the log-likelihood of the model M evaluated at its MLE.

19.14 Theorem. *The deviance is the likelihood ratio test statistic for*

$$H_0 : \text{the model is } M \quad \text{versus} \quad H_1 : \text{the model is } M_{sat}.$$

Under H_0, $dev(M) \overset{d}{\to} \chi^2_\nu$ with ν degrees of freedom equal to the difference in the number of parameters between the saturated model and M.

One way to find a good model is to use the deviance to test every sub-model. Every model that is not rejected by this test is then considered a plausible model. However, this is not a good strategy for two reasons. First, we will end up doing many tests which means that there is ample opportunity for making Type I and Type II errors. Second, we will end up using models where we failed to reject H_0. But we might fail to reject H_0 due to low power. The result is that we end up with a bad model just due to low power.

After finding a "best model" this way we can draw the corresponding graph.

19.15 Example. The following breast cancer data are from Morrison et al. (1973). The data are on diagnostic center (X_1), nuclear grade (X_2), and survival (X_3):

		malignant	malignant	benign	benign
	X_2				
	X_3	died	survived	died	survived
X_1	Boston	35	59	47	112
	Glamorgan	42	77	26	76

The saturated log-linear model is:

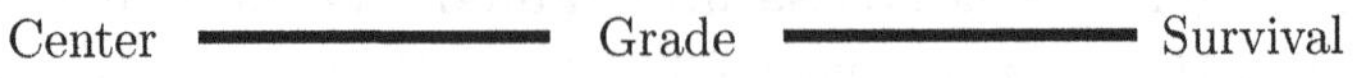

Center ━━━━━━━ Grade ━━━━━━━ Survival

FIGURE 19.5. The graph for Example 19.15.

Variable	$\widehat{\beta}_j$	$\widehat{se}$	W_j	p-value
(Intercept)	3.56	0.17	21.03	0.00 ***
center	0.18	0.22	0.79	0.42
grade	0.29	0.22	1.32	0.18
survival	0.52	0.21	2.44	0.01 *
center×grade	-0.77	0.33	-2.31	0.02 *
center×survival	0.08	0.28	0.29	0.76
grade×survival	0.34	0.27	1.25	0.20
center×grade×survival	0.12	0.40	0.29	0.76

The best sub-model, selected using AIC and backward searching is:

Variable	$\widehat{\beta}_j$	$\widehat{se}$	W_j	p-value
(Intercept)	3.52	0.13	25.62	< 0.00 ***
center	0.23	0.13	1.70	0.08
grade	0.26	0.18	1.43	0.15
survival	0.56	0.14	3.98	6.65e-05 ***
center×grade	-0.67	0.18	-3.62	0.00 ***
grade×survival	0.37	0.19	1.90	0.05

The graph for this model M is shown in Figure 19.5. To test the fit of this model, we compute the deviance of M which is 0.6. The appropriate χ^2 has $8 - 6 = 2$ degrees of freedom. The p-value is $\mathbb{P}(\chi_2^2 > .6) = .74$. So we have no evidence to suggest that the model is a poor fit. ∎

19.6 Bibliographic Remarks

For this chapter, I drew heavily on Whittaker (1990) which is an excellent text on log-linear models and graphical models. Some of the exercises are from Whittaker. A classic reference on log-linear models is Bishop et al. (1975).

19.7   Exercises

1. Solve for the $p'_{ij}s$ in terms of the β's in Example 19.3.

2. Prove Lemma 19.5.

3. Prove Lemma 19.9.

4. Consider random variables (X_1, X_2, X_3, X_4). Suppose the log-density is

$$\log f(x) = \psi_\emptyset(x) + \psi_{12}(x) + \psi_{13}(x) + \psi_{24}(x) + \psi_{34}(x).$$

   (a) Draw the graph G for these variables.

   (b) Write down all independence and conditional independence relations implied by the graph.

   (c) Is this model graphical? Is it hierarchical?

5. Suppose that parameters $p(x_1, x_2, x_3)$ are proportional to the following values:

	x_2	0	0	1	1
	x_3	0	1	0	1
x_1	0	2	8	4	16
	1	16	128	32	256

   Find the ψ-terms for the log-linear expansion. Comment on the model.

6. Let $X_1, \ldots, X_4$ be binary. Draw the independence graphs corresponding to the following log-linear models. Also, identify whether each is graphical and/or hierarchical (or neither).

   (a) $\log f = 7 + 11x_1 + 2x_2 + 1.5x_3 + 17x_4$

   (b) $\log f = 7 + 11x_1 + 2x_2 + 1.5x_3 + 17x_4 + 12x_2x_3 + 78x_2x_4 + 3x_3x_4 + 32x_2x_3x_4$

   (c) $\log f = 7 + 11x_1 + 2x_2 + 1.5x_3 + 17x_4 + 12x_2x_3 + 3x_3x_4 + x_1x_4 + 2x_1x_2$

   (d) $\log f = 7 + 5055x_1x_2x_3x_4$

20
Nonparametric Curve Estimation

In this Chapter we discuss nonparametric estimation of probability density functions and regression functions which we refer to as **curve estimation** or **smoothing**.

In Chapter 7 we saw that it is possible to consistently estimate a cumulative distribution function F without making any assumptions about F. If we want to estimate a probability density function $f(x)$ or a regression function $r(x) = \mathbb{E}(Y|X = x)$ the situation is different. We cannot estimate these functions consistently without making some smoothness assumptions. Correspondingly, we need to perform some sort of smoothing operation on the data.

An example of a density estimator is a **histogram**, which we discuss in detail in Section 20.2. To form a histogram estimator of a density f, we divide the real line to disjoint sets called **bins**. The histogram estimator is a piecewise constant function where the height of the function is proportional to number of observations in each bin; see Figure 20.3. The number of bins is an example of a **smoothing parameter**. If we smooth too much (large bins) we get a highly biased estimator while if we smooth too little (small bins) we get a highly variable estimator. Much of curve estimation is concerned with trying to optimally balance variance and bias.

This is a function of the data   This is the point at which we are
                                   evaluating $\widehat{g}(\cdot)$

FIGURE 20.1. A curve estimate $\widehat{g}$ is random because it is a function of the data. The point x at which we evaluate $\widehat{g}$ is not a random variable.

20.1   The Bias-Variance Tradeoff

Let g denote an unknown function such as a density function or a regression function. Let $\widehat{g}_n$ denote an estimator of g. Bear in mind that $\widehat{g}_n(x)$ is a random function evaluated at a point x. The estimator is random because it depends on the data. See Figure 20.1.

As a loss function, we will use the **integrated squared error (ISE)**: [1]

$$L(g, \widehat{g}_n) = \int (g(u) - \widehat{g}_n(u))^2 \, du. \tag{20.1}$$

The **risk** or **mean integrated squared error (MISE)** with respect to squared error loss is

$$R(f, \widehat{f}) = \mathbb{E}\left(L(g, \widehat{g}) \right). \tag{20.2}$$

20.1 Lemma. *The risk can be written as*

$$R(g, \widehat{g}_n) = \int b^2(x) \, dx + \int v(x) \, dx \tag{20.3}$$

where

$$b(x) = \mathbb{E}(\widehat{g}_n(x)) - g(x) \tag{20.4}$$

is the bias of $\widehat{g}_n(x)$ at a fixed x and

$$v(x) = \mathbb{V}(\widehat{g}_n(x)) = \mathbb{E}\left(\left(\widehat{g}_n(x) - \mathbb{E}(\widehat{g}_n(x))^2 \right) \right) \tag{20.5}$$

is the variance of $\widehat{g}_n(x)$ at a fixed x.

---

[1]We could use other loss functions. The results are similar but the analysis is much more complicated.

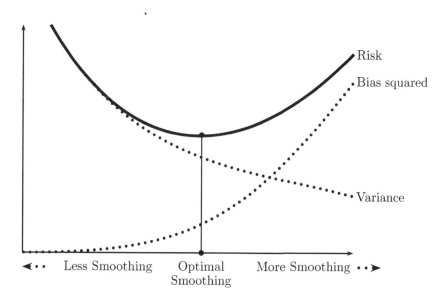

FIGURE 20.2. The Bias-Variance trade-off. The bias increases and the variance decreases with the amount of smoothing. The optimal amount of smoothing, indicated by the vertical line, minimizes the risk = bias^2 + variance.

In summary,

$$\text{RISK} = \text{BIAS}^2 + \text{VARIANCE}. \qquad (20.6)$$

When the data are oversmoothed, the bias term is large and the variance is small. When the data are undersmoothed the opposite is true; see Figure 20.2. This is called the **bias-variance tradeoff**. Minimizing risk corresponds to balancing bias and variance.

20.2   Histograms

Let $X_1, \ldots, X_n$ be IID on $[0, 1]$ with density f. The restriction to $[0, 1]$ is not crucial; we can always rescale the data to be on this interval. Let m be an integer and define **bins**

$$B_1 = \left[0, \frac{1}{m}\right), B_2 = \left[\frac{1}{m}, \frac{2}{m}\right), \quad \ldots, \quad B_m = \left[\frac{m-1}{m}, 1\right]. \qquad (20.7)$$

Define the **binwidth** $h = 1/m$, let ν_j be the number of observations in B_j, let $\widehat{p}_j = \nu_j/n$ and let $p_j = \int_{B_j} f(u)du$.

The **histogram estimator** is defined by

$$\widehat{f}_n(x) = \begin{cases} \widehat{p}_1/h & x \in B_1 \\ \widehat{p}_2/h & x \in B_2 \\ \vdots & \vdots \\ \widehat{p}_m/h & x \in B_m \end{cases}$$

which we can write more succinctly as

$$\widehat{f}_n(x) = \sum_{j=1}^n \frac{\widehat{p}_j}{h} I(x \in B_j). \qquad (20.8)$$

To understand the motivation for this estimator, let $p_j = \int_{B_j} f(u)du$ and note that, for $x \in B_j$ and h small,

$$\mathbb{E}(\widehat{f}_n(x)) = \frac{\mathbb{E}(\widehat{p}_j)}{h} = \frac{p_j}{h} = \frac{\int_{B_j} f(u)du}{h} \approx \frac{f(x)h}{f(x)} = f(x).$$

20.2 Example. Figure 20.3 shows three different histograms based on $n = 1,266$ data points from an astronomical sky survey. Each data point represents the distance from us to a galaxy. The galaxies lie on a "pencilbeam" pointing directly from the Earth out into space. Because of the finite speed of light, looking at galaxies farther and farther away corresponds to looking back in time. Choosing the right number of bins involves finding a good tradeoff between bias and variance. We shall see later that the top left histogram has too few bins resulting in oversmoothing and too much bias. The bottom left histogram has too many bins resulting in undersmoothing and too few bins. The top right histogram is just right. The histogram reveals the presence of clusters of galaxies. Seeing how the size and number of galaxy clusters varies with time, helps cosmologists understand the evolution of the universe. ∎

The mean and variance of $\widehat{f}_n(x)$ are given in the following Theorem.

20.3 Theorem. *Consider fixed x and fixed m, and let B_j be the bin containing x. Then,*

$$\mathbb{E}(\widehat{f}_n(x)) = \frac{p_j}{h} \quad \text{and} \quad \mathbb{V}(\widehat{f}_n(x)) = \frac{p_j(1-p_j)}{nh^2}. \qquad (20.9)$$

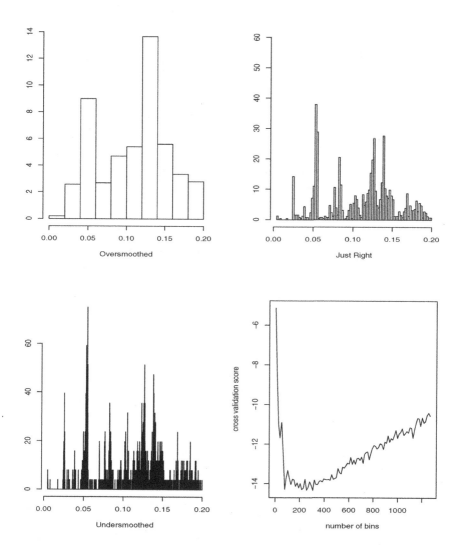

FIGURE 20.3. Three versions of a histogram for the astronomy data. The top left histogram has too few bins. The bottom left histogram has too many bins. The top right histogram is just right. The lower, right plot shows the estimated risk versus the number of bins.

Let's take a closer look at the bias-variance tradeoff using equation (20.9). Consider some $x \in B_j$. For any other $u \in B_j$,

$$f(u) \approx f(x) + (u - x)f'(x)$$

and so

$$
\begin{aligned}
p_j = \int_{B_j} f(u)du &\approx \int_{B_j} \left(f(x) + (u - x)f'(x) \right) du \\
&= f(x)h + hf'(x) \left(h \left(j - \frac{1}{2} \right) - x \right).
\end{aligned}
$$

Therefore, the bias $b(x)$ is

$$
\begin{aligned}
b(x) &= \mathbb{E}(\widehat{f}_n(x)) - f(x) = \frac{p_j}{h} - f(x) \\
&\approx \frac{f(x)h + hf'(x) \left(h \left(j - \frac{1}{2} \right) - x \right)}{h} - f(x) \\
&= f'(x) \left(h \left(j - \frac{1}{2} \right) - x \right).
\end{aligned}
$$

If $\widetilde{x}_j$ is the center of the bin, then

$$
\begin{aligned}
\int_{B_j} b^2(x)\, dx &\approx \int_{B_j} (f'(x))^2 \left(h \left(j - \frac{1}{2} \right) - x \right)^2 dx \\
&\approx (f'(\widetilde{x}_j))^2 \int_{B_j} \left(h \left(j - \frac{1}{2} \right) - x \right)^2 dx \\
&= (f'(\widetilde{x}_j))^2 \frac{h^3}{12}.
\end{aligned}
$$

Therefore,

$$
\begin{aligned}
\int_0^1 b^2(x)dx &= \sum_{j=1}^m \int_{B_j} b^2(x)dx \approx \sum_{j=1}^m (f'(\widetilde{x}_j))^2 \frac{h^3}{12} \\
&= \frac{h^2}{12} \sum_{j=1}^m h (f'(\widetilde{x}_j))^2 \approx \frac{h^2}{12} \int_0^1 (f'(x))^2 dx.
\end{aligned}
$$

Note that this increases as a function of h. Now consider the variance. For h small, $1 - p_j \approx 1$, so

$$
\begin{aligned}
v(x) &\approx \frac{p_j}{nh^2} \\
&= \frac{f(x)h + hf'(x) \left(h \left(j - \frac{1}{2} \right) - x \right)}{nh^2} \\
&\approx \frac{f(x)}{nh}
\end{aligned}
$$

where we have kept only the dominant term. So,

$$\int_0^1 v(x)dx \approx \frac{1}{nh}.$$

Note that this decreases with h. Putting all this together, we get:

20.4 Theorem. *Suppose that $\int (f'(u))^2 du < \infty$. Then*

$$R(\widehat{f}_n, f) \approx \frac{h^2}{12} \int (f'(u))^2 du + \frac{1}{nh}. \tag{20.10}$$

The value h^* that minimizes (20.10) is

$$h^* = \frac{1}{n^{1/3}} \left(\frac{6}{\int (f'(u))^2 du} \right)^{1/3}. \tag{20.11}$$

With this choice of binwidth,

$$R(\widehat{f}_n, f) \approx \frac{C}{n^{2/3}} \tag{20.12}$$

where $C = (3/4)^{2/3} \left(\int (f'(u))^2 du \right)^{1/3}$.

Theorem 20.4 is quite revealing. We see that with an optimally chosen binwidth, the MISE decreases to 0 at rate $n^{-2/3}$. By comparison, most parametric estimators converge at rate n^{-1}. The slower rate of convergence is the price we pay for being nonparametric. The formula for the optimal binwidth h^* is of theoretical interest but it is not useful in practice since it depends on the unknown function f.

A practical way to choose the binwidth is to estimate the risk function and minimize over h. Recall that the loss function, which we now write as a function of h, is

$$L(h) = \int (\widehat{f}_n(x) - f(x))^2 \, dx$$

$$= \int \widehat{f}_n^2(x) \, dx - 2 \int \widehat{f}_n(x) f(x) dx + \int f^2(x) \, dx.$$

The last term does not depend on the binwidth h so minimizing the risk is equivalent to minimizing the expected value of

$$J(h) = \int \widehat{f}_n^2(x) \, dx - 2 \int \widehat{f}_n(x) f(x) dx.$$

We shall refer to $\mathbb{E}(J(h))$ as the risk, although it differs from the true risk by the constant term $\int f^2(x)\,dx$.

20.5 Definition. *The* **cross-validation estimator of risk** *is*

$$\widehat{J}(h) = \int \left(\widehat{f}_n(x)\right)^2 dx - \frac{2}{n}\sum_{i=1}^{n}\widehat{f}_{(-i)}(X_i) \qquad (20.13)$$

where $\widehat{f}_{(-i)}$ is the histogram estimator obtained after removing the ith observation. We refer to $\widehat{J}(h)$ as the cross-validation score or estimated risk.

20.6 Theorem. *The cross-validation estimator is nearly unbiased:*

$$\mathbb{E}(\widehat{J}(x)) \approx \mathbb{E}(J(x)).$$

In principle, we need to recompute the histogram n times to compute $\widehat{J}(h)$. Moreover, this has to be done for all values of h. Fortunately, there is a shortcut formula.

20.7 Theorem. *The following identity holds:*

$$\widehat{J}(h) = \frac{2}{(n-1)h} - \frac{n+1}{(n-1)}\sum_{j=1}^{m}\widehat{p}_j^2. \qquad (20.14)$$

20.8 Example. We used cross-validation in the astronomy example. The cross-validation function is quite flat near its minimum. Any m in the range of 73 to 310 is an approximate minimizer but the resulting histogram does not change much over this range. The histogram in the top right plot in Figure 20.3 was constructed using $m = 73$ bins. The bottom right plot shows the estimated risk, or more precisely, $\widehat{A}$, plotted versus the number of bins. ∎

Next we want a confidence set for f. Suppose $\widehat{f}_n$ is a histogram with m bins and binwidth $h = 1/m$. We cannot realistically make confidence statements about the fine details of the true density f. Instead, we shall make confidence statements about f at the resolution of the histogram. To this end, define

$$\overline{f}_n(x) = \mathbb{E}(\widehat{f}_n(x)) = \frac{p_j}{h} \quad \text{for } x \in B_j \qquad (20.15)$$

where $p_j = \int_{B_j} f(u)du$. Think of $\overline{f}(x)$ as a "histogramized" version of f.

20.9 Definition. *A pair of functions* $(\ell_n(x), u_n(x))$ *is a* $1 - \alpha$ **confidence band** *(or* **confidence envelope***) if*

$$\mathbb{P}\left(\ell(x) \leq \overline{f}_n(x) \leq u(x) \ \text{for all } x\right) \geq 1 - \alpha. \qquad (20.16)$$

20.10 Theorem. *Let* $m = m(n)$ *be the number of bins in the histogram* $\widehat{f}_n$. *Assume that* $m(n) \to \infty$ *and* $m(n) \log n / n \to 0$ *as* $n \to \infty$. *Define*

$$\ell_n(x) = \left(\max\left\{\sqrt{\widehat{f}_n(x)} - c, 0\right\}\right)^2$$

$$u_n(x) = \left(\sqrt{\widehat{f}_n(x)} + c\right)^2 \qquad (20.17)$$

where

$$c = \frac{z_{\alpha/(2m)}}{2}\sqrt{\frac{m}{n}}. \qquad (20.18)$$

Then, $(\ell_n(x), u_n(x))$ *is an approximate* $1 - \alpha$ *confidence band.*

PROOF. Here is an outline of the proof. From the central limit theorem, $\widehat{p}_j \approx N(p_j, p_j(1 - p_j)/n)$. By the delta method, $\sqrt{\widehat{p}_j} \approx N(\sqrt{p_j}, 1/(4n))$. Moreover, it can be shown that the $\sqrt{\widehat{p}_j}$'s are approximately independent. Therefore,

$$2\sqrt{n}\left(\sqrt{\widehat{p}_j} - \sqrt{p_j}\right) \approx Z_j \qquad (20.19)$$

where $Z_1, \ldots, Z_m \sim N(0, 1)$. Let

$$A = \left\{\ell_n(x) \leq \overline{f}_n(x) \leq u_n(x) \ \text{for all } x\right\} = \left\{\max_x \left|\sqrt{\widehat{f}_n(x)} - \sqrt{\overline{f}(x)}\right| \leq c\right\}.$$

Then,

$$\mathbb{P}(A^c) = \mathbb{P}\left(\max_x \left|\sqrt{\widehat{f}_n(x)} - \sqrt{\overline{f}(x)}\right| > c\right) = \mathbb{P}\left(\max_j \left|\sqrt{\frac{\widehat{p}_j}{h}} - \sqrt{\frac{p_j}{h}}\right| > c\right)$$

$$= \mathbb{P}\left(\max_j 2\sqrt{n}\left|\sqrt{\widehat{p}_j} - \sqrt{p_j}\right| > z_{\alpha/(2m)}\right)$$

$$\approx \mathbb{P}\left(\max_j |Z_j| > z_{\alpha/(2m)}\right) \leq \sum_{j=1}^{m}\mathbb{P}\left(|Z_j| > z_{\alpha/(2m)}\right)$$

$$= \sum_{j=1}^{m}\frac{\alpha}{m} = \alpha. \quad \blacksquare$$

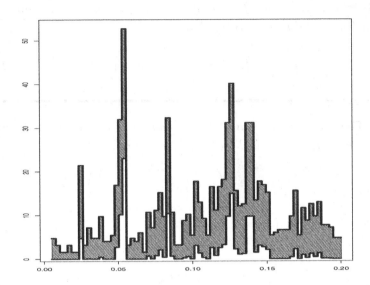

FIGURE 20.4. 95 percent confidence envelope for astronomy data using $m = 73$ bins.

20.11 Example. Figure 20.4 shows a 95 percent confidence envelope for the astronomy data. We see that even with over 1,000 data points, there is still substantial uncertainty. ∎

20.3   Kernel Density Estimation

Histograms are discontinuous. **Kernel density estimators** are smoother and they converge faster to the true density than histograms.

Let $X_1, \ldots, X_n$ denote the observed data, a sample from f. In this chapter, a **kernel** is defined to be any smooth function K such that $K(x) \geq 0$, $\int K(x)\, dx = 1$, $\int x K(x) dx = 0$ and $\sigma_K^2 \equiv \int x^2 K(x) dx > 0$. Two examples of kernels are the **Epanechnikov kernel**

$$K(x) = \begin{cases} \frac{3}{4}(1 - x^2/5)/\sqrt{5} & |x| < \sqrt{5} \\ 0 & otherwise \end{cases} \tag{20.20}$$

and the Gaussian (Normal) kernel $K(x) = (2\pi)^{-1/2} e^{-x^2/2}$.

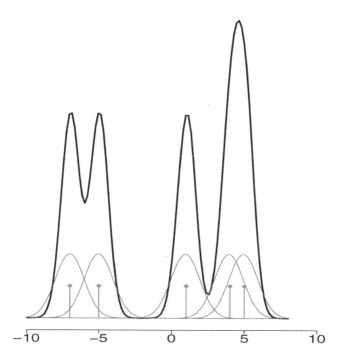

FIGURE 20.5. A kernel density estimator $\widehat{f}$. At each point x, $\widehat{f}(x)$ is the average of the kernels centered over the data points X_i. The data points are indicated by short vertical bars.

20.12 Definition. *Given a kernel K and a positive number h, called the* **bandwidth,** *the* **kernel density estimator** *is defined to be*

$$\widehat{f}(x) = \frac{1}{n} \sum_{i=1}^{n} \frac{1}{h} K\left(\frac{x - X_i}{h}\right). \tag{20.21}$$

An example of a kernel density estimator is show in Figure 20.5. The kernel estimator effectively puts a smoothed-out lump of mass of size $1/n$ over each data point X_i. The bandwidth h controls the amount of smoothing. When h is close to 0, $\widehat{f}_n$ consists of a set of spikes, one at each data point. The height of the spikes tends to infinity as $h \to 0$. When $h \to \infty$, $\widehat{f}_n$ tends to a uniform density.

20.13 Example. Figure 20.6 shows kernel density estimators for the astronomy data using three different bandwidths. In each case we used a Gaussian kernel. The properly smoothed kernel density estimator in the top right panel shows similar structure as the histogram. However, it is easier to see the clusters with the kernel estimator. ∎

To construct a kernel density estimator, we need to choose a kernel K and a bandwidth h. It can be shown theoretically and empirically that the choice of K is not crucial. [2] However, the choice of bandwidth h is very important. As with the histogram, we can make a theoretical statement about how the risk of the estimator depends on the bandwidth.

20.14 Theorem. *Under weak assumptions on f and K,*

$$R(f, \widehat{f}_n) \approx \frac{1}{4}\sigma_K^4 h^4 \int (f''(x))^2 + \frac{\int K^2(x)dx}{nh} \qquad (20.22)$$

where $\sigma_K^2 = \int x^2 K(x)dx$. The optimal bandwidth is

$$h^* = \frac{c_1^{2/5} c_2^{1/5} c_3^{-1/5}}{n^{1/5}} \qquad (20.23)$$

where $c_1 = \int x^2 K(x)dx$, $c_2 = \int K(x)^2 dx$ and $c_3 = \int (f''(x))^2 dx$. With this choice of bandwidth,

$$R(f, \widehat{f}_n) \approx \frac{c_4}{n^{4/5}}$$

for some constant $c_4 > 0$.

PROOF. Write $K_h(x, X) = h^{-1}K((x-X)/h)$ and $\widehat{f}_n(x) = n^{-1}\sum_i K_h(x, X_i)$. Thus, $\mathbb{E}[\widehat{f}_n(x)] = \mathbb{E}[K_h(x, X)]$ and $\mathbb{V}[\widehat{f}_n(x)] = n^{-1}\mathbb{V}[K_h(x, X)]$. Now,

$$
\begin{aligned}
\mathbb{E}[K_h(x, X)] &= \int \frac{1}{h}K\left(\frac{x-t}{h}\right)f(t)\,dt \\
&= \int K(u)f(x - hu)\,du \\
&= \int K(u)\left[f(x) - hf'(x) + \frac{1}{2}f''(x) + \cdots\right]du \\
&= f(x) + \frac{1}{2}h^2 f''(x)\int u^2 K(u)\,du \cdots
\end{aligned}
$$

since $\int K(x)\,dx = 1$ and $\int x K(x)\,dx = 0$. The bias is

$$\mathbb{E}[K_h(x, X)] - f(x) \approx \frac{1}{2}\sigma_k^2 h^2 f''(x).$$

---

[2] It can be shown that the Epanechnikov kernel is optimal in the sense of giving smallest asymptotic mean squared error, but it is really the choice of bandwidth which is crucial.

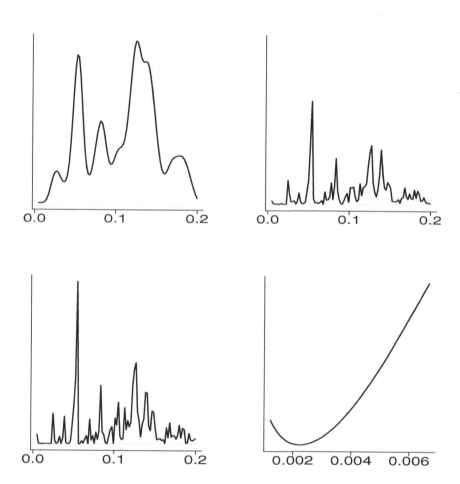

FIGURE 20.6. Kernel density estimators and estimated risk for the astronomy data. Top left: oversmoothed. Top right: just right (bandwidth chosen by cross-validation). Bottom left: undersmoothed. Bottom right: cross-validation curve as a function of bandwidth h. The bandwidth was chosen to be the value of h where the curve is a minimum.

By a similar calculation,

$$\mathbb{V}[\widehat{f}_n(x)] \approx \frac{f(x) \int K^2(x)\,dx}{n\,h_n}.$$

The result follows from integrating the squared bias plus the variance. ∎

We see that kernel estimators converge at rate $n^{-4/5}$ while histograms converge at the slower rate $n^{-2/3}$. It can be shown that, under weak assumptions, there does not exist a nonparametric estimator that converges faster than $n^{-4/5}$.

The expression for h^* depends on the unknown density f which makes the result of little practical use. As with the histograms, we shall use cross-validation to find a bandwidth. Thus, we estimate the risk (up to a constant) by

$$\widehat{J}(h) = \int \widehat{f}^2(x)dz - \frac{2}{n}\sum_{i=1}^{n}\widehat{f}_{-i}(X_i) \qquad (20.24)$$

where $\widehat{f}_{-i}$ is the kernel density estimator after omitting the ith observation.

20.15 Theorem. *For any $h > 0$,*

$$\mathbb{E}\left[\widehat{J}(h)\right] = \mathbb{E}\left[J(h)\right].$$

Also,

$$\widehat{J}(h) \approx \frac{1}{hn^2}\sum_i\sum_j K^*\left(\frac{X_i - X_j}{h}\right) + \frac{2}{nh}K(0) \qquad (20.25)$$

where $K^*(x) = K^{(2)}(x) - 2K(x)$ and $K^{(2)}(z) = \int K(z-y)K(y)dy$. In particular, if K is a $N(0,1)$ Gaussian kernel then $K^{(2)}(z)$ is the $N(0,2)$ density.

We then choose the bandwidth h_n that minimizes $\widehat{J}(h)$.[3] A justification for this method is given by the following remarkable theorem due to Stone.

20.16 Theorem (Stone's Theorem). *Suppose that f is bounded. Let $\widehat{f}_h$ denote the kernel estimator with bandwidth h and let h_n denote the bandwidth chosen by cross-validation. Then,*

$$\frac{\int \left(f(x) - \widehat{f}_{h_n}(x)\right)^2 dx}{\inf_h \int \left(f(x) - \widehat{f}_h(x)\right)^2 dx} \xrightarrow{\text{P}} 1. \qquad (20.26)$$

---

[3]For large data sets, $\widehat{f}$ and (20.25) can be computed quickly using the fast Fourier transform.

20.17 Example. The top right panel of Figure 20.6 is based on cross-validation. These data are rounded which problems for cross-validation. Specifically, it causes the minimizer to be $h = 0$. To overcome this problem, we added a small amount of random Normal noise to the data. The result is that $\widehat{J}(h)$ is very smooth with a well defined minimum. ∎

20.18 Remark. Do not assume that, if the estimator $\widehat{f}$ is wiggly, then cross-validation has let you down. The eye is not a good judge of risk.

To construct confidence bands, we use something similar to histograms. Again, the confidence band is for the smoothed version,

$$\overline{f}_n = \mathbb{E}(\widehat{f}_n(x)) = \int \frac{1}{h} K\left(\frac{x-u}{h}\right) f(u)\, du,$$

of the true density f. [4] Assume the density is on an interval (a, b). The band is

$$\ell_n(x) = \widehat{f}_n(x) - q\ \mathsf{se}(x), \quad u_n(x) = \widehat{f}_n(x) + q\ \mathsf{se}(x) \tag{20.27}$$

where

$$\mathsf{se}(x) = \frac{s(x)}{\sqrt{n}},$$

$$s^2(x) = \frac{1}{n-1}\sum_{i=1}^{n}(Y_i(x) - \overline{Y}_n(x))^2,$$

$$Y_i(x) = \frac{1}{h}K\left(\frac{x-X_i}{h}\right),$$

$$q = \Phi^{-1}\left(\frac{1+(1-\alpha)^{1/m}}{2}\right),$$

$$m = \frac{b-a}{\omega}$$

where ω is the width of the kernel. In case the kernel does not have finite width then we take ω to be the effective width, that is, the range over which the kernel is non-negligible. In particular, we take $\omega = 3h$ for the Normal kernel.

20.19 Example. Figure 20.7 shows approximate 95 percent confidence bands for the astronomy data. ∎

---

[4] This is a modified version of the band described in Chaudhuri and Marron (1999).

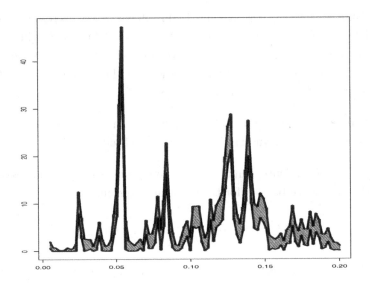

FIGURE 20.7. 95 percent confidence bands for kernel density estimate for the astronomy data.

Suppose now that the data $X_i = (X_{i1}, \ldots, X_{id})$ are d-dimensional. The kernel estimator can easily be generalized to d dimensions. Let $h = (h_1, \ldots, h_d)$ be a vector of bandwidths and define

$$\widehat{f}_n(x) = \frac{1}{n} \sum_{i=1}^{n} K_h(x - X_i) \tag{20.28}$$

where

$$K_h(x - X_i) = \frac{1}{nh_1 \cdots h_d} \left\{ \prod_{j=1}^{d} K\left(\frac{x_i - X_{ij}}{h_j}\right) \right\} \tag{20.29}$$

where $h_1, \ldots, h_d$ are bandwidths. For simplicity, we might take $h_j = s_j h$ where s_j is the standard deviation of the jth variable. There is now only a single bandwidth h to choose. Using calculations like those in the one-dimensional case, the risk is given by

$$
R(f, \widehat{f}_n) \approx \frac{1}{4}\sigma_K^4 \left[\sum_{j=1}^{d} h_j^4 \int f_{jj}^2(x)dx + \sum_{j \neq k} h_j^2 h_k^2 \int f_{jj}f_{kk}dx \right]
$$
$$
+ \frac{\left(\int K^2(x)dx\right)^d}{nh_1 \cdots h_d}
$$

where f_{jj} is the second partial derivative of f. The optimal bandwidth satisfies $h_i \approx c_1 n^{-1/(4+d)}$, leading to a risk of order $n^{-4/(4+d)}$. From this fact, we see that the risk increases quickly with dimension, a problem usually called the **curse of dimensionality**. To get a sense of how serious this problem is, consider the following table from Silverman (1986) which shows the sample size required to ensure a relative mean squared error less than 0.1 at 0 when the density is multivariate normal and the optimal bandwidth is selected:

Dimension	Sample Size
1	4
2	19
3	67
4	223
5	768
6	2790
7	10,700
8	43,700
9	187,000
10	842,000

This is bad news indeed. It says that having 842,000 observations in a ten-dimensional problem is really like having 4 observations in a one-dimensional problem.

20.4 Nonparametric Regression

Consider pairs of points $(x_1, Y_1), \ldots, (x_n, Y_n)$ related by

$$Y_i = r(x_i) + \epsilon_i \tag{20.30}$$

where $\mathbb{E}(\epsilon_i) = 0$. We have written the x_i's in lower case since we will treat them as fixed. We can do this since, in regression, it is only the mean of Y conditional on x that we are interested in. We want to estimate the regression function $r(x) = \mathbb{E}(Y|X = x)$.

There are many nonparametric regression estimators. Most involve estimating $r(x)$ by taking some sort of weighted average of the Y_i's, giving higher weight to those points near x. A popular version is the Nadaraya-Watson kernel estimator.

20.20 Definition. *The* **Nadaraya-Watson kernel estimator** *is defined by*

$$\widehat{r}(x) = \sum_{i=1}^{n} w_i(x) Y_i \tag{20.31}$$

where K is a kernel and the weights $w_i(x)$ are given by

$$w_i(x) = \frac{K\left(\frac{x-x_i}{h}\right)}{\sum_{j=1}^{n} K\left(\frac{x-x_j}{h}\right)}. \tag{20.32}$$

The form of this estimator comes from first estimating the joint density $f(x, y)$ using kernel density estimation and then inserting the estimate into the formula,

$$r(x) = \mathbb{E}(Y|X = x) = \int y f(y|x) dy = \frac{\int y f(x, y) dy}{\int f(x, y) dy}.$$

20.21 Theorem. *Suppose that* $\mathbb{V}(\epsilon_i) = \sigma^2$. *The risk of the Nadaraya-Watson kernel estimator is*

$$R(\widehat{r}_n, r) \approx \frac{h^4}{4} \left(\int x^2 K^2(x) dx \right)^4 \int \left(r''(x) + 2r'(x)\frac{f'(x)}{f(x)} \right)^2 dx$$

$$+ \int \frac{\sigma^2 \int K^2(x) dx}{n h f(x)} dx. \tag{20.33}$$

The optimal bandwidth decreases at rate $n^{-1/5}$ *and with this choice the risk decreases at rate* $n^{-4/5}$.

In practice, to choose the bandwidth h we minimize the cross validation score

$$\widehat{J}(h) = \sum_{i=1}^{n} (Y_i - \widehat{r}_{-i}(x_i))^2 \tag{20.34}$$

where $\widehat{r}_{-i}$ is the estimator we get by omitting the ith variable. Fortunately, there is a shortcut formula for computing $\widehat{J}$.

20.22 Theorem. $\widehat{J}$ *can be written as*

$$\widehat{J}(h) = \sum_{i=1}^{n} (Y_i - \widehat{r}(x_i))^2 \frac{1}{\left(1 - \frac{K(0)}{\sum_{j=1}^{n} K\left(\frac{x_i - x_j}{h}\right)} \right)^2}. \tag{20.35}$$

20.23 Example. Figures 20.8 shows cosmic microwave background (CMB) data from BOOMERaNG (Netterfield et al. (2002)), Maxima (Lee et al. (2001)), and DASI (Halverson et al. (2002))). The data consist of n pairs $(x_1, Y_1), \ldots, (x_n, Y_n)$ where x_i is called the multipole moment and Y_i is the

estimated power spectrum of the temperature fluctuations. What you are seeing are sound waves in the cosmic microwave background radiation which is
the heat, left over from the big bang. If $r(x)$ denotes the true power spectrum,
then

$$Y_i = r(x_i) + \epsilon_i$$

where ϵ_i is a random error with mean 0. The location and size of peaks in
$r(x)$ provides valuable clues about the behavior of the early universe. Figure
20.8 shows the fit based on cross-validation as well as an undersmoothed and
oversmoothed fit. The cross-validation fit shows the presence of three well-
defined peaks, as predicted by the physics of the big bang. ∎

The procedure for finding confidence bands is similar to that for density
estimation. However, we first need to estimate σ^2. Suppose that the x_i's are
ordered. Assuming $r(x)$ is smooth, we have $r(x_{i+1}) - r(x_i) \approx 0$ and hence

$$Y_{i+1} - Y_i = \left[r(x_{i+1}) + \epsilon_{i+1} \right] - \left[r(x_i) + \epsilon_i \right] \approx \epsilon_{i+1} - \epsilon_i$$

and hence

$$\mathbb{V}(Y_{i+1} - Y_i) \approx \mathbb{V}(\epsilon_{i+1} - \epsilon_i) = \mathbb{V}(\epsilon_{i+1}) + \mathbb{V}(\epsilon_i) = 2\sigma^2.$$

We can thus use the average of the $n-1$ differences $Y_{i+1} - Y_i$ to estimate σ^2.
Hence, define

$$\widehat{\sigma}^2 = \frac{1}{2(n-1)} \sum_{i=1}^{n-1} (Y_{i+1} - Y_i)^2. \tag{20.36}$$

As with density estimate, the confidence band is for the smoothed version
$\overline{r}_n(x) = \mathbb{E}(\widehat{r}_n(x))$ of the true regression function r.

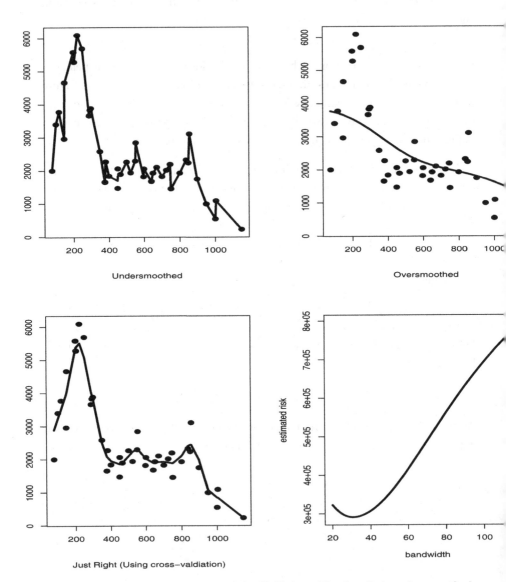

FIGURE 20.8. Regression analysis of the CMB data. The first fit is undersmoothed, the second is oversmoothed, and the third is based on cross-validation. The last panel shows the estimated risk versus the bandwidth of the smoother. The data are from BOOMERaNG, Maxima, and DASI.

Confidence Bands for Kernel Regression

An approximate $1 - \alpha$ confidence band for $\bar{r}_n(x)$ is

$$\ell_n(x) = \widehat{r}_n(x) - q \ \widehat{se}(x), \quad u_n(x) = \widehat{r}_n(x) + q \ \widehat{se}(x) \qquad (20.37)$$

where

$$\widehat{se}(x) = \widehat{\sigma} \sqrt{\sum_{i=1}^{n} w_i^2(x)},$$

$$q = \Phi^{-1}\left(\frac{1 + (1 - \alpha)^{1/m}}{2}\right),$$

$$m = \frac{b - a}{\omega},$$

$\widehat{\sigma}$ is defined in (20.36) and ω is the width of the kernel. In case the kernel does not have finite width then we take ω to be the effective width, that is, the range over which the kernel is non-negligible. In particular, we take $\omega = 3h$ for the Normal kernel.

20.24 Example. Figure 20.9 shows a 95 percent confidence envelope for the CMB data. We see that we are highly confident of the existence and position of the first peak. We are more uncertain about the second and third peak. At the time of this writing, more accurate data are becoming available that apparently provide sharper estimates of the second and third peak. ∎

The extension to multiple regressors $X = (X_1, \ldots, X_p)$ is straightforward. As with kernel density estimation we just replace the kernel with a multivariate kernel. However, the same caveats about the curse of dimensionality apply. In some cases, we might consider putting some restrictions on the regression function which will then reduce the curse of dimensionality. For example, **additive regression** is based on the model

$$Y = \sum_{j=1}^{p} r_j(X_j) + \epsilon. \qquad (20.38)$$

Now we only need to fit p one-dimensional functions. The model can be enriched by adding various interactions, for example,

$$Y = \sum_{j=1}^{p} r_j(X_j) + \sum_{j<k} r_{jk}(X_j X_k) + \epsilon. \qquad (20.39)$$

Additive models are usually fit by an algorithm called **backfitting**.

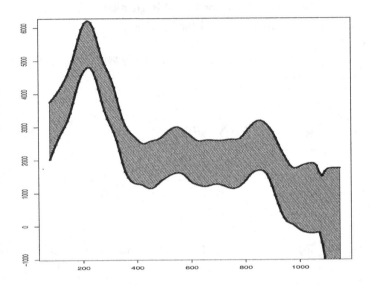

FIGURE 20.9. 95 percent confidence envelope for the CMB data.

Backfitting

1. Initialize $r_1(x_1), \ldots, r_p(x_p)$.

2. For $j = 1, \ldots, p$:

    (a) Let $\epsilon_i = Y_i - \sum_{s \neq j} r_s(x_i)$.

    (b) Let r_j be the function estimate obtained by regressing the ϵ_i's on the jth covariate.

3. If converged STOP. Else, go back to step 2.

Additive models have the advantage that they avoid the curse of dimensionality and they can be fit quickly, but they have one disadvantage: the model is not fully nonparametric. In other words, the true regression function $r(x)$ may not be of the form (20.38).

20.5 Appendix

CONFIDENCE SETS AND BIAS. The confidence bands we computed are not for the density function or regression function but rather for the smoothed function. For example, the confidence band for a kernel density estimate with bandwidth h is a band for the function one gets by smoothing the true function with a kernel with the same bandwidth. Getting a confidence set for the true function is complicated for reasons we now explain.

Let $\widehat{f}_n(x)$ denote an estimate of the function $f(x)$. Denote the mean and standard deviation of $\widehat{f}_n(x)$ by $\overline{f}_n(x)$ and $s_n(x)$. Then,

$$\frac{\widehat{f}_n(x) - f(x)}{s_n(x)} = \frac{\widehat{f}_n(x) - \overline{f}_n(x)}{s_n(x)} + \frac{\overline{f}_n(x) - f(x)}{s_n(x)}.$$

Typically, the first term converges to a standard Normal from which one derives confidence bands. The second term is the bias divided by the standard deviation. In parametric inference, the bias is usually smaller than the standard deviation of the estimator so this term goes to 0 as the sample size increases. In nonparametric inference, optimal smoothing leads us to balance the bias and the standard deviation. Thus the second term does not vanish even with large sample sizes. This means that the confidence interval will not be centered around the true function f.

20.6 Bibliographic Remarks

Two very good books on density estimation are Scott (1992) and Silverman (1986). The literature on nonparametric regression is very large. Two good starting points are Hardle (1990) and Loader (1999). The latter emphasizes a class of techniques called local likelihood methods.

20.7 Exercises

1. Let $X_1, \ldots, X_n \sim f$ and let $\widehat{f}_n$ be the kernel density estimator using the boxcar kernel:

$$K(x) = \begin{cases} 1 & -\frac{1}{2} < x < \frac{1}{2} \\ 0 & \text{otherwise.} \end{cases}$$

(a) Show that

$$\mathbb{E}(\widehat{f}(x)) = \frac{1}{h} \int_{x-(h/2)}^{x+(h/2)} f(y)dy$$

and

$$\mathbb{V}(\widehat{f}(x)) = \frac{1}{nh^2} \left[\int_{x-(h/2)}^{x+(h/2)} f(y)dy - \left(\int_{x-(h/2)}^{x+(h/2)} f(y)dy \right)^2 \right].$$

(b) Show that if $h \to 0$ and $nh \to \infty$ as $n \to \infty$, then $\widehat{f}_n(x) \xrightarrow{P} f(x)$.

2. Get the data on fragments of glass collected in forensic work from the book website. Estimate the density of the first variable (refractive index) using a histogram and use a kernel density estimator. Use cross-validation to choose the amount of smoothing. Experiment with different binwidths and bandwidths. Comment on the similarities and differences. Construct 95 percent confidence bands for your estimators.

3. Consider the data from question 2. Let Y be refractive index and let x be aluminum content (the fourth variable). Perform a nonparametric regression to fit the model $Y = f(x) + \epsilon$. Use cross-validation to estimate the bandwidth. Construct 95 percent confidence bands for your estimate.

4. Prove Lemma 20.1.

5. Prove Theorem 20.3.

6. Prove Theorem 20.7.

7. Prove Theorem 20.15.

8. Consider regression data $(x_1, Y_1), \ldots, (x_n, Y_n)$. Suppose that $0 \le x_i \le 1$ for all i. Define bins B_j as in equation (20.7). For $x \in B_j$ define

$$\widehat{r}_n(x) = \overline{Y}_j$$

where $\overline{Y}_j$ is the mean of all the Y_i's corresponding to those x_i's in B_j. Find the approximate risk of this estimator. From this expression for the risk, find the optimal bandwidth. At what rate does the risk go to zero?

9. Show that with suitable smoothness assumptions on $r(x)$, $\widehat{\sigma}^2$ in equation (20.36) is a consistent estimator of σ^2.

10. Prove Theorem 20.22.

21
Smoothing Using Orthogonal Functions

In this chapter we will study an approach to nonparametric curve estimation based on **orthogonal functions**. We begin with a brief introduction to
the theory of orthogonal functions, then we turn to density estimation and
regression.

21.1 Orthogonal Functions and L_2 Spaces

Let $v = (v_1, v_2, v_3)$ denote a three-dimensional vector, that is, a list of three
real numbers. Let $\mathcal{V}$ denote the set of all such vectors. If a is a scalar (a
number) and v is a vector, we define $av = (av_1, av_2, av_3)$. The sum of vectors
v and w is defined by $v + w = (v_1 + w_1, v_2 + w_2, v_3 + w_3)$. The **inner product**
between two vectors v and w is defined by $\langle v, w \rangle = \sum_{i=1}^{3} v_i w_i$. The **norm**
(or length) of a vector v is defined by

$$||v|| = \sqrt{\langle v, v \rangle} = \sqrt{\sum_{i=1}^{3} v_i^2}. \tag{21.1}$$

Two vectors are **orthogonal (or perpendicular)** if $\langle v, w \rangle = 0$. A set of
vectors are orthogonal if each pair in the set is orthogonal. A vector is **normal**
if $||v|| = 1$.

Let $\phi_1 = (1, 0, 0)$, $\phi_2 = (0, 1, 0)$, $\phi_3 = (0, 0, 1)$. These vectors are said to be an **orthonormal basis** for $\mathcal{V}$ since they have the following properties:
(i) they are orthogonal;
(ii) they are normal;
(iii) they form a basis for $\mathcal{V}$, which means that any $v \in \mathcal{V}$ can be written as a linear combination of ϕ_1, ϕ_2, ϕ_3:

$$v = \sum_{j=1}^{3} \beta_j \phi_j \quad \text{where} \quad \beta_j = \langle \phi_j, v \rangle. \tag{21.2}$$

For example, if $v = (12, 3, 4)$ then $v = 12\phi_1 + 3\phi_2 + 4\phi_3$. There are other orthonormal bases for $\mathcal{V}$, for example,

$$\psi_1 = \left(\frac{1}{\sqrt{3}}, \frac{1}{\sqrt{3}}, \frac{1}{\sqrt{3}} \right), \ \psi_2 = \left(\frac{1}{\sqrt{2}}, -\frac{1}{\sqrt{2}}, 0 \right), \ \psi_3 = \left(\frac{1}{\sqrt{6}}, \frac{1}{\sqrt{6}}, -\frac{2}{\sqrt{6}} \right).$$

You can check that these three vectors also form an orthonormal basis for $\mathcal{V}$. Again, if v is any vector then we can write

$$v = \sum_{j=1}^{3} \beta_j \psi_j \quad \text{where} \quad \beta_j = \langle \psi_j, v \rangle.$$

For example, if $v = (12, 3, 4)$ then

$$v = 10.97\psi_1 + 6.36\psi_2 + 2.86\psi_3.$$

Now we make the leap from vectors to functions. Basically, we just replace vectors with functions and sums with integrals. Let $L_2(a, b)$ denote all functions defined on the interval $[a, b]$ such that $\int_a^b f(x)^2 dx < \infty$:

$$L_2(a, b) = \left\{ f : [a, b] \to \mathbb{R}, \ \int_a^b f(x)^2 dx < \infty \right\}. \tag{21.3}$$

We sometimes write L_2 instead of $L_2(a, b)$. The inner product between two functions $f, g \in L_2$ is defined by $\int f(x)g(x)dx$. The norm of f is

$$\|f\| = \sqrt{\int f(x)^2 dx}. \tag{21.4}$$

Two functions are orthogonal if $\int f(x)g(x)dx = 0$. A function is normal if $\|f\| = 1$.

A sequence of functions $\phi_1, \phi_2, \phi_3, \ldots$ is **orthonormal** if $\int \phi_j^2(x)dx = 1$ for each j and $\int \phi_i(x)\phi_j(x)dx = 0$ for $i \neq j$. An orthonormal sequence is **complete** if the only function that is orthogonal to each ϕ_j is the zero function.

In this case, the functions $\phi_1, \phi_2, \phi_3, \ldots$ form in basis, meaning that if $f \in L_2$ then f can be written as[1]

$$f(x) = \sum_{j=1}^{\infty} \beta_j \phi_j(x), \quad \text{where} \quad \beta_j = \int_a^b f(x)\phi_j(x)dx. \tag{21.5}$$

A useful result is **Parseval's relation** which says that

$$||f||^2 \equiv \int f^2(x)\,dx = \sum_{j=1}^{\infty} \beta_j^2 \equiv ||\beta||^2 \tag{21.6}$$

where $\beta = (\beta_1, \beta_2, \ldots)$.

21.1 Example. An example of an orthonormal basis for $L_2(0,1)$ is the **cosine basis** defined as follows. Let $\phi_0(x) = 1$ and for $j \geq 1$ define

$$\phi_j(x) = \sqrt{2}\cos(j\pi x). \tag{21.7}$$

The first six functions are plotted in Figure 21.1. ∎

21.2 Example. Let

$$f(x) = \sqrt{x(1-x)}\,\sin\left(\frac{2.1\pi}{(x+.05)}\right)$$

which is called the "doppler function." Figure 21.2 shows f (top left) and its approximation

$$f_J(x) = \sum_{j=1}^{J} \beta_j \phi_j(x)$$

with J equal to 5 (top right), 20 (bottom left), and 200 (bottom right). As J increases we see that $f_J(x)$ gets closer to $f(x)$. The coefficients $\beta_j = \int_0^1 f(x)\phi_j(x)dx$ were computed numerically. ∎

21.3 Example. The **Legendre polynomials** on $[-1,1]$ are defined by

$$P_j(x) = \frac{1}{2^j j!}\frac{d^j}{dx^j}(x^2-1)^j, \quad j = 0,1,2,\ldots \tag{21.8}$$

It can be shown that these functions are complete and orthogonal and that

$$\int_{-1}^{1} P_j^2(x)dx = \frac{2}{2j+1}. \tag{21.9}$$

---

[1]The equality in the displayed equation means that $\int (f(x) - f_n(x))^2 dx \to 0$ where $f_n(x) = \sum_{j=1}^{n} \beta_j \phi_j(x)$.

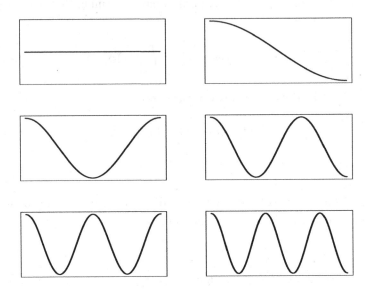

FIGURE 21.1. The first six functions in the cosine basis.

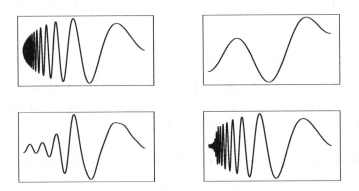

FIGURE 21.2. Approximating the doppler function with its expansion
in the cosine basis. The function f (top left) and its approximation
$f_J(x) = \sum_{j=1}^{J} \beta_j \phi_j(x)$ with J equal to 5 (top right), 20 (bottom left),
and 200 (bottom right). The coefficients $\beta_j = \int_0^1 f(x)\phi_j(x)dx$ were
computed numerically.

It follows that the functions $\phi_j(x) = \sqrt{(2j+1)/2}P_j(x)$, $j = 0, 1, \ldots$ form an orthonormal basis for $L_2(-1, 1)$. The first few Legendre polynomials are:

$$
\begin{aligned}
P_0(x) &= 1, \\
P_1(x) &= x, \\
P_2(x) &= \frac{1}{2}\left(3x^2 - 1\right), \text{ and} \\
P_3(x) &= \frac{1}{2}\left(5x^3 - 3x\right).
\end{aligned}
$$

These polynomials may be constructed explicitly using the following recursive relation:

$$P_{j+1}(x) = \frac{(2j+1)xP_j(x) - jP_{j-1}(x)}{j+1}. \quad \blacksquare \tag{21.10}$$

The coefficients $\beta_1, \beta_2, \ldots$ are related to the smoothness of the function f. To see why, note that if f is smooth, then its derivatives will be finite. Thus we expect that, for some k, $\int_0^1 (f^{(k)}(x))^2 dx < \infty$ where $f^{(k)}$ is the kth derivative of f. Now consider the cosine basis (21.7) and let $f(x) = \sum_{j=0}^{\infty} \beta_j \phi_j(x)$. Then,

$$\int_0^1 (f^{(k)}(x))^2 dx = 2\sum_{j=1}^{\infty} \beta_j^2 (\pi j)^{2k}.$$

The only way that $\sum_{j=1}^{\infty} \beta_j^2 (\pi j)^{2k}$ can be finite is if the β_j's get small when j gets large. To summarize:

If the function f is smooth, then the coefficients β_j will be small when j is large.

For the rest of this chapter, assume we are using the cosine basis unless otherwise specified.

21.2 Density Estimation

Let $X_1, \ldots, X_n$ be IID observations from a distribution on $[0, 1]$ with density f. Assuming $f \in L_2$ we can write

$$f(x) = \sum_{j=0}^{\infty} \beta_j \phi_j(x)$$

where $\phi_1, \phi_2, \ldots$ is an orthonormal basis. Define

$$\widehat{\beta}_j = \frac{1}{n}\sum_{i=1}^{n} \phi_j(X_i). \tag{21.11}$$

21.4 Theorem. *The mean and variance of $\widehat{\beta}_j$ are*

$$\mathbb{E}\left(\widehat{\beta}_j\right) = \beta_j, \quad \mathbb{V}\left(\widehat{\beta}_j\right) = \frac{\sigma_j^2}{n} \tag{21.12}$$

where

$$\sigma_j^2 = \mathbb{V}(\phi_j(X_i)) = \int (\phi_j(x) - \beta_j)^2 f(x) dx. \tag{21.13}$$

PROOF. The mean is

$$
\begin{aligned}
\mathbb{E}\left(\widehat{\beta}_j\right) &= \frac{1}{n}\sum_{i=1}^n \mathbb{E}\left(\phi_j(X_i)\right) \\
&= \mathbb{E}\left(\phi_j(X_1)\right) \\
&= \int \phi_j(x) f(x) dx = \beta_j.
\end{aligned}
$$

The calculation for the variance is similar. ∎

Hence, $\widehat{\beta}_j$ is an unbiased estimate of β_j. It is tempting to estimate f by $\sum_{j=1}^{\infty} \widehat{\beta}_j \phi_j(x)$ but this turns out to have a very high variance. Instead, consider the estimator

$$\widehat{f}(x) = \sum_{j=1}^{J} \widehat{\beta}_j \phi_j(x). \tag{21.14}$$

The number of terms J is a smoothing parameter. Increasing J will decrease bias while increasing variance. For technical reasons, we restrict J to lie in the range

$$1 \leq J \leq p$$

where $p = p(n) = \sqrt{n}$. To emphasize the dependence of the risk function on J, we write the risk function as $R(J)$.

21.5 Theorem. *The risk of $\widehat{f}$ is*

$$R(J) = \sum_{j=1}^{J} \frac{\sigma_j^2}{n} + \sum_{j=J+1}^{\infty} \beta_j^2. \tag{21.15}$$

An estimate of the risk is

$$\widehat{R}(J) = \sum_{j=1}^{J} \frac{\widehat{\sigma}_j^2}{n} + \sum_{j=J+1}^{p} \left(\widehat{\beta}_j^2 - \frac{\widehat{\sigma}_j^2}{n}\right)_+ \tag{21.16}$$

where $a_+ = \max\{a, 0\}$ and

$$\widehat{\sigma}_j^2 = \frac{1}{n-1}\sum_{i=1}^n \left(\phi_j(X_i) - \widehat{\beta}_j\right)^2. \tag{21.17}$$

To motivate this estimator, note that $\widehat{\sigma}_j^2$ is an unbiased estimate of σ_j^2 and $\widehat{\beta}_j^2 - \widehat{\sigma}_j^2$ is an unbiased estimator of β_j^2. We take the positive part of the latter term since we know that β_j^2 cannot be negative. We now choose $1 \le \widehat{J} \le p$ to minimize $\widehat{R}(\widehat{f}, f)$. Here is a summary:

Summary of Orthogonal Function Density Estimation

1. Let

$$\widehat{\beta}_j = \frac{1}{n} \sum_{i=1}^{n} \phi_j(X_i).$$

2. Choose $\widehat{J}$ to minimize $\widehat{R}(J)$ over $1 \le J \le p = \sqrt{n}$ where $\widehat{R}$ is given in equation (21.16).

3. Let

$$\widehat{f}(x) = \sum_{j=1}^{\widehat{J}} \widehat{\beta}_j \phi_j(x).$$

The estimator $\widehat{f}_n$ can be negative. If we are interested in exploring the shape of f, this is not a problem. However, if we need our estimate to be a probability density function, we can truncate the estimate and then normalize it. That is, we take $\widehat{f}^* = \max\{\widehat{f}_n(x), 0\} / \int_0^1 \max\{\widehat{f}_n(u), 0\} du$.

Now let us construct a confidence band for f. Suppose we estimate f using J orthogonal functions. We are essentially estimating $f_J(x) = \sum_{j=1}^{J} \beta_j \phi_j(x)$ not the true density $f(x) = \sum_{j=1}^{\infty} \beta_j \phi_j(x)$. Thus, the confidence band should be regarded as a band for $f_J(x)$.

21.6 Theorem. *An approximate $1 - \alpha$ confidence band for f_J is $(\ell(x), u(x))$ where*

$$\ell(x) = \widehat{f}_n(x) - c, \quad u(x) = \widehat{f}_n(x) + c \tag{21.18}$$

where

$$c = K^2 \sqrt{\frac{J\chi_{J,\alpha}^2}{n}} \tag{21.19}$$

and

$$K = \max_{1 \le j \le J} \max_{x} |\phi_j(x)|.$$

For the cosine basis, $K = \sqrt{2}$.

PROOF. Here is an outline of the proof. Let $L = \sum_{j=1}^{J} (\widehat{\beta}_j - \beta_j)^2$. By the central limit theorem, $\widehat{\beta}_j \approx N(\beta_j, \sigma_j^2/n)$. Hence, $\widehat{\beta}_j \approx \beta_j + \sigma_j \epsilon_j / \sqrt{n}$ where

$\epsilon_j \sim N(0,1)$, and therefore

$$L \approx \frac{1}{n} \sum_{j=1}^{J} \sigma_j^2 \epsilon_j^2 \leq \frac{K^2}{n} \sum_{j=1}^{J} \epsilon_j^2 \overset{d}{=} \frac{K^2}{n} \chi_J^2. \tag{21.20}$$

Thus we have, approximately, that

$$\mathbb{P}\left(L > \frac{K^2}{n} \chi_{J,\alpha}^2\right) \leq \mathbb{P}\left(\frac{K^2}{n} \chi_J^2 > \frac{K^2}{n} \chi_{J,\alpha}^2\right) = \alpha.$$

Also,

$$\begin{aligned}
\max_x |\widehat{f}_J(x) - f_J(x)| &\leq \max_x \sum_{j=1}^{J} |\phi_j(x)| \, |\widehat{\beta}_j - \beta_j| \\
&\leq K \sum_{j=1}^{J} |\widehat{\beta}_j - \beta_j| \\
&\leq \sqrt{J} K \sqrt{\sum_{j=1}^{J} (\widehat{\beta}_j - \beta_j)^2} \\
&= \sqrt{J} K \sqrt{L}
\end{aligned}$$

where the third inequality is from the Cauchy-Schwartz inequality (Theorem 4.8). So,

$$\begin{aligned}
\mathbb{P}\left(\max_x |\widehat{f}_J(x) - f_J(x)| > K^2 \sqrt{\frac{J\chi_{J,\alpha}^2}{n}}\right) &\leq \mathbb{P}\left(\sqrt{J} K \sqrt{L} > K^2 \sqrt{\frac{J\chi_{J,\alpha}^2}{n}}\right) \\
&= \mathbb{P}\left(\sqrt{L} > K \sqrt{\frac{\chi_{J,\alpha}^2}{n}}\right) \\
&= \mathbb{P}\left(L > \frac{K^2 \chi_{J,\alpha}^2}{n}\right) \\
&\leq \alpha. \quad \blacksquare
\end{aligned}$$

21.7 Example. Let

$$f(x) = \frac{5}{6}\phi(x;0,1) + \frac{1}{6}\sum_{j=1}^{5} \phi(x;\mu_j,.1)$$

where $\phi(x;\mu,\sigma)$ denotes a Normal density with mean μ and standard deviation σ, and $(\mu_1,\ldots,\mu_5) = (-1,-1/2,0,1/2,1)$. Marron and Wand (1992) call this

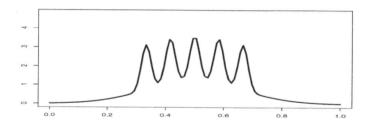

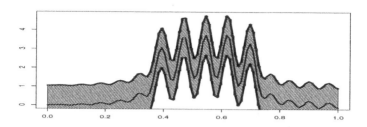

FIGURE 21.3. The top plot is the true density for the Bart Simpson distribution (rescaled to have most of its mass between 0 and 1). The bottom plot is the orthogonal function density estimate and 95 percent confidence band.

"the claw" although the "Bart Simpson" might be more appropriate. Figure 21.3 shows the true density as well as the estimated density based on $n = 5,000$ observations and a 95 percent confidence band. The density has been rescaled to have most of its mass between 0 and 1 using the transformation $y = (x + 3)/6$. ∎

21.3 Regression

Consider the regression model

$$Y_i = r(x_i) + \epsilon_i, \quad i = 1, \dots, n \tag{21.21}$$

where the ϵ_i are independent with mean 0 and variance σ^2. We will initially focus on the special case where $x_i = i/n$. We assume that $r \in L_2(0,1)$ and hence we can write

$$r(x) = \sum_{j=1}^{\infty} \beta_j \phi_j(x) \quad \text{where} \quad \beta_j = \int_0^1 r(x)\phi_j(x)dx \tag{21.22}$$

where $\phi_1, \phi_2, \dots$ where is an orthonormal basis for $[0, 1]$.

Define

$$\widehat{\beta}_j = \frac{1}{n} \sum_{i=1}^{n} Y_i \, \phi_j(x_i), \quad j = 1, 2, \ldots \tag{21.23}$$

Since $\widehat{\beta}_j$ is an average, the central limit theorem tells us that $\widehat{\beta}_j$ will be approximately Normally distributed.

21.8 Theorem.

$$\widehat{\beta}_j \approx N \left(\beta_j, \frac{\sigma^2}{n} \right). \tag{21.24}$$

PROOF. The mean of $\widehat{\beta}_j$ is

$$\begin{aligned}
\mathbb{E}(\widehat{\beta}_j) &= \frac{1}{n} \sum_{i=1}^{n} \mathbb{E}(Y_i)\phi_j(x_i) = \frac{1}{n} \sum_{i=1}^{n} r(x_i)\phi_j(x_i) \\
&\approx \int r(x)\phi_j(x)dx = \beta_j
\end{aligned}$$

where the approximate equality follows from the definition of a Riemann integral: $\sum_i \Delta_n h(x_i) \to \int_0^1 h(x)dx$ where $\Delta_n = 1/n$. The variance is

$$\begin{aligned}
\mathbb{V}(\widehat{\beta}_j) &= \frac{1}{n^2} \sum_{i=1}^{n} \mathbb{V}(Y_i)\phi_j^2(x_i) \\
&= \frac{\sigma^2}{n^2} \sum_{i=1}^{n} \phi_j^2(x_i) = \frac{\sigma^2}{n} \frac{1}{n} \sum_{i=1}^{n} \phi_j^2(x_i) \\
&\approx \frac{\sigma^2}{n} \int \phi_j^2(x)dx = \frac{\sigma^2}{n}
\end{aligned}$$

since $\int \phi_j^2(x)dx = 1$. ∎

Let

$$\widehat{r}(x) = \sum_{j=1}^{J} \widehat{\beta}_j \phi_j(x),$$

and let

$$R(J) = \mathbb{E} \int (r(x) - \widehat{r}(x))^2 \, dx$$

be the risk of the estimator.

21.9 Theorem. *The risk $R(J)$ of the estimator $\widehat{r}_n(x) = \sum_{j=1}^{J} \widehat{\beta}_j \phi_j(x)$ is*

$$R(J) = \frac{J\sigma^2}{n} + \sum_{j=J+1}^{\infty} \beta_j^2. \tag{21.25}$$

To estimate for $\sigma^2 = \mathbb{V}(\epsilon_i)$ we use

$$\widehat{\sigma}^2 = \frac{n}{k} \sum_{i=n-k+1}^{n} \widehat{\beta}_j^2 \tag{21.26}$$

where $k = n/4$. To motivate this estimator, recall that if f is smooth, then $\beta_j \approx 0$ for large j. So, for $j \geq k$, $\widehat{\beta}_j \approx N(0, \sigma^2/n)$ and thus, $\widehat{\beta}_j \approx \sigma Z_j/\sqrt{n}$ for for $j \geq k$, where $Z_j \sim N(0,1)$. Therefore,

$$
\begin{aligned}
\widehat{\sigma}^2 &= \frac{n}{k} \sum_{i=n-k+1}^{n} \widehat{\beta}_j^2 \approx \frac{n}{k} \sum_{i=n-k+1}^{n} \left(\frac{\sigma}{\sqrt{n}} \widehat{\beta}_j \right)^2 \\
&= \frac{\sigma^2}{k} \sum_{i=n-k+1}^{n} \widehat{\beta}_j^2 = \frac{\sigma^2}{k} \chi_k^2
\end{aligned}
$$

since a sum of k Normals has a χ_k^2 distribution. Now $\mathbb{E}(\chi_k^2) = k$ and hence $\mathbb{E}(\widehat{\sigma}^2) \approx \sigma^2$. Also, $\mathbb{V}(\chi_k^2) = 2k$ and hence $\mathbb{V}(\widehat{\sigma}^2) \approx (\sigma^4/k^2)(2k) = (2\sigma^4/k) \to 0$ as $n \to \infty$. Thus we expect $\widehat{\sigma}^2$ to be a consistent estimator of σ^2. There is nothing special about the choice $k = n/4$. Any k that increases with n at an appropriate rate will suffice.

We estimate the risk with

$$\widehat{R}(J) = J\frac{\widehat{\sigma}^2}{n} + \sum_{j=J+1}^{n} \left(\widehat{\beta}_j^2 - \frac{\widehat{\sigma}^2}{n} \right)_{+}. \tag{21.27}$$

21.10 Example. Figure 21.4 shows the doppler function f and $n = 2,048$ observations generated from the model

$$Y_i = r(x_i) + \epsilon_i$$

where $x_i = i/n$, $\epsilon_i \sim N(0, (.1)^2)$. The figure shows the data and the estimated function. The estimate was based on $\widehat{J} = 234$ terms. ∎

We are now ready to give a complete description of the method.

Orthogonal Series Regression Estimator

1. Let

$$\widehat{\beta}_j = \frac{1}{n} \sum_{i=1}^{n} Y_i \phi_j(x_i), \quad j = 1, \ldots, n.$$

2. Let

$$\widehat{\sigma}^2 = \frac{n}{k} \sum_{i=n-k+1}^{n} \widehat{\beta}_j^2 \tag{21.28}$$

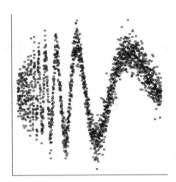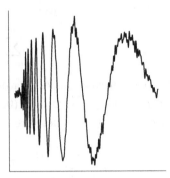

FIGURE 21.4. Data from the doppler test function and the estimated function. See Example 21.10.

where $k \approx n/4$.

3. For $1 \leq J \leq n$, compute the risk estimate

$$\widehat{R}(J) = J\frac{\widehat{\sigma}^2}{n} + \sum_{j=J+1}^{n} \left(\widehat{\beta}_j^2 - \frac{\widehat{\sigma}^2}{n}\right)_+ .$$

4. Choose $\widehat{J} \in \{1, \ldots n\}$ to minimize $\widehat{R}(J)$.

5. Let

$$\widehat{r}(x) = \sum_{j=1}^{\widehat{J}} \widehat{\beta}_j \phi_j(x).$$

Finally, we turn to confidence bands. As before, these bands are not really for the true function $r(x)$ but rather for the smoothed version of the function $r_J(x) = \sum_{j=1}^{\widehat{J}} \beta_j \phi_j(x)$.

21.11 Theorem. *Suppose the estimate $\widehat{r}$ is based on J terms and $\widehat{\sigma}$ is defined as in equation (21.28). Assume that $J < n - k + 1$. An approximate $1 - \alpha$ confidence band for r_J is (ℓ, u) where*

$$\ell(x) = \widehat{r}_n(x) - c, \quad u(x) = \widehat{r}_n(x) + c, \tag{21.29}$$

where

$$c = \frac{a(x)\, \widehat{\sigma}\, \chi_{J,\alpha}}{\sqrt{n}}, \quad a(x) = \sqrt{\sum_{j=1}^{J} \phi_j^2(x)},$$

and $\widehat{\sigma}$ is given in equation (21.28).

PROOF. Let $L = \sum_{j=1}^{J}(\widehat{\beta}_j - \beta_j)^2$. By the central limit theorem, $\widehat{\beta}_j \approx N(\beta_j, \sigma^2/n)$. Hence, $\widehat{\beta}_j \approx \beta_j + \sigma\epsilon_j/\sqrt{n}$ where $\epsilon_j \sim N(0,1)$ and therefore

$$L \approx \frac{\sigma^2}{n} \sum_{j=1}^{J} \epsilon_j^2 \stackrel{d}{=} \frac{\sigma^2}{n} \chi_J^2.$$

Thus,

$$\mathbb{P}\left(L > \frac{\sigma^2}{n}\chi_{J,\alpha}^2\right) = \mathbb{P}\left(\frac{\sigma^2}{n}\chi_J^2 > \frac{\sigma^2}{n}\chi_{J,\alpha}^2\right) = \alpha.$$

Also,

$$
\begin{aligned}
|\widehat{r}(x) - r_J(x)| &\leq \sum_{j=1}^{J} |\phi_j(x)|\, |\widehat{\beta}_j - \beta_j| \\
&\leq \sqrt{\sum_{j=1}^{J} \phi_j^2(x)} \sqrt{\sum_{j=1}^{J}(\widehat{\beta}_j - \beta_j^2)} \\
&\leq a(x)\,\sqrt{L}
\end{aligned}
$$

by the Cauchy-Schwartz inequality (Theorem 4.8). So,

$$
\begin{aligned}
\mathbb{P}\left(\max_x \frac{|\widehat{f}_J(x) - \overline{f}(x)|}{a(x)} > \frac{\widehat{\sigma}\chi_{J,\alpha}}{\sqrt{n}}\right) &\leq \mathbb{P}\left(\sqrt{L} > \frac{\widehat{\sigma}\chi_{J,\alpha}}{\sqrt{n}}\right) \\
&= \alpha
\end{aligned}
$$

and the result follows. ∎

21.12 Example. Figure 21.5 shows the confidence envelope for the doppler signal. The first plot is based on $J = 234$ (the value of J that minimizes the estimated risk). The second is based on $J = 45 \approx \sqrt{n}$. Larger J yields a higher resolution estimator at the cost of large confidence bands. Smaller J yields a lower resolution estimator but has tighter confidence bands. ∎

So far, we have assumed that the x_i's are of the form $\{1/n, 2/n, \ldots, 1\}$. If the x_i's are on interval $[a, b]$, then we can rescale them so that are in the interval $[0, 1]$. If the x_i's are not equally spaced, the methods we have discussed still apply so long as the x_i's "fill out" the interval $[0,1]$ in such a way so as to not be too clumped together. If we want to treat the x_i's as random instead of fixed, then the method needs significant modifications which we shall not deal with here.

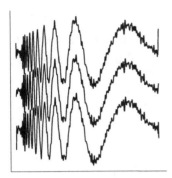 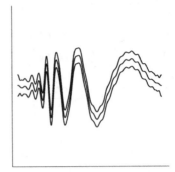

FIGURE 21.5. Estimates and confidence bands for the doppler test function using $n = 2,048$ observations. First plot: $J = 234$ terms. Second plot: $J = 45$ terms.

21.4 Wavelets

Suppose there is a sharp jump in a regression function f at some point x but that f is otherwise very smooth. Such a function f is said to be **spatially inhomogeneous**. The doppler function is an example of a spatially inhomogeneous function; it is smooth for large x and unsmooth for small x.

It is hard to estimate f using the methods we have discussed so far. If we use a cosine basis and only keep low order terms, we will miss the peak; if we allow higher order terms we will find the peak but we will make the rest of the curve very wiggly. Similar comments apply to kernel regression. If we use a large bandwidth, then we will smooth out the peak; if we use a small bandwidth, then we will find the peak but we will make the rest of the curve very wiggly.

One way to estimate inhomogeneous functions is to use a more carefully chosen basis that allows us to place a "blip" in some small region without adding wiggles elsewhere. In this section, we describe a special class of bases called **wavelets**, that are aimed at fixing this problem. Statistical inference using wavelets is a large and active area. We will just discuss a few of the main ideas to get a flavor of this approach.

We start with a particular wavelet called the **Haar wavelet**. The **Haar father wavelet** or **Haar scaling function** is defined by

$$\phi(x) = \begin{cases} 1 & \text{if } 0 \le x < 1 \\ 0 & \text{otherwise.} \end{cases} \tag{21.30}$$

The **mother Haar wavelet** is defined by

$$\psi(x) = \begin{cases} -1 & \text{if } 0 \le x \le \frac{1}{2}, \\ 1 & \text{if } \frac{1}{2} < x \le 1. \end{cases} \tag{21.31}$$

For any integers j and k define

$$\psi_{j,k}(x) = 2^{j/2}\psi(2^j x - k). \tag{21.32}$$

The function $\psi_{j,k}$ has the same shape as ψ but it has been rescaled by a factor of $2^{j/2}$ and shifted by a factor of k.

See Figure 21.6 for some examples of Haar wavelets. Notice that for large j, $\psi_{j,k}$ is a very localized function. This makes it possible to add a blip to a function in one place without adding wiggles elsewhere. Increasing j is like looking in a microscope at increasing degrees of resolution. In technical terms, we say that wavelets provide a **multiresolution analysis** of $L_2(0,1)$.

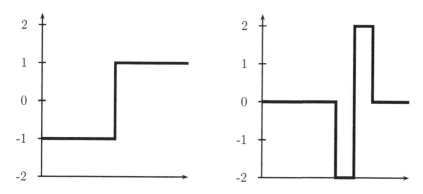

FIGURE 21.6. Some Haar wavelets. Left: the mother wavelet $\psi(x)$; Right: $\psi_{2,2}(x)$.

Let

$$W_j = \{\psi_{jk}, \ k = 0, 1, \ldots, 2^j - 1\}$$

be the set of rescaled and shifted mother wavelets at resolution j.

21.13 Theorem. *The set of functions*

$$\left\{ \phi, W_0, W_1, W_2, \ldots, \right\}$$

is an orthonormal basis for $L_2(0,1)$.

It follows from this theorem that we can expand any function $f \in L_2(0, 1)$ in this basis. Because each W_j is itself a set of functions, we write the expansion as a double sum:

$$f(x) = \alpha\,\phi(x) + \sum_{j=0}^{\infty} \sum_{k=0}^{2^j-1} \beta_{j,k}\psi_{j,k}(x) \qquad (21.33)$$

where

$$\alpha = \int_0^1 f(x)\phi(x)\,dx, \quad \beta_{j,k} = \int_0^1 f(x)\psi_{j,k}(x)\,dx.$$

We call α the **scaling coefficient** and the $\beta_{j,k}$'s are called the **detail coefficients**. We call the finite sum

$$f_J(x) = \alpha\phi(x) + \sum_{j=0}^{J-1} \sum_{k=0}^{2^j-1} \beta_{j,k}\psi_{j,k}(x) \qquad (21.34)$$

the **resolution** J approximation to f. The total number of terms in this sum is

$$1 + \sum_{j=0}^{J-1} 2^j = 1 + 2^J - 1 = 2^J.$$

21.14 Example. Figure 21.7 shows the doppler signal, and its reconstruction using $J = 3, 5$ and $J = 8$. ∎

Haar wavelets are localized, meaning that they are zero outside an interval. But they are not smooth. This raises the question of whether there exist smooth, localized wavelets that from an orthonormal basis. In 1988, Ingrid Daubechie showed that such wavelets do exist. These smooth wavelets are difficult to describe. They can be constructed numerically but there is no closed form formula for the smoother wavelets. To keep things simple, we will continue to use Haar wavelets.

Consider the regression model $Y_i = r(x_i) + \sigma\epsilon_i$ where $\epsilon_i \sim N(0, 1)$ and $x_i = i/n$. To simplify the discussion we assume that $n = 2^J$ for some J.

There is one major difference between estimation using wavelets instead of a cosine (or polynomial) basis. With the cosine basis, we used all the terms $1 \le j \le J$ for some J. The number of terms J acted as a smoothing parameter. With wavelets, we control smoothing using a method called **thresholding** where we keep a term in the function approximation if its coefficient is large,

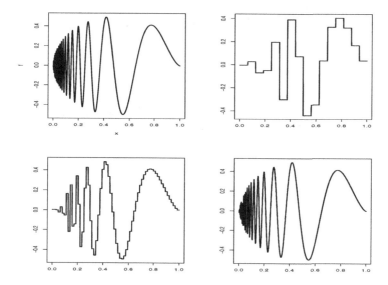

FIGURE   21.7.   The   doppler   signal   and   its   reconstruction
$f_J(x) = \alpha\phi(x) + \sum_{j=0}^{J-1} \sum_k \beta_{j,k}\psi_{j,k}(x)$ based on $J = 3$, $J = 5$, and $J = 8$.

otherwise, we throw out that term. There are many versions of thresholding.
The simplest is called hard, universal thresholding. Let $J = \log_2(n)$ and define

$$\widehat{\alpha} = \frac{1}{n}\sum_i \phi_k(x_i)Y_i \quad \text{and} \quad D_{j,k} = \frac{1}{n}\sum_i \psi_{j,k}(x_i)Y_i \tag{21.35}$$

for $0 \le j \le J - 1$.

Haar Wavelet Regression

1. Compute $\widehat{\alpha}$ and $D_{j,k}$ as in (21.35), for $0 \le j \le J - 1$.

2. Estimate σ; see (21.37).

3. Apply universal thresholding:

$$\widehat{\beta}_{j,k} = \left\{ \begin{array}{ll} D_{j,k} & \text{if } |D_{j,k}| > \widehat{\sigma}\sqrt{\frac{2\log n}{n}} \\ 0 & \text{otherwise.} \end{array} \right\} \tag{21.36}$$

4. Set $\widehat{f}(x) = \widehat{\alpha}\phi(x) + \sum_{j=j_0}^{J-1} \sum_{k=0}^{2^j - 1} \widehat{\beta}_{j,k}\psi_{j,k}(x)$.

In practice, we do not compute S_k and $D_{j,k}$ using (21.35). Instead, we use the **discrete wavelet transform (DWT)** which is very fast. The DWT for Haar wavelets is described in the appendix. The estimate of σ is

$$\widehat{\sigma} = \sqrt{n} \times \frac{\text{median}\left(|D_{J-1,k}| :\ k = 0, \ldots, 2^{J-1} - 1\right)}{0.6745}. \tag{21.37}$$

The estimate for σ may look strange. It is similar to the estimate we used for the cosine basis but it is designed to be insensitive to sharp peaks in the function.

To understand the intuition behind universal thresholding, consider what happens when there is no signal, that is, when $\beta_{j,k} = 0$ for all j and k.

21.15 Theorem. *Suppose that $\beta_{j,k} = 0$ for all j and k and let $\widehat{\beta}_{j,k}$ be the universal threshold estimator. Then*

$$\mathbb{P}(\widehat{\beta}_{j,k} = 0 \text{ for all } j, k) \to 1$$

as $n \to \infty$.

PROOF. To simplify the proof, assume that σ is known. Now $D_{j,k} \approx N(0, \sigma^2/n)$. We will need Mill's inequality (Theorem 4.7): if $Z \sim N(0,1)$ then $\mathbb{P}(|Z| > t) \le (c/t)e^{-t^2/2}$ where $c = \sqrt{2/\pi}$ is a constant. Thus,

$$
\begin{aligned}
\mathbb{P}(\max |D_{j,k}| > \lambda) \ &\le\ \sum_{j,k} \mathbb{P}(|D_{j,k}| > \lambda) = \sum_{j,k} \mathbb{P}\left(\frac{\sqrt{n}|D_{j,k}|}{\sigma} > \frac{\sqrt{n}\lambda}{\sigma}\right) \\
&\le\ \sum_{j,k} \frac{c\sigma}{\lambda\sqrt{n}} \exp\left\{-\frac{1}{2}\frac{n\lambda^2}{\sigma^2}\right\} \\
&=\ \frac{c}{\sqrt{2\log n}} \to 0. \ \blacksquare
\end{aligned}
$$

21.16 Example. Consider $Y_i = r(x_i) + \sigma\epsilon_i$ where f is the doppler signal, $\sigma = .1$ and $n = 2,048$. Figure 21.8 shows the data and the estimated function using universal thresholding. Of course, the estimate is not smooth since Haar wavelets are not smooth. Nonetheless, the estimate is quite accurate. $\blacksquare$

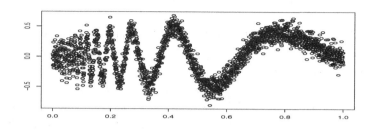

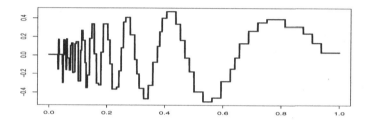

FIGURE 21.8. Estimate of the Doppler function using Haar wavelets and universal thresholding.

21.5   Appendix

THE DWT FOR HAAR WAVELETS. Let y be the vector of Y_i's (length n) and let $J = \log_2(n)$. Create a list D with elements

$$D[[0]], \ldots, D[[J-1]].$$

Set:

$$temp \leftarrow y/\sqrt{n}.$$

Then do:

$$
\begin{aligned}
for(j \quad &in \quad (J-1):0)\{ \\
m \quad &\leftarrow \quad 2^j \\
I \quad &\leftarrow \quad (1:m) \\
D[[j]] \quad &\leftarrow \quad \left(temp[2*I] - temp[(2*I)-1]\right)/\sqrt{2} \\
temp \quad &\leftarrow \quad \left(temp[2*I] + temp[(2*I)-1]\right)/\sqrt{2} \\
&\}
\end{aligned}
$$

21.6   Bibliographic Remarks

Efromovich (1999) is a reference for orthogonal function methods. See also Beran (2000) and Beran and Dümbgen (1998). An introduction to wavelets is given in Ogden (1997). A more advanced treatment can be found in Härdle et al. (1998). The theory of statistical estimation using wavelets has been developed by many authors, especially David Donoho and Ian Johnstone. See Donoho and Johnstone (1994), Donoho and Johnstone (1995), Donoho et al. (1995), and Donoho and Johnstone (1998).

21.7   Exercises

1. Prove Theorem 21.5.

2. Prove Theorem 21.9.

3. Let

$$\psi_1 = \left(\frac{1}{\sqrt{3}}, \frac{1}{\sqrt{3}}, \frac{1}{\sqrt{3}}\right), \ \psi_2 = \left(\frac{1}{\sqrt{2}}, -\frac{1}{\sqrt{2}}, 0\right), \ \psi_3 = \left(\frac{1}{\sqrt{6}}, \frac{1}{\sqrt{6}}, -\frac{2}{\sqrt{6}}\right).$$

   Show that these vectors have norm 1 and are orthogonal.

4. Prove Parseval's relation equation (21.6).

5. Plot the first five Legendre polynomials. Verify, numerically, that they are orthonormal.

6. Expand the following functions in the cosine basis on $[0, 1]$. For (a) and (b), find the coefficients β_j analytically. For (c) and (d), find the coefficients β_j numerically, i.e.

$$\beta_j = \int_0^1 f(x)\phi_j(x) \approx \frac{1}{N} \sum_{r=1}^N f\left(\frac{r}{N}\right) \phi_j\left(\frac{r}{N}\right)$$

   for some large integer N. Then plot the partial sum $\sum_{j=1}^n \beta_j \phi_j(x)$ for increasing values of n.

   (a) $f(x) = \sqrt{2}\cos(3\pi x)$.

   (b) $f(x) = \sin(\pi x)$.

   (c) $f(x) = \sum_{j=1}^{11} h_j K(x - t_j)$ where $K(t) = (1 + \text{sign}(t))/2$, $\text{sign}(x) = -1$ if $x < 0$, $\text{sign}(x) = 0$ if $x = 0$, $\text{sign}(x) = 1$ if $x > 0$,

$(t_j) = (.1, .13, .15, .23, .25, .40, .44, .65, .76, .78, .81)$,

$(h_j) = (4, -5, 3, -4, 5, -4.2, 2.1, 4.3, -3.1, 2.1, -4.2)$.

(d) $f = \sqrt{x(1-x)} \sin\left(\frac{2.1\pi}{(x+.05)}\right)$.

7. Consider the glass fragments data from the book's website. Let Y be refractive index and let X be aluminum content (the fourth variable).

(a) Do a nonparametric regression to fit the model $Y = f(x) + \epsilon$ using the cosine basis method. The data are not on a regular grid. Ignore this when estimating the function. (But do sort the data first according to x.) Provide a function estimate, an estimate of the risk, and a confidence band.

(b) Use the wavelet method to estimate f.

8. Show that the Haar wavelets are orthonormal.

9. Consider again the doppler signal:

$$f(x) = \sqrt{x(1-x)} \sin\left(\frac{2.1\pi}{x + 0.05}\right).$$

Let $n = 1,024$, $\sigma = 0.1$, and let $(x_1, \ldots, x_n) = (1/n, \ldots, 1)$. Generate data

$$Y_i = f(x_i) + \sigma\epsilon_i$$

where $\epsilon_i \sim N(0, 1)$.

(a) Fit the curve using the cosine basis method. Plot the function estimate and confidence band for $J = 10, 20, \ldots, 100$.

(b) Use Haar wavelets to fit the curve.

10. (Haar density Estimation.) Let $X_1, \ldots, X_n \sim f$ for some density f on $[0, 1]$. Let's consider constructing a wavelet histogram. Let ϕ and ψ be the Haar father and mother wavelet. Write

$$f(x) \approx \phi(x) + \sum_{j=0}^{J-1} \sum_{k=0}^{2^j - 1} \beta_{j,k} \psi_{j,k}(x)$$

where $J \approx \log_2(n)$. Let

$$\widehat{\beta}_{j,k} = \frac{1}{n} \sum_{i=1}^{n} \psi_{j,k}(X_i).$$

(a) Show that $\widehat{\beta}_{j,k}$ is an unbiased estimate of $\beta_{j,k}$.

(b) Define the Haar histogram

$$\widehat{f}(x) = \phi(x) + \sum_{j=0}^{B} \sum_{k=0}^{2^j - 1} \widehat{\beta}_{j,k} \psi_{j,k}(x)$$

for $0 \leq B \leq J - 1$.

(c) Find an approximate expression for the MSE as a function of B.

(d) Generate $n = 1{,}000$ observations from a Beta (15,4) density. Estimate the density using the Haar histogram. Use leave-one-out cross validation to choose B.

11. In this question, we will explore the motivation for equation (21.37). Let $X_1, \ldots, X_n \sim N(0, \sigma^2)$. Let

$$\widehat{\sigma} = \sqrt{n} \times \frac{\text{median}\,(|X_1|, \ldots, |X_n|)}{0.6745}.$$

(a) Show that $\mathbb{E}(\widehat{\sigma}) = \sigma$.

(b) Simulate $n = 100$ observations from a N(0,1) distribution. Compute $\widehat{\sigma}$ as well as the usual estimate of σ. Repeat 1,000 times and compare the MSE.

(c) Repeat (b) but add some outliers to the data. To do this, simulate each observation from a N(0,1) with probability .95 and simulate each observation from a N(0,10) with probability .95.

12. Repeat question 6 using the Haar basis.

22
Classification

22.1 Introduction

The problem of predicting a discrete random variable Y from another random variable X is called **classification, supervised learning, discrimination,** or **pattern recognition.**

Consider IID data $(X_1, Y_1), \ldots, (X_n, Y_n)$ where

$$X_i = (X_{i1}, \ldots, X_{id}) \in \mathcal{X} \subset \mathbb{R}^d$$

is a d-dimensional vector and Y_i takes values in some finite set $\mathcal{Y}$. A **classification rule** is a function $h : \mathcal{X} \to \mathcal{Y}$. When we observe a new X, we predict Y to be $h(X)$.

22.1 Example. Here is a an example with fake data. Figure 22.1 shows 100 data points. The covariate $X = (X_1, X_2)$ is 2-dimensional and the outcome $Y \in \mathcal{Y} = \{0, 1\}$. The Y values are indicated on the plot with the triangles representing $Y = 1$ and the squares representing $Y = 0$. Also shown is a linear classification rule represented by the solid line. This is a rule of the form

$$h(x) = \begin{cases} 1 & \text{if } a + b_1 x_1 + b_2 x_2 > 0 \\ 0 & \text{otherwise.} \end{cases}$$

Everything above the line is classified as a 0 and everything below the line is classified as a 1. ∎

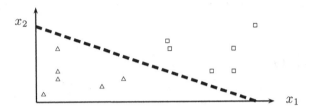

FIGURE 22.1. Two covariates and a linear decision boundary. $\triangle$ means $Y = 1$. $\square$ means $Y = 0$. These two groups are perfectly separated by the linear decision boundary; you probably won't see real data like this.

22.2 Example. Recall the the Coronary Risk-Factor Study (CORIS) data from Example 13.17. There are 462 males between the ages of 15 and 64 from three rural areas in South Africa. The outcome Y is the presence $(Y = 1)$ or absence $(Y = 0)$ of coronary heart disease and there are 9 covariates: systolic blood pressure, cumulative tobacco (kg), ldl (low density lipoprotein choles-terol), adiposity, famhist (family history of heart disease), typea (type-A be-havior), obesity, alcohol (current alcohol consumption), and age. I computed a linear decision boundary using the LDA method based on two of the co-variates, systolic blood pressure and tobacco consumption. The LDA method will be explained shortly. In this example, the groups are hard to tell apart. In fact, 141 of the 462 subjects are misclassified using this classification rule. ∎

At this point, it is worth revisiting the Statistics/Data Mining dictionary:

Statistics	Computer Science	Meaning
classification	supervised learning	predicting a discrete Y from X
data	training sample	$(X_1, Y_1), \ldots, (X_n, Y_n)$
covariates	features	the X_i's
classifier	hypothesis	map $h : \mathcal{X} \to \mathcal{Y}$
estimation	learning	finding a good classifier

22.2   Error Rates and the Bayes Classifier

Our goal is to find a classification rule h that makes accurate predictions. We start with the following definitions:

22.3 Definition. *The* **true error rate**[1] *of a classifier h is*

$$L(h) = \mathbb{P}(\{h(X) \neq Y\}) \tag{22.1}$$

and the **empirical error rate** *or* **training error rate** *is*

$$\widehat{L}_n(h) = \frac{1}{n} \sum_{i=1}^{n} I(h(X_i) \neq Y_i). \tag{22.2}$$

First we consider the special case where $\mathcal{Y} = \{0, 1\}$. Let

$$r(x) = \mathbb{E}(Y|X = x) = \mathbb{P}(Y = 1|X = x)$$

denote the **regression function**. From Bayes' theorem we have that

$$
\begin{aligned}
r(x) &= \mathbb{P}(Y = 1|X = x) \\
&= \frac{f(x|Y = 1)\mathbb{P}(Y = 1)}{f(x|Y = 1)\mathbb{P}(Y = 1) + f(x|Y = 0)\mathbb{P}(Y = 0)} \\
&= \frac{\pi f_1(x)}{\pi f_1(x) + (1 - \pi)f_0(x)}
\end{aligned}
\tag{22.3}
$$

where

$$
\begin{aligned}
f_0(x) &= f(x|Y = 0) \\
f_1(x) &= f(x|Y = 1) \\
\pi &= \mathbb{P}(Y = 1).
\end{aligned}
$$

22.4 Definition. *The* **Bayes classification rule** h^* *is*

$$h^*(x) = \begin{cases} 1 & \text{if } r(x) > \frac{1}{2} \\ 0 & \text{otherwise.} \end{cases} \tag{22.4}$$

The set $\mathcal{D}(h) = \{x : \mathbb{P}(Y = 1|X = x) = \mathbb{P}(Y = 0|X = x)\}$ *is called the* **decision boundary.**

Warning! The Bayes rule has nothing to do with Bayesian inference. We could estimate the Bayes rule using either frequentist or Bayesian methods. The Bayes rule may be written in several equivalent forms:

---

[1] One can use other loss functions. For simplicity we will use the error rate as our loss function.

$$h^*(x) = \begin{cases} 1 & \text{if } \mathbb{P}(Y = 1|X = x) > \mathbb{P}(Y = 0|X = x) \\ 0 & \text{otherwise} \end{cases} \qquad (22.5)$$

and

$$h^*(x) = \begin{cases} 1 & \text{if } \pi f_1(x) > (1 - \pi)f_0(x) \\ 0 & \text{otherwise.} \end{cases} \qquad (22.6)$$

22.5 Theorem. *The Bayes rule is optimal, that is, if h is any other classification rule then $L(h^*) \le L(h)$.*

The Bayes rule depends on unknown quantities so we need to use the data to find some approximation to the Bayes rule. At the risk of oversimplifying, there are three main approaches:

1. **Empirical Risk Minimization.** Choose a set of classifiers $\mathcal{H}$ and find $\widehat{h} \in \mathcal{H}$ that minimizes some estimate of $L(h)$.

2. **Regression.** Find an estimate $\widehat{r}$ of the regression function r and define

$$\widehat{h}(x) = \begin{cases} 1 & \text{if } \widehat{r}(x) > \frac{1}{2} \\ 0 & \text{otherwise.} \end{cases}$$

3. **Density Estimation.** Estimate f_0 from the X_i's for which $Y_i = 0$, estimate f_1 from the X_i's for which $Y_i = 1$ and let $\widehat{\pi} = n^{-1} \sum_{i=1}^n Y_i$. Define

$$\widehat{r}(x) = \widehat{\mathbb{P}}(Y = 1|X = x) = \frac{\widehat{\pi}\widehat{f_1}(x)}{\widehat{\pi}\widehat{f_1}(x) + (1 - \widehat{\pi})\widehat{f_0}(x)}$$

and

$$\widehat{h}(x) = \begin{cases} 1 & \text{if } \widehat{r}(x) > \frac{1}{2} \\ 0 & \text{otherwise.} \end{cases}$$

Now let us generalize to the case where Y takes on more than two values as follows.

22.6 Theorem. *Suppose that $Y \in \mathcal{Y} = \{1, \ldots, K\}$. The optimal rule is*

$$\begin{aligned} h(x) &= \operatorname{argmax}_k \mathbb{P}(Y = k|X = x) & (22.7) \\ &= \operatorname{argmax}_k \pi_k f_k(x) & (22.8) \end{aligned}$$

where

$$\mathbb{P}(Y = k|X = x) = \frac{f_k(x)\pi_k}{\sum_r f_r(x)\pi_r}, \qquad (22.9)$$

$\pi_r = P(Y = r)$, $f_r(x) = f(x|Y = r)$ *and* argmax_k *means "the value of k that maximizes that expression."*

22.3   Gaussian and Linear Classifiers

Perhaps the simplest approach to classification is to use the density estimation strategy and assume a parametric model for the densities. Suppose that
$\mathcal{Y} = \{0,1\}$ and that $f_0(x) = f(x|Y = 0)$ and $f_1(x) = f(x|Y = 1)$ are both
multivariate Gaussians:

$$f_k(x) = \frac{1}{(2\pi)^{d/2}|\Sigma_k|^{1/2}} \exp\left\{-\frac{1}{2}(x - \mu_k)^T \Sigma_k^{-1}(x - \mu_k)\right\}, \quad k = 0,1.$$

Thus, $X|Y = 0 \sim N(\mu_0, \Sigma_0)$ and $X|Y = 1 \sim N(\mu_1, \Sigma_1)$.

22.7 Theorem. *If $X|Y = 0 \sim N(\mu_0, \Sigma_0)$ and $X|Y = 1 \sim N(\mu_1, \Sigma_1)$, then the
Bayes rule is*

$$h^*(x) = \begin{cases} 1 & \text{if } r_1^2 < r_0^2 + 2\log\left(\frac{\pi_1}{\pi_0}\right) + \log\left(\frac{|\Sigma_0|}{|\Sigma_1|}\right) \\ 0 & \text{otherwise} \end{cases} \qquad (22.10)$$

where

$$r_i^2 = (x - \mu_i)^T \Sigma_i^{-1}(x - \mu_i), \quad i = 1, 2 \qquad (22.11)$$

*is the **Manalahobis distance**. An equivalent way of expressing the Bayes'
rule is*

$$h^*(x) = \text{argmax}_k \delta_k(x)$$

where

$$\delta_k(x) = -\frac{1}{2}\log|\Sigma_k| - \frac{1}{2}(x - \mu_k)^T \Sigma_k^{-1}(x - \mu_k) + \log \pi_k \qquad (22.12)$$

and $|A|$ denotes the determinant of a matrix A.

The decision boundary of the above classifier is quadratic so this procedure
is called **quadratic discriminant analysis (QDA)**. In practice, we use
sample estimates of $\pi, \mu_1, \mu_2, \Sigma_0, \Sigma_1$ in place of the true value, namely:

$$\widehat{\pi}_0 = \frac{1}{n}\sum_{i=1}^{n}(1 - Y_i), \quad \widehat{\pi}_1 = \frac{1}{n}\sum_{i=1}^{n}Y_i$$

$$\widehat{\mu}_0 = \frac{1}{n_0}\sum_{i:\ Y_i=0} X_i, \quad \widehat{\mu}_1 = \frac{1}{n_1}\sum_{i:\ Y_i=1} X_i$$

$$S_0 = \frac{1}{n_0}\sum_{i:\ Y_i=0}(X_i - \widehat{\mu}_0)(X_i - \widehat{\mu}_0)^T, \quad S_1 = \frac{1}{n_1}\sum_{i:\ Y_i=1}(X_i - \widehat{\mu}_1)(X_i - \widehat{\mu}_1)^T$$

where $n_0 = \sum_i(1 - Y_i)$ and $n_1 = \sum_i Y_i$.

A simplification occurs if we assume that $\Sigma_0 = \Sigma_0 = \Sigma$. In that case, the Bayes rule is

$$h^*(x) = \mathrm{argmax}_k \delta_k(x) \tag{22.13}$$

where now

$$\delta_k(x) = x^T \Sigma^{-1} \mu_k - \frac{1}{2} \mu_k^T \Sigma^{-1} + \log \pi_k. \tag{22.14}$$

The parameters are estimated as before, except that the MLE of Σ is

$$S = \frac{n_0 S_0 + n_1 S_1}{n_0 + n_1}.$$

The classification rule is

$$h^*(x) = \begin{cases} 1 & \text{if } \delta_1(x) > \delta_0(x) \\ 0 & \text{otherwise} \end{cases} \tag{22.15}$$

where

$$\delta_j(x) = x^T S^{-1} \widehat{\mu}_j - \frac{1}{2} \widehat{\mu}_j^T S^{-1} \widehat{\mu}_j + \log \widehat{\pi}_j$$

is called the **discriminant function**. The decision boundary $\{x : \delta_0(x) = \delta_1(x)\}$ is linear so this method is called **linear discrimination analysis (LDA)**.

22.8 Example. Let us return to the South African heart disease data. The decision rule in in Example 22.2 was obtained by linear discrimination. The outcome was

	classified as 0	classified as 1
$y = 0$	277	25
$y = 1$	116	44

The observed misclassification rate is $141/462 = .31$. Including all the covariates reduces the error rate to .27. The results from quadratic discrimination are

	classified as 0	classified as 1
$y = 0$	272	30
$y = 1$	113	47

which has about the same error rate $143/462 = .31$. Including all the covariates reduces the error rate to .26. In this example, there is little advantage to QDA over LDA. ∎

Now we generalize to the case where Y takes on more than two values.

22.9 Theorem. *Suppose that $Y \in \{1, \ldots, K\}$. If $f_k(x) = f(x|Y = k)$ is Gaussian, the Bayes rule is*

$$h(x) = \operatorname{argmax}_k \delta_k(x)$$

where

$$\delta_k(x) = -\frac{1}{2} \log |\Sigma_k| - \frac{1}{2}(x - \mu_k)^T \Sigma_k^{-1}(x - \mu_k) + \log \pi_k. \qquad (22.16)$$

If the variances of the Gaussians are equal, then

$$\delta_k(x) = x^T \Sigma^{-1} \mu_k - \frac{1}{2} \mu_k^T \Sigma^{-1} + \log \pi_k. \qquad (22.17)$$

We estimate $\delta_k(x)$ by by inserting estimates of μ_k, Σ_k and π_k. There is another version of linear discriminant analysis due to Fisher. The idea is to first reduce the dimension of covariates to one dimension by projecting the data onto a line. Algebraically, this means replacing the covariate $X = (X_1, \ldots, X_d)$ with a linear combination $U = w^T X = \sum_{j=1}^{d} w_j X_j$. The goal is to choose the vector $w = (w_1, \ldots, w_d)$ that "best separates the data." Then we perform classification with the one-dimensional covariate Z instead of X.

We need define what we mean by separation of the groups. We would like the two groups to have means that are far apart relative to their spread. Let μ_j denote the mean of X for Y_j and let Σ be the variance matrix of X. Then $\mathbb{E}(U|Y = j) = \mathbb{E}(w^T X|Y = j) = w^T \mu_j$ and $\mathbb{V}(U) = w^T \Sigma w$. [2] Define the separation by

$$
\begin{aligned}
J(w) &= \frac{(\mathbb{E}(U|Y = 0) - \mathbb{E}(U|Y = 1))^2}{w^T \Sigma w} \\
&= \frac{(w^T \mu_0 - w^T \mu_1)^2}{w^T \Sigma w} \\
&= \frac{w^T (\mu_0 - \mu_1)(\mu_0 - \mu_1)^T w}{w^T \Sigma w}.
\end{aligned}
$$

We estimate J as follows. Let $n_j = \sum_{i=1}^{n} I(Y_i = j)$ be the number of observations in group j, let $\overline{X}_j$ be the sample mean vector of the X's for group j, and let S_j be the sample covariance matrix in group j. Define

$$\widehat{J}(w) = \frac{w^T S_B w}{w^T S_W w} \qquad (22.18)$$

---

[2] The quantity J arises in physics, where it is called the Rayleigh coefficient.

where

$$S_B = (\overline{X}_0 - \overline{X}_1)(\overline{X}_0 - \overline{X}_1)^T$$
$$S_W = \frac{(n_0 - 1)S_0 + (n_1 - 1)S_1}{(n_0 - 1) + (n_1 - 1)}.$$

22.10 Theorem. *The vector*

$$w = S_W^{-1}(\overline{X}_0 - \overline{X}_1) \qquad (22.19)$$

is a minimizer of $\widehat{J}(w)$. *We call*

$$U = w^T X = (\overline{X}_0 - \overline{X}_1)^T S_W^{-1} X \qquad (22.20)$$

the **Fisher linear discriminant function.** *The midpoint* m *between* $\overline{X}_0$ *and* $\overline{X}_1$ *is*

$$m = \frac{1}{2}(\overline{X}_0 + \overline{X}_1) = \frac{1}{2}(\overline{X}_0 - \overline{X}_1)^T S_B^{-1}(\overline{X}_0 + \overline{X}_1) \qquad (22.21)$$

Fisher's classification rule is

$$h(x) = \begin{cases} 0 & \text{if } w^T X \geq m \\ 1 & \text{if } w^T X < m. \end{cases}$$

Fisher's rule is the same as the Bayes linear classifier in equation (22.14) when $\widehat{\pi} = 1/2$.

22.4 Linear Regression and Logistic Regression

A more direct approach to classification is to estimate the regression function $r(x) = \mathbb{E}(Y|X = x)$ without bothering to estimate the densities f_k. For the rest of this section, we will only consider the case where $\mathcal{Y} = \{0, 1\}$. Thus, $r(x) = \mathbb{P}(Y = 1|X = x)$ and once we have an estimate $\widehat{r}$, we will use the classification rule

$$\widehat{h}(x) = \begin{cases} 1 & \text{if } \widehat{r}(x) > \frac{1}{2} \\ 0 & \text{otherwise.} \end{cases} \qquad (22.22)$$

The simplest regression model is the linear regression model

$$Y = r(x) + \epsilon = \beta_0 + \sum_{j=1}^{d} \beta_j X_j + \epsilon \qquad (22.23)$$

where $\mathbb{E}(\epsilon) = 0$. This model can't be correct since it does not force $Y = 0$ or 1. Nonetheless, it can sometimes lead to a decent classifier.

Recall that the least squares estimate of $\beta = (\beta_0, \beta_1, \ldots, \beta_d)^T$ minimizes the residual sums of squares

$$\text{RSS}(\beta) = \sum_{i=1}^{n} \left(Y_i - \beta_0 - \sum_{j=1}^{d} X_{ij}\beta_j \right)^2.$$

Let $\mathbf{X}$ denote the $N \times (d+1)$ matrix of the form

$$\mathbf{X} = \begin{bmatrix} 1 & X_{11} & \cdots & X_{1d} \\ 1 & X_{21} & \cdots & X_{2d} \\ \vdots & \vdots & \vdots & \vdots \\ 1 & X_{n1} & \cdots & X_{nd} \end{bmatrix}.$$

Also let $\mathbf{Y} = (Y_1, \ldots, Y_n)^T$. Then,

$$RSS(\beta) = (\mathbf{Y} - \mathbf{X}\beta)^T (\mathbf{Y} - \mathbf{X}\beta)$$

and the model can be written as

$$\mathbf{Y} = \mathbf{X}\beta + \epsilon$$

where $\epsilon = (\epsilon_1, \ldots, \epsilon_n)^T$. From Theorem 13.13,

$$\widehat{\beta} = (\mathbf{X}^T\mathbf{X})^{-1}\mathbf{X}^T Y.$$

The predicted values are

$$\widehat{\mathbf{Y}} = \mathbf{X}\widehat{\beta}.$$

Now we use (22.22) to classify, where $\widehat{r}(x) = \widehat{\beta}_0 + \sum_j \widehat{\beta}_j x_j$.

An alternative is to use logistic regression which was also discussed in Chapter 13. The model is

$$r(x) = \mathbb{P}(Y = 1 | X = x) = \frac{e^{\beta_0 + \sum_j \beta_j x_j}}{1 + e^{\beta_0 + \sum_j \beta_j x_j}} \qquad (22.24)$$

and the MLE $\widehat{\beta}$ is obtained numerically.

22.11 Example. Let us return to the heart disease data. The MLE is given in Example 13.17. The error rate, using this model for classification, is .27. The error rate from a linear regression is .26.

We can get a better classifier by fitting a richer model. For example, we could fit

$$\text{logit } \mathbb{P}(Y = 1 | X = x) = \beta_0 + \sum_j \beta_j x_j + \sum_{j,k} \beta_{jk} x_j x_k. \qquad (22.25)$$

More generally, we could add terms of up to order r for some integer r. Large values of r give a more complicated model which should fit the data better. But there is a bias–variance tradeoff which we'll discuss later.

22.12 Example. If we use model (22.25) for the heart disease data with $r = 2$, the error rate is reduced to .22. ∎

22.5   Relationship Between Logistic Regression and LDA

LDA and logistic regression are almost the same thing. If we assume that each group is Gaussian with the same covariance matrix, then we saw earlier that

$$
\begin{aligned}
\log\left(\frac{\mathbb{P}(Y = 1|X = x)}{\mathbb{P}(Y = 0|X = x)}\right) &= \log\left(\frac{\pi_0}{\pi_1}\right) - \frac{1}{2}(\mu_0 + \mu_1)^T \Sigma^{-1}(\mu_1 - \mu_0) \\
&\quad + x^T \Sigma^{-1}(\mu_1 - \mu_0) \\
&\equiv \alpha_0 + \alpha^T x.
\end{aligned}
$$

On the other hand, the logistic model is, by assumption,

$$
\log\left(\frac{\mathbb{P}(Y = 1|X = x)}{\mathbb{P}(Y = 0|X = x)}\right) = \beta_0 + \beta^T x.
$$

These are the same model since they both lead to classification rules that are linear in x. The difference is in how we estimate the parameters.

The joint density of a single observation is $f(x, y) = f(x|y)f(y) = f(y|x)f(x)$. In LDA we estimated the whole joint distribution by maximizing the likelihood

$$
\prod_i f(x_i, y_i) = \underbrace{\prod_i f(x_i|y_i)}_{\text{Gaussian}} \underbrace{\prod_i f(y_i)}_{\text{Bernoulli}}. \tag{22.26}
$$

In logistic regression we maximized the conditional likelihood $\prod_i f(y_i|x_i)$ but we ignored the second term $f(x_i)$:

$$
\prod_i f(x_i, y_i) = \underbrace{\prod_i f(y_i|x_i)}_{\text{logistic}} \underbrace{\prod_i f(x_i)}_{\text{ignored}}. \tag{22.27}
$$

Since classification only requires knowing $f(y|x)$, we don't really need to estimate the whole joint distribution. Logistic regression leaves the marginal distribution $f(x)$ unspecified so it is more nonparametric than LDA. This is an advantage of the logistic regression approach over LDA.

To summarize: LDA and logistic regression both lead to a linear classification rule. In LDA we estimate the entire joint distribution $f(x, y) = f(x|y)f(y)$. In logistic regression we only estimate $f(y|x)$ and we don't bother estimating $f(x)$.

22.6   Density Estimation and Naive Bayes

The Bayes rule is $h(x) = \text{argmax}_k \, \pi_k \, f_k(x)$. If we can estimate π_k and f_k then we can estimate the Bayes classification rule. Estimating π_k is easy but what about f_k? We did this previously by assuming f_k was Gaussian. Another strategy is to estimate f_k with some nonparametric density estimator $\widehat{f}_k$ such as a kernel estimator. But if $x = (x_1, \ldots, x_d)$ is high-dimensional, nonparametric density estimation is not very reliable. This problem is ameliorated if we assume that $X_1, \ldots, X_d$ are independent, for then, $f_k(x_1, \ldots, x_d) = \prod_{j=1}^{d} f_{kj}(x_j)$. This reduces the problem to d one-dimensional density estimation problems, within each of the k groups. The resulting classifier is called **the naive Bayes classifier.** The assumption that the components of X are independent is usually wrong yet the resulting classifier might still be accurate. Here is a summary of the steps in the naive Bayes classifier:

The Naive Bayes Classifier

1. For each group k, compute an estimate $\widehat{f}_{kj}$ of the density f_{kj} for X_j, using the data for which $Y_i = k$.

2. Let
$$\widehat{f}_k(x) = \widehat{f}_k(x_1, \ldots, x_d) = \prod_{j=1}^{d} \widehat{f}_{kj}(x_j).$$

3. Let
$$\widehat{\pi}_k = \frac{1}{n} \sum_{i=1}^{n} I(Y_i = k)$$
where $I(Y_i = k) = 1$ if $Y_i = k$ and $I(Y_i = k) = 0$ if $Y_i \neq k$.

4. Let
$$h(x) = \text{argmax}_k \, \widehat{\pi}_k \, \widehat{f}_k(x).$$

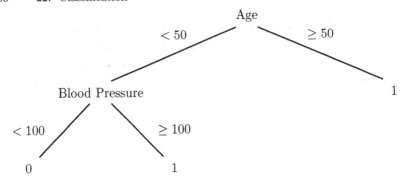

FIGURE 22.2. A simple classification tree.

The naive Bayes classifier is popular when x is high-dimensional and discrete. In that case, $\widehat{f}_{kj}(x_j)$ is especially simple.

22.7   Trees

Trees are classification methods that partition the covariate space $\mathcal{X}$ into disjoint pieces and then classify the observations according to which partition element they fall in. As the name implies, the classifier can be represented as a tree.

For illustration, suppose there are two covariates, $X_1 =$ age and $X_2 =$ blood pressure. Figure 22.2 shows a classification tree using these variables.

The tree is used in the following way. If a subject has Age ≥ 50 then we classify him as $Y = 1$. If a subject has Age < 50 then we check his blood pressure. If systolic blood pressure is < 100 then we classify him as $Y = 1$, otherwise we classify him as $Y = 0$. Figure 22.3 shows the same classifier as a partition of the covariate space.

Here is how a tree is constructed. First, suppose that $y \in \mathcal{Y} = \{0, 1\}$ and that there is only a single covariate X. We choose a split point t that divides the real line into two sets $A_1 = (-\infty, t]$ and $A_2 = (t, \infty)$. Let $\widehat{p}_s(j)$ be the proportion of observations in A_s such that $Y_i = j$:

$$\widehat{p}_s(j) = \frac{\sum_{i=1}^n I(Y_i = j, X_i \in A_s)}{\sum_{i=1}^n I(X_i \in A_s)} \tag{22.28}$$

for $s = 1, 2$ and $j = 0, 1$. The **impurity** of the split t is defined to be

$$I(t) = \sum_{s=1}^2 \gamma_s \tag{22.29}$$

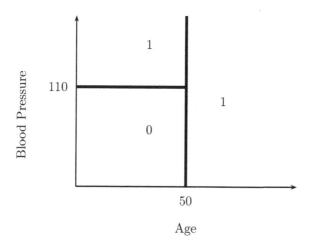

FIGURE 22.3. Partition representation of classification tree.

where

$$\gamma_s = 1 - \sum_{j=0}^{1} \widehat{p}_s(j)^2. \tag{22.30}$$

This particular measure of impurity is known as the **Gini index**. If a partition element A_s contains all 0's or all 1's, then $\gamma_s = 0$. Otherwise, $\gamma_s > 0$. We choose the split point t to minimize the impurity. (Other indices of impurity besides can be used besides the Gini index.)

When there are several covariates, we choose whichever covariate and split that leads to the lowest impurity. This process is continued until some stopping criterion is met. For example, we might stop when every partition element has fewer than n_0 data points, where n_0 is some fixed number. The bottom nodes of the tree are called the **leaves**. Each leaf is assigned a 0 or 1 depending on whether there are more data points with $Y = 0$ or $Y = 1$ in that partition element.

This procedure is easily generalized to the case where $Y \in \{1, \ldots, K\}$. We simply define the impurity by

$$\gamma_s = 1 - \sum_{j=1}^{k} \widehat{p}_s(j)^2 \tag{22.31}$$

where $\widehat{p}_i(j)$ is the proportion of observations in the partition element for which $Y = j$.

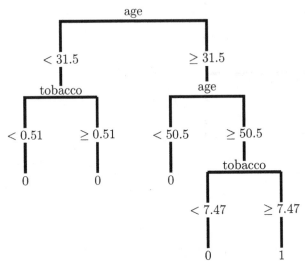

FIGURE 22.4. A classification tree for the heart disease data using two covariates.

22.13 Example. A classification tree for the heart disease data yields a mis-classification rate of .21. If we build a tree using only tobacco and age, the misclassification rate is then .29. The tree is shown in Figure 22.4. ∎

Our description of how to build trees is incomplete. If we keep splitting until there are few cases in each leaf of the tree, we are likely to overfit the data. We should choose the complexity of the tree in such a way that the estimated true error rate is low. In the next section, we discuss estimation of the error rate.

22.8   Assessing Error Rates and Choosing a Good Classifier

How do we choose a good classifier? We would like to have a classifier h with a low true error rate $L(h)$. Usually, we can't use the training error rate $\widehat{L}_n(h)$ as an estimate of the true error rate because it is biased downward.

22.14 Example. Consider the heart disease data again. Suppose we fit a se-quence of logistic regression models. In the first model we include one co-variate. In the second model we include two covariates, and so on. The ninth model includes all the covariates. We can go even further. Let's also fit a tenth model that includes all nine covariates plus the first covariate squared. Then we fit an eleventh model that includes all nine covariates plus the first covariate squared and the second covariate squared. Continuing this way we will get a sequence of 18 classifiers of increasing complexity. The solid line in Figure 22.5 shows the observed classification error which steadily decreases as we make the model more complex. If we keep going, we can make a model with zero observed classification error. The dotted line shows the **10-fold cross-validation estimate** of the error rate (to be explained shortly) which is a better estimate of the true error rate than the observed classification error. The estimated error decreases for a while then increases. This is essentially the bias–variance tradeoff phenomenon we have seen in Chapter 20. ∎

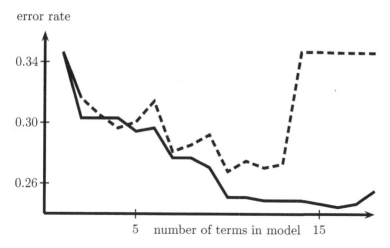

FIGURE 22.5. The solid line is the observed error rate and dashed line is the cross-validation estimate of true error rate.

There are many ways to estimate the error rate. We'll consider two: **cross-validation** and **probability inequalities**.

CROSS-VALIDATION. The basic idea of cross-validation, which we have already encountered in curve estimation, is to leave out some of the data when fitting a model. The simplest version of cross-validation involves randomly splitting the data into two pieces: the **training set** $\mathcal{T}$ and the **validation set** $\mathcal{V}$. Often, about 10 per cent of the data might be set aside as the validation set. The classifier h is constructed from the training set. We then estimate

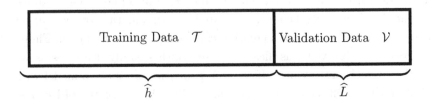

FIGURE 22.6. Cross-validation. The data are divided into two groups: the training data and the validation data. The training data are used to produce an estimated classifier $\widehat{h}$. Then, $\widehat{h}$ is applied to the validation data to obtain an estimate $\widehat{L}$ of the error rate of $\widehat{h}$.

the error by

$$\widehat{L}(h) = \frac{1}{m} \sum_{X_i \in \mathcal{V}} I(h(X_i) \neq Y_I). \qquad (22.32)$$

where m is the size of the validation set. See Figure 22.6.

Another approach to cross-validation is **K-fold cross-validation** which is obtained from the following algorithm.

K-fold cross-validation.

1. Randomly divide the data into K chunks of approximately equal size. A common choice is $K = 10$.

2. For k = 1 to K, do the following:

   (a) Delete chunk k from the data.

   (b) Compute the classifier $\widehat{h}_{(k)}$ from the rest of the data.

   (c) Use $\widehat{h}_{(k)}$ to the predict the data in chunk k. Let $\widehat{L}_{(k)}$ denote the observed error rate.

3. Let

$$\widehat{L}(h) = \frac{1}{K} \sum_{k=1}^{K} \widehat{L}_{(k)}. \qquad (22.33)$$

22.15 Example. We applied 10-fold cross-validation to the heart disease data. The minimum cross-validation error as a function of the number of leaves occurred at six. Figure 22.7 shows the tree with six leaves. ∎

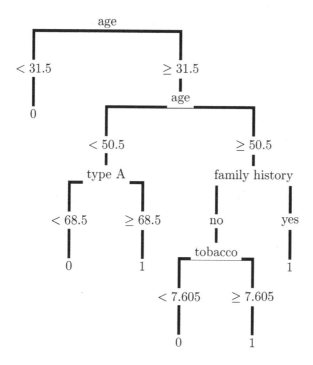

FIGURE 22.7. Smaller classification tree with size chosen by cross-validation.

PROBABILITY INEQUALITIES. Another approach to estimating the error rate is to find a confidence interval for $\widehat{L}_n(h)$ using probability inequalities. This method is useful in the context of **empirical risk minimization**.

Let $\mathcal{H}$ be a set of classifiers, for example, all linear classifiers. Empirical risk minimization means choosing the classifier $\widehat{h} \in \mathcal{H}$ to minimize the training error $\widehat{L}_n(h)$, also called the empirical risk. Thus,

$$\widehat{h} = \operatorname{argmin}_{h \in \mathcal{H}} \widehat{L}_n(h) = \operatorname{argmin}_{h \in \mathcal{H}} \left(\frac{1}{n} \sum_i I(h(X_i) \neq Y_i) \right). \qquad (22.34)$$

Typically, $\widehat{L}_n(\widehat{h})$ underestimates the true error rate $L(\widehat{h})$ because $\widehat{h}$ was chosen to make $\widehat{L}_n(\widehat{h})$ small. Our goal is to assess how much underestimation is taking place. Our main tool for this analysis is **Hoeffding's inequality** (Theorem 4.5). Recall that if $X_1, \ldots, X_n \sim \text{Bernoulli}(p)$, then, for any $\epsilon > 0$,

$$\mathbb{P}\left(|\widehat{p} - p| > \epsilon\right) \leq 2e^{-2n\epsilon^2} \qquad (22.35)$$

where $\widehat{p} = n^{-1} \sum_{i=1}^n X_i$.

First, suppose that $\mathcal{H} = \{h_1, \ldots, h_m\}$ consists of finitely many classifiers. For any fixed h, $\widehat{L}_n(h)$ converges in almost surely to $L(h)$ by the law of large numbers. We will now establish a stronger result.

22.16 Theorem (Uniform Convergence). *Assume $\mathcal{H}$ is finite and has m elements. Then,*

$$\mathbb{P}\left(\max_{h \in \mathcal{H}} |\widehat{L}_n(h) - L(h)| > \epsilon \right) \leq 2m e^{-2n\epsilon^2}.$$

PROOF. We will use Hoeffding's inequality and we will also use the fact that if $A_1, \ldots, A_m$ is a set of events then $\mathbb{P}(\bigcup_{i=1}^m A_i) \leq \sum_{i=1}^m \mathbb{P}(A_i)$. Now,

$$
\begin{aligned}
\mathbb{P}\left(\max_{h \in \mathcal{H}} |\widehat{L}_n(h) - L(h)| > \epsilon \right) &= \mathbb{P}\left(\bigcup_{h \in \mathcal{H}} |\widehat{L}_n(h) - L(h)| > \epsilon \right) \\
&\leq \sum_{H \in \mathcal{H}} \mathbb{P}\left(|\widehat{L}_n(h) - L(h)| > \epsilon \right) \\
&\leq \sum_{H \in \mathcal{H}} 2e^{-2n\epsilon^2} = 2m e^{-2n\epsilon^2}. \quad \blacksquare
\end{aligned}
$$

22.17 Theorem. *Let*

$$\epsilon = \sqrt{\frac{2}{n} \log\left(\frac{2m}{\alpha} \right)}.$$

Then $\widehat{L}_n(\widehat{h}) \pm \epsilon$ is a $1 - \alpha$ confidence interval for $L(\widehat{h})$.

PROOF. This follows from the fact that

$$
\begin{aligned}
\mathbb{P}(|\widehat{L}_n(\widehat{h}) - L(\widehat{h})| > \epsilon) &\leq \mathbb{P}\left(\max_{h \in \mathcal{H}} |\widehat{L}_n(\widehat{h}) - L(\widehat{h})| > \epsilon \right) \\
&\leq 2m e^{-2n\epsilon^2} = \alpha. \quad \blacksquare
\end{aligned}
$$

When $\mathcal{H}$ is large the confidence interval for $L(\widehat{h})$ is large. The more functions there are in $\mathcal{H}$ the more likely it is we have "overfit" which we compensate for by having a larger confidence interval.

In practice we usually use sets $\mathcal{H}$ that are infinite, such as the set of linear classifiers. To extend our analysis to these cases we want to be able to say something like

$$\mathbb{P}\left(\sup_{h \in \mathcal{H}} |\widehat{L}_n(h) - L(h)| > \epsilon \right) \leq \text{something not too big.}$$

One way to develop such a generalization is by way of the **Vapnik-Chervonenkis** or **VC dimension**.

Let $\mathcal{A}$ be a class of sets. Give a finite set $F = \{x_1, \ldots, x_n\}$ let

$$N_{\mathcal{A}}(F) = \#\left\{F \bigcap A : A \in \mathcal{A}\right\} \tag{22.36}$$

be the number of subsets of F "picked out" by $\mathcal{A}$. Here $\#(B)$ denotes the number of elements of a set B. The **shatter coefficient** is defined by

$$s(\mathcal{A}, n) = \max_{F \in \mathcal{F}_n} N_{\mathcal{A}}(F) \tag{22.37}$$

where $\mathcal{F}_n$ consists of all finite sets of size n. Now let $X_1, \ldots, X_n \sim \mathbb{P}$ and let

$$\mathbb{P}_n(A) = \frac{1}{n} \sum_i I(X_i \in A)$$

denote the **empirical probability measure**. The following remarkable theorem bounds the distance between $\mathbb{P}$ and $\mathbb{P}_n$.

22.18 Theorem (Vapnik and Chervonenkis (1971)). *For any $\mathbb{P}$, n and $\epsilon > 0$,*

$$\mathbb{P}\left\{\sup_{A \in \mathcal{A}} |\mathbb{P}_n(A) - \mathbb{P}(A)| > \epsilon\right\} \leq 8s(\mathcal{A}, n)e^{-n\epsilon^2/32}. \tag{22.38}$$

The proof, though very elegant, is long and we omit it. If $\mathcal{H}$ is a set of classifiers, define $\mathcal{A}$ to be the class of sets of the form $\{x : h(x) = 1\}$. We then define $s(\mathcal{H}, n) = s(\mathcal{A}, n)$.

22.19 Theorem.

$$\mathbb{P}\left\{\sup_{h \in \mathcal{H}} |\widehat{L}_n(h) - L(h)| > \epsilon\right\} \leq 8s(\mathcal{H}, n)e^{-n\epsilon^2/32}.$$

A $1 - \alpha$ confidence interval for $L(\widehat{h})$ is $\widehat{L}_n(\widehat{h}) \pm \epsilon_n$ where

$$\epsilon_n^2 = \frac{32}{n} \log\left(\frac{8s(\mathcal{H}, n)}{\alpha}\right).$$

These theorems are only useful if the shatter coefficients do not grow too quickly with n. This is where VC dimension enters.

22.20 Definition. *The VC (Vapnik-Chervonenkis) dimension of a class of sets $\mathcal{A}$ is defined as follows. If $s(\mathcal{A}, n) = 2^n$ for all n, set $VC(\mathcal{A}) = \infty$. Otherwise, define $VC(\mathcal{A})$ to be the largest k for which $s(\mathcal{A}, n) = 2^k$.*

Thus, the VC-dimension is the size of the largest finite set F that can be **shattered** by $\mathcal{A}$ meaning that $\mathcal{A}$ picks out each subset of F. If $\mathcal{H}$ is a set of classifiers we define $VC(\mathcal{H}) = VC(\mathcal{A})$ where $\mathcal{A}$ is the class of sets of the form $\{x : h(x) = 1\}$ as h varies in $\mathcal{H}$. The following theorem shows that if $\mathcal{A}$ has finite VC-dimension, then the shatter coefficients grow as a polynomial in n.

22.21 Theorem. *If $\mathcal{A}$ has finite VC-dimension v, then*

$$s(\mathcal{A}, n) \leq n^v + 1.$$

22.22 Example. Let $\mathcal{A} = \{(-\infty, a]; \ a \in \mathcal{R}\}$. The $\mathcal{A}$ shatters every 1-point set $\{x\}$ but it shatters no set of the form $\{x, y\}$. Therefore, $VC(\mathcal{A}) = 1$. ∎

22.23 Example. Let $\mathcal{A}$ be the set of closed intervals on the real line. Then $\mathcal{A}$ shatters $S = \{x, y\}$ but it cannot shatter sets with 3 points. Consider $S = \{x, y, z\}$ where $x < y < z$. One cannot find an interval A such that $A \bigcap S = \{x, z\}$. So, $VC(\mathcal{A}) = 2$. ∎

22.24 Example. Let $\mathcal{A}$ be all linear half-spaces on the plane. Any 3-point set (not all on a line) can be shattered. No 4 point set can be shattered. Consider, for example, 4 points forming a diamond. Let T be the left and rightmost points. This can't be picked out. Other configurations can also be seen to be unshatterable. So $VC(\mathcal{A}) = 3$. In general, halfspaces in $\mathcal{R}^d$ have VC dimension $d + 1$. ∎

22.25 Example. Let $\mathcal{A}$ be all rectangles on the plane with sides parallel to the axes. Any 4 point set can be shattered. Let S be a 5 point set. There is one point that is not leftmost, rightmost, uppermost, or lowermost. Let T be all points in S except this point. Then T can't be picked out. So $VC(\mathcal{A}) = 4$. ∎

22.26 Theorem. *Let x have dimension d and let $\mathcal{H}$ be th set of linear classifiers. The VC-dimension of $\mathcal{H}$ is $d + 1$. Hence, a $1 - \alpha$ confidence interval for the true error rate is $\widehat{L}(\widehat{h}) \pm \epsilon$ where*

$$\epsilon_n^2 = \frac{32}{n} \log \left(\frac{8(n^{d+1} + 1)}{\alpha} \right).$$

22.9   Support Vector Machines

In this section we consider a class of linear classifiers called **support vector machines**. Throughout this section, we assume that Y is binary. It will be convenient to label the outcomes as -1 and $+1$ instead of 0 and 1. A linear classifier can then be written as

$$h(x) = \text{sign}\Big(H(x)\Big)$$

where $x = (x_1, \ldots, x_d)$,

$$H(x) = a_0 + \sum_{i=1}^{d} a_i x_i$$

and

$$\text{sign}(z) = \begin{cases} -1 & \text{if } z < 0 \\ 0 & \text{if } z = 0 \\ 1 & \text{if } z > 0. \end{cases}$$

First, suppose that the data are **linearly separable**, that is, there exists a hyperplane that perfectly separates the two classes.

22.27 Lemma. *The data can be separated by some hyperplane if and only if there exists a hyperplane $H(x) = a_0 + \sum_{i=1}^{d} a_i x_i$ such that*

$$Y_i H(x_i) \geq 1, \quad i = 1, \ldots, n. \tag{22.39}$$

PROOF. Suppose the data can be separated by a hyperplane $W(x) = b_0 + \sum_{i=1}^{d} b_i x_i$. It follows that there exists some constant c such that $Y_i = 1$ implies $W(X_i) \geq c$ and $Y_i = -1$ implies $W(X_i) \leq -c$. Therefore, $Y_i W(X_i) \geq c$ for all i. Let $H(x) = a_0 + \sum_{i=1}^{d} a_i x_i$ where $a_j = b_j / c$. Then $Y_i H(X_i) \geq 1$ for all i. The reverse direction is straightforward. ∎

In the separable case, there will be many separating hyperplanes. How should we choose one? Intuitively, it seems reasonable to choose the hyperplane "furthest" from the data in the sense that it separates the +1s and -1s and maximizes the distance to the closest point. This hyperplane is called the **maximum margin hyperplane.** The margin is the distance to from the hyperplane to the nearest point. Points on the boundary of the margin are called **support vectors.** See Figure 22.8.

22.28 Theorem. *The hyperplane $\widehat{H}(x) = \widehat{a}_0 + \sum_{i=1}^{d} \widehat{a}_i x_i$ that separates the data and maximizes the margin is given by minimizing $(1/2) \sum_{j=1}^{d} a_j^2$ subject to (22.39).*

It turns out that this problem can be recast as a quadratic programming problem. Let $\langle X_i, X_k \rangle = X_i^T X_k$ denote the inner product of X_i and X_k.

22.29 Theorem. *Let $\widehat{H}(x) = \widehat{a}_0 + \sum_{i=1}^{d} \widehat{a}_i x_i$ denote the optimal (largest margin) hyperplane. Then, for $j = 1, \ldots, d$,*

$$\widehat{a}_j = \sum_{i=1}^{n} \widehat{\alpha}_i Y_i X_j(i)$$

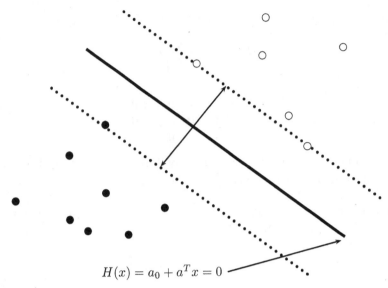

$$H(x) = a_0 + a^T x = 0$$

FIGURE 22.8. The hyperplane $H(x)$ has the largest margin of all hyperplanes that separate the two classes.

where $X_j(i)$ is the value of the covariate X_j for the ith data point, and $\widehat{\alpha} = (\widehat{\alpha}_1, \ldots, \widehat{\alpha}_n)$ is the vector that maximizes

$$\sum_{i=1}^{n} \alpha_i - \frac{1}{2} \sum_{i=1}^{n} \sum_{k=1}^{n} \alpha_i \alpha_k Y_i Y_k \langle X_i, X_k \rangle \qquad (22.40)$$

subject to

$$\alpha_i \geq 0$$

and

$$0 = \sum_i \alpha_i Y_i.$$

The points X_i for which $\widehat{\alpha} \neq 0$ are called **support vectors**. $\widehat{a}_0$ can be found by solving

$$\widehat{\alpha}_i \left(Y_i (X_i^T \widehat{a} + \widehat{\beta}_0) \right) = 0$$

for any support point X_i. $\widehat{H}$ may be written as

$$\widehat{H}(x) = \widehat{a}_0 + \sum_{i=1}^{n} \widehat{\alpha}_i Y_i \langle x, X_i \rangle.$$

There are many software packages that will solve this problem quickly. If there is no perfect linear classifier, then one allows overlap between the groups by replacing the condition (22.39) with

$$Y_i H(x_i) \geq 1 - \xi_i, \quad \xi_i \geq 0, \quad i = 1, \ldots, n. \tag{22.41}$$

The variables $\xi_1, \ldots, \xi_n$ are called **slack variables**.

We now maximize (22.40) subject to

$$0 \leq \xi_i \leq c, \quad i = 1, \ldots, n$$

and

$$\sum_{i=1}^{n} \alpha_i Y_i = 0.$$

The constant c is a tuning parameter that controls the amount of overlap.

22.10 Kernelization

There is a trick called **kernelization** for improving a computationally simple classifier h. The idea is to map the covariate X — which takes values in $\mathcal{X}$ — into a higher dimensional space $\mathcal{Z}$ and apply the classifier in the bigger space $\mathcal{Z}$. This can yield a more flexible classifier while retaining computationally simplicity.

The standard example of this idea is illustrated in Figure 22.9. The covariate $x = (x_1, x_2)$. The Y_is can be separated into two groups using an ellipse. Define a mapping ϕ by

$$z = (z_1, z_2, z_3) = \phi(x) = (x_1^2, \sqrt{2} x_1 x_2, x_2^2).$$

Thus, ϕ maps $\mathcal{X} = \mathbb{R}^2$ into $\mathcal{Z} = \mathbb{R}^3$. In the higher-dimensional space $\mathcal{Z}$, the Y_i's are separable by a linear decision boundary. In other words,

a linear classifier in a higher-dimensional space corresponds to a nonlinear classifier in the original space.

The point is that to get a richer set of classifiers we do not need to give up the convenience of linear classifiers. We simply map the covariates to a higher-dimensional space. This is akin to making linear regression more flexible by using polynomials.

There is a potential drawback. If we significantly expand the dimension of the problem, we might increase the computational burden. For example, if x has dimension $d = 256$ and we wanted to use all fourth-order terms, then $z = \phi(x)$ has dimension 183,181,376. We are spared this computational

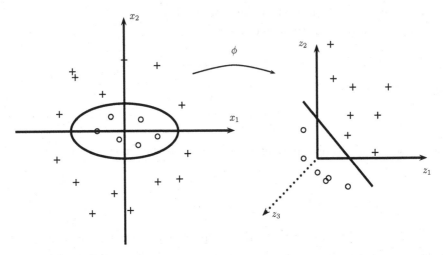

FIGURE 22.9. Kernelization. Mapping the covariates into a higher-dimensional space can make a complicated decision boundary into a simpler decision boundary.

nightmare by the following two facts. First, many classifiers do not require that we know the values of the individual points but, rather, just the inner product between pairs of points. Second, notice in our example that the inner product in $\mathcal{Z}$ can be written

$$
\begin{aligned}
\langle z, \tilde{z} \rangle &= \langle \phi(x), \phi(\tilde{x}) \rangle \\
&= x_1^2 \tilde{x}_1^2 + 2x_1 \tilde{x}_1 x_2 \tilde{x}_2 + x_2^2 \tilde{x}_2^2 \\
&= (\langle x, \tilde{x} \rangle)^2 \equiv K(x, \tilde{x}).
\end{aligned}
$$

Thus, we can compute $\langle z, \tilde{z} \rangle$ without ever computing $Z_i = \phi(X_i)$.

To summarize, kernelization involves finding a mapping $\phi : \mathcal{X} \to \mathcal{Z}$ and a classifier such that:

1. $\mathcal{Z}$ has higher dimension than $\mathcal{X}$ and so leads a richer set of classifiers.

2. The classifier only requires computing inner products.

3. There is a function K, called a kernel, such that $\langle \phi(x), \phi(\tilde{x}) \rangle = K(x, \tilde{x})$.

4. Everywhere the term $\langle x, \tilde{x} \rangle$ appears in the algorithm, replace it with $K(x, \tilde{x})$.

In fact, we never need to construct the mapping ϕ at all. We only need to specify a kernel $K(x, \tilde{x})$ that corresponds to $\langle \phi(x), \phi(\tilde{x}) \rangle$ for some ϕ. This raises an interesting question: given a function of two variables $K(x, y)$, does there exist a function $\phi(x)$ such that $K(x, y) = \langle \phi(x), \phi(y) \rangle$? The answer is provided by **Mercer's theorem** which says, roughly, that if K is positive definite — meaning that

$$\int \int K(x, y) f(x) f(y) dx dy \geq 0$$

for square integrable functions f — then such a ϕ exists. Examples of commonly used kernels are:

$$\begin{aligned}
\text{polynomial} \quad K(x, \tilde{x}) &= \left(\langle x, \tilde{x} \rangle + a \right)^r \\
\text{sigmoid} \quad K(x, \tilde{x}) &= \tanh(a \langle x, \tilde{x} \rangle + b) \\
\text{Gaussian} \quad K(x, \tilde{x}) &= \exp\left(-||x - \tilde{x}||^2 / (2\sigma^2) \right)
\end{aligned}$$

Let us now see how we can use this trick in LDA and in support vector machines.

Recall that the Fisher linear discriminant method replaces X with $U = w^T X$ where w is chosen to maximize the Rayleigh coefficient

$$J(w) = \frac{w^T S_B w}{w^T S_W w},$$

$$S_B = (\overline{X}_0 - \overline{X}_1)(\overline{X}_0 - \overline{X}_1)^T$$

and

$$S_W = \left(\frac{(n_0 - 1) S_0}{(n_0 - 1) + (n_1 - 1)} \right) + \left(\frac{(n_1 - 1) S_1}{(n_0 - 1) + (n_1 - 1)} \right).$$

In the kernelized version, we replace X_i with $Z_i = \phi(X_i)$ and we find w to maximize

$$J(w) = \frac{w^T \widetilde{S}_B w}{w^T \widetilde{S}_W w} \tag{22.42}$$

where

$$\widetilde{S}_B = (\overline{Z}_0 - \overline{Z}_1)(\overline{Z}_0 - \overline{Z}_1)^T$$

and

$$\widetilde{S}_W = \left(\frac{(n_0 - 1) \widetilde{S}_0}{(n_0 - 1) + (n_1 - 1)} \right) + \left(\frac{(n_1 - 1) \widetilde{S}_1}{(n_0 - 1) + (n_1 - 1)} \right).$$

Here, $\widetilde{S}_j$ is the sample of covariance of the Z_i's for which $Y = j$. However, to take advantage of kernelization, we need to re-express this in terms of inner products and then replace the inner products with kernels.

It can be shown that the maximizing vector w is a linear combination of the Z_i's. Hence we can write

$$w = \sum_{i=1}^{n} \alpha_i Z_i.$$

Also,

$$\overline{Z}_j = \frac{1}{n_j} \sum_{i=1}^{n} \phi(X_i) I(Y_i = j).$$

Therefore,

$$
\begin{aligned}
w^T \overline{Z}_j &= \left(\sum_{i=1}^{n} \alpha_i Z_i \right)^T \left(\frac{1}{n_j} \sum_{i=1}^{n} \phi(X_i) I(Y_i = j) \right) \\
&= \frac{1}{n_j} \sum_{i=1}^{n} \sum_{s=1}^{n} \alpha_i I(Y_s = j) Z_i^T \phi(X_s) \\
&= \frac{1}{n_j} \sum_{i=1}^{n} \alpha_i \sum_{s=1}^{n} I(Y_s = j) \phi(X_i)^T \phi(X_s) \\
&= \frac{1}{n_j} \sum_{i=1}^{n} \alpha_i \sum_{s=1}^{n} I(Y_s = j) K(X_i, X_s) \\
&= \alpha^T M_j
\end{aligned}
$$

where M_j is a vector whose i^{th} component is

$$M_j(i) = \frac{1}{n_j} \sum_{s=1}^{n} K(X_i, X_s) I(Y_i = j).$$

It follows that

$$w^T \widetilde{S}_B w = \alpha^T M \alpha$$

where $M = (M_0 - M_1)(M_0 - M_1)^T$. By similar calculations, we can write

$$w^T \widetilde{S}_W w = \alpha^T N \alpha$$

where

$$N = K_0 \left(I - \frac{1}{n_0} \mathbf{1} \right) K_0^T + K_1 \left(I - \frac{1}{n_1} \mathbf{1} \right) K_1^T,$$

I is the identity matrix, $\mathbf{1}$ is a matrix of all one's, and K_j is the $n \times n_j$ matrix with entries $(K_j)_{rs} = K(x_r, x_s)$ with x_s varying over the observations in group j. Hence, we now find α to maximize

$$J(\alpha) = \frac{\alpha^T M \alpha}{\alpha^T N \alpha}.$$

All the quantities are expressed in terms of the kernel. Formally, the solution is $\alpha = N^{-1}(M_0 - M_1)$. However, N might be non-invertible. In this case one replaces N by $N + bI$, for some constant b. Finally, the projection onto the new subspace can be written as

$$U = w^T \phi(x) = \sum_{i=1}^{n} \alpha_i K(x_i, x).$$

The support vector machine can similarly be kernelized. We simply replace $\langle X_i, X_j \rangle$ with $K(X_i, X_j)$. For example, instead of maximizing (22.40), we now maximize

$$\sum_{i=1}^{n} \alpha_i - \frac{1}{2} \sum_{i=1}^{n} \sum_{k=1}^{n} \alpha_i \alpha_k Y_i Y_k K(X_i, X_j). \tag{22.43}$$

The hyperplane can be written as $\widehat{H}(x) = \widehat{a}_0 + \sum_{i=1}^{n} \widehat{a}_i Y_i K(X, X_i)$.

22.11   Other Classifiers

There are many other classifiers and space precludes a full discussion of all of them. Let us briefly mention a few.

The **k-nearest-neighbors** classifier is very simple. Given a point x, find the k data points closest to x. Classify x using the majority vote of these k neighbors. Ties can be broken randomly. The parameter k can be chosen by cross-validation.

Bagging is a method for reducing the variability of a classifier. It is most helpful for highly nonlinear classifiers such as trees. We draw B bootstrap samples from the data. The bth bootstrap sample yields a classifier h_b. The final classifier is

$$\widehat{h}(x) = \begin{cases} 1 & \text{if } \frac{1}{B} \sum_{b=1}^{B} h_b(x) \geq \frac{1}{2} \\ 0 & \text{otherwise.} \end{cases}$$

Boosting is a method for starting with a simple classifier and gradually improving it by refitting the data giving higher weight to misclassified samples. Suppose that $\mathcal{H}$ is a collection of classifiers, for example, trees with only one split. Assume that $Y_i \in \{-1, 1\}$ and that each h is such that $h(x) \in \{-1, 1\}$. We usually give equal weight to all data points in the methods we have discussed. But one can incorporate unequal weights quite easily in most algorithms. For example, in constructing a tree, we could replace the impurity measure with a weighted impurity measure. The original version of boosting, called AdaBoost, is as follows.

1. Set the weights $w_i = 1/n$, $i = 1, \ldots, n$.

2. For $j = 1, \ldots, J$, do the following steps:

    (a) Constructing a classifier h_j from the data using the weights $w_1, \ldots, w_n$.

    (b) Compute the weighted error estimate:

    $$\widehat{L}_j = \frac{\sum_{i=1}^n w_i I(Y_i \neq h_j(X_i))}{\sum_{i=1}^n w_i}.$$

    (c) Let $\alpha_j = \log((1 - \widehat{L}_j)/\widehat{L}_j)$.

    (d) Update the weights:

    $$w_i \longleftarrow w_i e^{\alpha_j I(Y_i \neq h_j(X_i))}$$

3. The final classifier is

$$\widehat{h}(x) = \text{sign}\left(\sum_{j=1}^J \alpha_j h_j(x)\right).$$

There is now an enormous literature trying to explain and improve on boosting. Whereas bagging is a variance reduction technique, boosting can be thought of as a bias reduction technique. We starting with a simple — and hence highly-biased — classifier, and we gradually reduce the bias. The disadvantage of boosting is that the final classifier is quite complicated.

Neural Networks are regression models of the form [3]

$$Y = \beta_0 + \sum_{j=1}^p \beta_j \sigma(\alpha_0 + \alpha^T X)$$

where σ is a smooth function, often taken to be $\sigma(v) = e^v/(1 + e^v)$. This is really nothing more than a nonlinear regression model. Neural nets were fashionable for some time but they pose great computational difficulties. In particular, one often encounters multiple minima when trying to find the least squares estimates of the parameters. Also, the number of terms p is essentially a smoothing parameter and there is the usual problem of trying to choose p to find a good balance between bias and variance.

---

[3] This is the simplest version of a neural net. There are more complex versions of the model.

22.12   Bibliographic Remarks

The literature on classification is vast and is growing quickly. An excellent reference is Hastie et al. (2001). For more on the theory, see Devroye et al. (1996) and Vapnik (1998). Two recent books on kernels are Scholkopf and Smola (2002) and Herbich (2002).

22.13   Exercises

1. Prove Theorem 22.5.

2. Prove Theorem 22.7.

3. Download the spam data from:

   http://www-stat.stanford.edu/~tibs/ElemStatLearn/index.html

   The data file can also be found on the course web page. The data contain 57 covariates relating to email messages. Each email message was classified as spam (Y=1) or not spam (Y=0). The outcome Y is the last column in the file. The goal is to predict whether an email is spam or not.

   (a) Construct classification rules using (i) LDA, (ii) QDA, (iii) logistic regression, and (iv) a classification tree. For each, report the observed misclassification error rate and construct a 2-by-2 table of the form

	$\widehat{h}(x) = 0$	$\widehat{h}(x) = 1$
$Y = 0$	??	??
$Y = 1$	??	??

   (b) Use 5-fold cross-validation to estimate the prediction accuracy of LDA and logistic regression.

   (c) Sometimes it helps to reduce the number of covariates. One strategy is to compare X_i for the spam and email group. For each of the 57 covariates, test whether the mean of the covariate is the same or different between the two groups. Keep the 10 covariates with the smallest p-values. Try LDA and logistic regression using only these 10 variables.

4. Let $\mathcal{A}$ be the set of two-dimensional spheres. That is, $A \in \mathcal{A}$ if $A = \{(x, y) : (x-a)^2 + (y-b)^2 \leq c^2\}$ for some a, b, c. Find the VC-dimension of $\mathcal{A}$.

5. Classify the spam data using support vector machines. Free software for the support vector machine is at http://svmlight.joachims.org/

6. Use VC theory to get a confidence interval on the true error rate of the LDA classifier for the iris data (from the book web site).

7. Suppose that $X_i \in \mathbb{R}$ and that $Y_i = 1$ whenever $|X_i| \leq 1$ and $Y_i = 0$ whenever $|X_i| > 1$. Show that no linear classifier can perfectly classify these data. Show that the kernelized data $Z_i = (X_i, X_i^2)$ can be linearly separated.

8. Repeat question 5 using the kernel $K(x, \tilde{x}) = (1 + x^T \tilde{x})^p$. Choose p by cross-validation.

9. Apply the k nearest neighbors classifier to the "iris data." Choose k by cross-validation.

10. (Curse of Dimensionality.) Suppose that X has a uniform distribution on the d-dimensional cube $[-1/2, 1/2]^d$. Let R be the distance from the origin to the closest neighbor. Show that the median of R is

$$\left(\frac{\left(1 - \left(\frac{1}{2}\right)^{1/n}\right)^{1/d}}{v_d(1)} \right)^{1/d}$$

where

$$v_d(r) = r^d \frac{\pi^{d/2}}{\Gamma((d/2) + 1)}$$

is the volume of a sphere of radius r. For what dimension d does the median of R exceed the edge of the cube when $n = 100$, $n = 1,000$, $n = 10,000$? (Hastie et al. (2001), p. 22–27.)

11. Fit a tree to the data in question 3. Now apply bagging and report your results.

12. Fit a tree that uses only one split on one variable to the data in question 3. Now apply boosting.

13. Let $r(x) = \mathbb{P}(Y = 1|X = x)$ and let $\widehat{r}(x)$ be an estimate of $r(x)$. Consider the classifier

$$h(x) = \begin{cases} 1 & \text{if } \widehat{r}(x) \geq 1/2 \\ 0 & \text{otherwise.} \end{cases}$$

Assume that $\widehat{r}(x) \approx N(\overline{r}(x), \sigma^2(x))$ for some functions $\overline{r}(x)$ and $\sigma^2(x)$. Show that, for fixed x,

$$\mathbb{P}(Y \neq h(x)) \approx \mathbb{P}(Y \neq h^*(x))$$

$$+ \left| 2r(x) - 1 \right| \times \left[1 - \Phi\left(\frac{\text{sign}\left(r(x) - (1/2) \right)\left(\overline{r}(x) - (1/2)\right)}{\sigma(x)} \right) \right]$$

where Φ is the standard Normal CDF and h^* is the Bayes rule. Regard $\text{sign}\left((r(x) - (1/2))(\overline{r}(x) - (1/2)) \right)$ as a type of bias term. Explain the implications for the bias–variance tradeoff in classification (Friedman (1997)).

Hint: first show that

$$\mathbb{P}(Y \neq h(x)) = |2r(x) - 1|\mathbb{P}(h(x) \neq h^*(x)) + \mathbb{P}(Y \neq h^*(x)).$$

23

Probability Redux: Stochastic Processes

23.1 Introduction

Most of this book has focused on IID sequences of random variables. Now we consider sequences of dependent random variables. For example, daily temperatures will form a sequence of time-ordered random variables and clearly the temperature on one day is not independent of the temperature on the previous day.

A **stochastic process** $\{X_t : t \in T\}$ is a collection of random variables. We shall sometimes write $X(t)$ instead of X_t. The variables X_t take values in some set $\mathcal{X}$ called the **state space**. The set T is called the **index set** and for our purposes can be thought of as time. The index set can be discrete $T = \{0, 1, 2, \ldots\}$ or continuous $T = [0, \infty)$ depending on the application.

23.1 Example (IID observations). A sequence of IID random variables can be written as $\{X_t : t \in T\}$ where $T = \{1, 2, 3, \ldots, \}$. Thus, a sequence of IID random variables is an example of a stochastic process. ∎

23.2 Example (The Weather). Let $\mathcal{X} = \{\text{sunny}, \text{cloudy}\}$. A typical sequence (depending on where you live) might be

$$\text{sunny}, \text{sunny}, \text{cloudy}, \text{sunny}, \text{cloudy}, \text{cloudy}, \cdots$$

This process has a discrete state space and a discrete index set. ∎

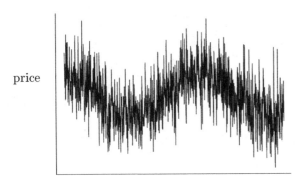

time

FIGURE 23.1. Stock price over ten week period.

23.3 Example (Stock Prices). Figure 23.1 shows the price of a fictitious stock over time. The price is monitored continuously so the index set T is continuous. Price is discrete but for all practical purposes we can treat it as a continuous variable. ∎

23.4 Example (Empirical Distribution Function). Let $X_1, \ldots, X_n \sim F$ where F is some CDF on $[0,1]$. Let

$$\widehat{F}_n(t) = \frac{1}{n} \sum_{i=1}^{n} I(X_i \le t)$$

be the empirical CDF. For any fixed value t, $\widehat{F}_n(t)$ is a random variable. But the whole empirical CDF

$$\left\{ \widehat{F}_n(t) : \ t \in [0,1] \right\}$$

is a stochastic process with a continuous state space and a continuous index set. ∎

We end this section by recalling a basic fact. If $X_1, \ldots, X_n$ are random variables, then we can write the joint density as

$$
\begin{aligned}
f(x_1, \ldots, x_n) &= f(x_1) f(x_2 | x_1) \cdots f(x_n | x_1, \ldots, x_{n-1}) \\
&= \prod_{i=1}^{n} f(x_i | \text{past}_i) \qquad (23.1)
\end{aligned}
$$

where $past_i = (X_1, \ldots, X_{i-1})$.

23.2   Markov Chains

A Markov chain is a stochastic process for which the distribution of X_t depends only on X_{t-1}. In this section we assume that the state space is discrete, either $\mathcal{X} = \{1, \ldots, N\}$ or $\mathcal{X} = \{1, 2, \ldots, \}$ and that the index set is $T = \{0, 1, 2, \ldots\}$. Typically, most authors write X_n instead of X_t when discussing Markov chains and I will do so as well.

23.5 Definition. *The process* $\{X_n : n \in T\}$ *is a* **Markov chain** *if*

$$\mathbb{P}(X_n = x \mid X_0, \ldots, X_{n-1}) = \mathbb{P}(X_n = x \mid X_{n-1}) \tag{23.2}$$

for all n and for all $x \in \mathcal{X}$.

For a Markov chain, equation (23.1) simplifies to

$$f(x_1, \ldots, x_n) = f(x_1)f(x_2|x_1)f(x_3|x_2)\cdots f(x_n|x_{n-1}).$$

A Markov chain can be represented by the following DAG:

$$X_0 \longrightarrow X_1 \longrightarrow X_2 \longrightarrow \quad \cdots \quad \longrightarrow X_n \longrightarrow \quad \cdots$$

Each variable has a single parent, namely, the previous observation.

The theory of Markov chains is a very rich and complex. We have to get through many definitions before we can do anything interesting. Our goal is to answer the following questions:

1. When does a Markov chain "settle down" into some sort of equilibrium?

2. How do we estimate the parameters of a Markov chain?

3. How can we construct Markov chains that converge to a given equilibrium distribution and why would we want to do that?

We will answer questions 1 and 2 in this chapter. We will answer question 3 in the next chapter. To understand question 1, look at the two chains in Figure 23.2. The first chain oscillates all over the place and will continue to do so forever. The second chain eventually settles into an equilibrium. If we constructed a histogram of the first process, it would keep changing as we got more and more observations. But a histogram from the second chain would eventually converge to some fixed distribution.

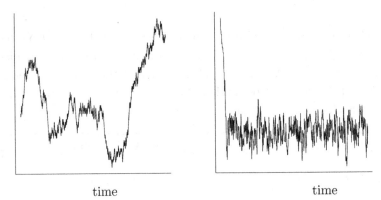

time   time

FIGURE 23.2. Two Markov chains. The first chain does not settle down into an equilibrium. The second does.

TRANSITION PROBABILITIES. The key quantities of a Markov chain are the probabilities of jumping from one state into another state. A Markov chain is **homogeneous** if $\mathbb{P}(X_{n+1} = j | X_n = i)$ does not change with time. Thus, for a homogeneous Markov chain, $\mathbb{P}(X_{n+1} = j | X_n = i) = \mathbb{P}(X_1 = j | X_0 = i)$. We shall only deal with homogeneous Markov chains.

23.6 Definition. *We call*

$$p_{ij} \equiv \mathbb{P}(X_{n+1} = j | X_n = i) \qquad (23.3)$$

the **transition probabilities.** *The matrix* $\mathbf{P}$ *whose* (i, j) *element is* p_{ij} *is called the* **transition matrix.**

We will only consider homogeneous chains. Notice that $\mathbf{P}$ has two properties: (i) $p_{ij} \geq 0$ and (ii) $\sum_i p_{ij} = 1$. Each row can be regarded as a probability mass function.

23.7 Example (Random Walk With Absorbing Barriers). Let $\mathcal{X} = \{1, \ldots, N\}$. Suppose you are standing at one of these points. Flip a coin with $\mathbb{P}(\text{Heads}) = p$ and $\mathbb{P}(\text{Tails}) = q = 1 - p$. If it is heads, take one step to the right. If it is tails, take one step to the left. If you hit one of the endpoints, stay there. The transition matrix is

$$\mathbf{P} = \begin{bmatrix} 1 & 0 & 0 & 0 & \cdots & 0 & 0 \\ q & 0 & p & 0 & \cdots & 0 & 0 \\ 0 & q & 0 & p & \cdots & 0 & 0 \\ \cdots & \cdots & \cdots & \cdots & \cdots & \cdots & \cdots \\ 0 & 0 & 0 & 0 & q & 0 & p \\ 0 & 0 & 0 & 0 & 0 & 0 & 1 \end{bmatrix}. \quad \blacksquare$$

23.8 Example. Suppose the state space is $\mathcal{X} = \{\text{sunny}, \text{cloudy}\}$. Then X_1, $X_2, \ldots$ represents the weather for a sequence of days. The weather today clearly depends on yesterday's weather. It might also depend on the weather two days ago but as a first approximation we might assume that the dependence is only one day back. In that case the weather is a Markov chain and a typical transition matrix might be

	Sunny	Cloudy
Sunny	0.4	0.6
Cloudy	0.8	0.2

For example, if it is sunny today, there is a 60 per cent chance it will be cloudy tomorrow. $\blacksquare$

Let

$$p_{ij}(n) = \mathbb{P}(X_{m+n} = j | X_m = i) \quad (23.4)$$

be the probability of of going from state i to state j in n steps. Let $\mathbf{P}_n$ be the matrix whose (i, j) element is $p_{ij}(n)$. These are called the **n-step transition probabilities.**

23.9 Theorem (The Chapman-Kolmogorov equations). *The n-step probabilities satisfy*

$$p_{ij}(m + n) = \sum_k p_{ik}(m) p_{kj}(n). \quad (23.5)$$

PROOF. Recall that, in general,

$$\mathbb{P}(X = x, Y = y) = \mathbb{P}(X = x)\mathbb{P}(Y = y | X = x).$$

This fact is true in the more general form

$$\mathbb{P}(X = x, Y = y | Z = z) = \mathbb{P}(X = x | Z = z)\mathbb{P}(Y = y | X = x, Z = z).$$

Also, recall the law of total probability:

$$\mathbb{P}(X = x) = \sum_y \mathbb{P}(X = x, Y = y).$$

Using these facts and the Markov property we have

$$
\begin{aligned}
p_{ij}(m+n) &= \mathbb{P}(X_{m+n} = j | X_0 = i) \\
&= \sum_k \mathbb{P}(X_{m+n} = j, X_m = k | X_0 = i) \\
&= \sum_k \mathbb{P}(X_{m+n} = j | X_m = k, X_0 = i) \mathbb{P}(X_m = k | X_0 = i) \\
&= \sum_k \mathbb{P}(X_{m+n} = j | X_m = k) \mathbb{P}(X_m = k | X_0 = i) \\
&= \sum_k p_{ik}(m) p_{kj}(n). \quad \blacksquare
\end{aligned}
$$

Look closely at equation (23.5). This is nothing more than the equation for matrix multiplication. Hence we have shown that

$$
\mathbf{P}_{m+n} = \mathbf{P}_m \mathbf{P}_n. \tag{23.6}
$$

By definition, $\mathbf{P}_1 = \mathbf{P}$. Using the above theorem, $\mathbf{P}_2 = \mathbf{P}_{1+1} = \mathbf{P}_1 \mathbf{P}_1 = \mathbf{PP} = \mathbf{P}^2$. Continuing this way, we see that

$$
\mathbf{P}_n = \mathbf{P}^n \equiv \underbrace{\mathbf{P} \times \mathbf{P} \times \cdots \times \mathbf{P}}_{\text{multiply the matrix n times}}. \tag{23.7}
$$

Let $\mu_n = (\mu_n(1), \ldots, \mu_n(N))$ be a row vector where

$$
\mu_n(i) = \mathbb{P}(X_n = i) \tag{23.8}
$$

is the marginal probability that the chain is in state i at time n. In particular, μ_0 is called the **initial distribution**. To simulate a Markov chain, all you need to know is μ_0 and $\mathbf{P}$. The simulation would look like this:

Step 1: Draw $X_0 \sim \mu_0$. Thus, $\mathbb{P}(X_0 = i) = \mu_0(i)$.
Step 2: Denote the outcome of step 1 by i. Draw $X_1 \sim \mathbf{P}$. In other words, $\mathbb{P}(X_1 = j | X_0 = i) = p_{ij}$.
Step 3: Suppose the outcome of step 2 is j. Draw $X_2 \sim \mathbf{P}$. In other words, $\mathbb{P}(X_2 = k | X_1 = j) = p_{jk}$.
    And so on.

It might be difficult to understand the meaning of μ_n. Imagine simulating the chain many times. Collect all the outcomes at time n from all the chains. This histogram would look approximately like μ_n. A consequence of theorem 23.9 is the following:

23.10 Lemma. *The marginal probabilities are given by*

$$\mu_n = \mu_0 \mathbf{P}^n.$$

PROOF.

$$
\begin{aligned}
\mu_n(j) &= \mathbb{P}(X_n = j) \\
&= \sum_i \mathbb{P}(X_n = j | X_0 = i) P(X_0 = i) \\
&= \sum_i \mu_0(i) p_{ij}(n) = \mu_0 \mathbf{P}^n. \quad \blacksquare
\end{aligned}
$$

Summary of Terminology

1. Transition matrix: $\mathbf{P}(i,j) = \mathbb{P}(X_{n+1} = j | X_n = i) = p_{ij}$.

2. n-step matrix: $\mathbf{P}_n(i,j) = \mathbb{P}(X_{n+m} = j | X_m = i)$.

3. $\mathbf{P}_n = \mathbf{P}^n$.

4. Marginal: $\mu_n(i) = \mathbb{P}(X_n = i)$.

5. $\mu_n = \mu_0 \mathbf{P}^n$.

STATES. The states of a Markov chain can be classified according to various properties.

23.11 Definition. *We say that i reaches j (or j is accessible from i) if $p_{ij}(n) > 0$ for some n, and we write $i \to j$. If $i \to j$ and $j \to i$ then we write $i \leftrightarrow j$ and we say that i and j communicate.*

23.12 Theorem. *The communication relation satisfies the following properties:*

1. $i \leftrightarrow i$.

2. *If $i \leftrightarrow j$ then $j \leftrightarrow i$.*

3. *If $i \leftrightarrow j$ and $j \leftrightarrow k$ then $i \leftrightarrow k$.*

4. *The set of states $\mathcal{X}$ can be written as a disjoint union of **classes** $\mathcal{X} = \mathcal{X}_1 \bigcup \mathcal{X}_2 \bigcup \cdots$ where two states i and j communicate with each other if and only if they are in the same class.*

If all states communicate with each other, then the chain is called **irreducible**. A set of states is **closed** if, once you enter that set of states you never leave. A closed set consisting of a single state is called an **absorbing state**.

23.13 Example. Let $\mathcal{X} = \{1, 2, 3, 4\}$ and

$$\mathbf{P} = \begin{pmatrix} \frac{1}{3} & \frac{2}{3} & 0 & 0 \\ \frac{2}{3} & \frac{1}{3} & 0 & 0 \\ \frac{1}{4} & \frac{1}{4} & \frac{1}{4} & \frac{1}{4} \\ 0 & 0 & 0 & 1 \end{pmatrix}$$

The classes are $\{1, 2\}, \{3\}$ and $\{4\}$. State 4 is an absorbing state. ∎

Suppose we start a chain in state i. Will the chain ever return to state i? If so, that state is called persistent or recurrent.

23.14 Definition. *State i is* **recurrent** *or* **persistent** *if*

$$\mathbb{P}(X_n = i \text{ for some } n \geq 1 \mid X_0 = i) = 1.$$

Otherwise, state i is **transient**.

23.15 Theorem. *A state i is recurrent if and only if*

$$\sum_n p_{ii}(n) = \infty. \tag{23.9}$$

A state i is transient if and only if

$$\sum_n p_{ii}(n) < \infty. \tag{23.10}$$

PROOF. Define

$$I_n = \begin{cases} 1 & \text{if } X_n = i \\ 0 & \text{if } X_n \neq i. \end{cases}$$

The number of times that the chain is in state i is $Y = \sum_{n=0}^{\infty} I_n$. The mean of Y, given that the chain starts in state i, is

$$\mathbb{E}(Y \mid X_0 = i) = \sum_{n=0}^{\infty} \mathbb{E}(I_n \mid X_0 = i) = \sum_{n=0}^{\infty} \mathbb{P}(X_n = i \mid X_0 = i) = \sum_{n=0}^{\infty} p_{ii}(n).$$

Define $a_i = \mathbb{P}(X_n = i \text{ for some } n \geq 1 \mid X_0 = i)$. If i is recurrent, $a_i = 1$. Thus, the chain will eventually return to i. Once it does return to i, we argue again that since $a_i = 1$, the chain will return to state i again. By repeating this argument, we conclude that $\mathbb{E}(Y|X_0 = i) = \infty$. If i is transient, then $a_i < 1$. When the chain is in state i, there is a probability $1 - a_i > 0$ that it will never return to state i. Thus, the probability that the chain is in state i exactly n times is $a_i^{n-1}(1 - a_i)$. This is a geometric distribution which has finite mean. ∎

23.16 Theorem. *Facts about recurrence.*

1. *If state i is recurrent and $i \leftrightarrow j$, then j is recurrent.*

2. *If state i is transient and $i \leftrightarrow j$, then j is transient.*

3. *A finite Markov chain must have at least one recurrent state.*

4. *The states of a finite, irreducible Markov chain are all recurrent.*

23.17 Theorem (Decomposition Theorem). *The state space $\mathcal{X}$ can be written as the disjoint union*

$$\mathcal{X} = \mathcal{X}_T \bigcup \mathcal{X}_1 \bigcup \mathcal{X}_2 \cdots$$

where $\mathcal{X}_T$ are the transient states and each $\mathcal{X}_i$ is a closed, irreducible set of recurrent states.

23.18 Example (Random Walk). *Let $\mathcal{X} = \{\ldots, -2, -1, 0, 1, 2, \ldots, \}$ and suppose that $p_{i,i+1} = p$, $p_{i,i-1} = q = 1 - p$. All states communicate, hence either all the states are recurrent or all are transient. To see which, suppose we start at $X_0 = 0$. Note that*

$$p_{00}(2n) = \binom{2n}{n} p^n q^n \tag{23.11}$$

since the only way to get back to 0 is to have n heads (steps to the right) and n tails (steps to the left). We can approximate this expression using Stirling's formula which says that

$$n! \sim n^n \sqrt{n} e^{-n} \sqrt{2\pi}.$$

Inserting this approximation into (23.11) shows that

$$p_{00}(2n) \sim \frac{(4pq)^n}{\sqrt{n\pi}}.$$

It is easy to check that $\sum_n p_{00}(n) < \infty$ if and only if $\sum_n p_{00}(2n) < \infty$. Moreover, $\sum_n p_{00}(2n) = \infty$ if and only if $p = q = 1/2$. By Theorem (23.15), the chain is recurrent if $p = 1/2$ otherwise it is transient. ∎

CONVERGENCE OF MARKOV CHAINS. To discuss the convergence of chains, we need a few more definitions. Suppose that $X_0 = i$. Define the **recurrence time**

$$T_{ij} = \min\{n > 0 : X_n = j\} \tag{23.12}$$

assuming X_n ever returns to state i, otherwise define $T_{ij} = \infty$. The **mean recurrence time** of a recurrent state i is

$$m_i = \mathbb{E}(T_{ii}) = \sum_n n f_{ii}(n) \tag{23.13}$$

where

$$f_{ij}(n) = \mathbb{P}(X_1 \neq j, X_2 \neq j, \dots, X_{n-1} \neq j, X_n = j | X_0 = i).$$

A recurrent state is **null** if $m_i = \infty$ otherwise it is called **non-null** or **positive**.

23.19 Lemma. *If a state is null and recurrent, then $p_{ii}^n \to 0$.*

23.20 Lemma. *In a finite state Markov chain, all recurrent states are positive.*

Consider a three-state chain with transition matrix

$$\begin{bmatrix} 0 & 1 & 0 \\ 0 & 0 & 1 \\ 1 & 0 & 0 \end{bmatrix}.$$

Suppose we start the chain in state 1. Then we will be in state 3 at times 3, 6, 9, ... This is an example of a periodic chain. Formally, the **period** of state i is d if $p_{ii}(n) = 0$ whenever n is not divisible by d and d is the largest integer with this property. Thus, $d = \gcd\{n : p_{ii}(n) > 0\}$ where gcd means "greater common divisor." State i is **periodic** if $d(i) > 1$ and **aperiodic** if $d(i) = 1$. A state with period 1 is called **aperiodic**.

23.21 Lemma. *If state i has period d and $i \leftrightarrow j$ then j has period d.*

23.22 Definition. *A state is **ergodic** if it is recurrent, non-null and aperiodic. A chain is ergodic if all its states are ergodic.*

Let $\pi = (\pi_i : i \in \mathcal{X})$ be a vector of non-negative numbers that sum to one. Thus π can be thought of as a probability mass function.

23.23 Definition. *We say that π is a **stationary** (or **invariant**) distribution if $\pi = \pi \mathbf{P}$.*

Here is the intuition. Draw X_0 from distribution π and suppose that π is a stationary distribution. Now draw X_1 according to the transition probability of the chain. The distribution of X_1 is then $\mu_1 = \mu_0 \mathbf{P} = \pi \mathbf{P} = \pi$. The distribution of X_2 is $\pi \mathbf{P}^2 = (\pi \mathbf{P})\mathbf{P} = \pi \mathbf{P} = \pi$. Continuing this way, we see that the distribution of X_n is $\pi \mathbf{P}^n = \pi$. In other words:

If at any time the chain has distribution π, then it will continue to have distribution π forever.

23.24 Definition. *We say that a chain has* **limiting distribution** *if*

$$\mathbf{P}^n \to \begin{bmatrix} \pi \\ \pi \\ \vdots \\ \pi \end{bmatrix}$$

for some π, that is, $\pi_j = \lim_{n\to\infty} \mathbf{P}^n_{ij}$ exists and is independent of i.

Here is the main theorem about convergence. The theorem says that an ergodic chain converges to its stationary distribution. Also, sample averages converge to their theoretical expectations under the stationary distribution.

23.25 Theorem. *An irreducible, ergodic Markov chain has a unique stationary distribution π. The limiting distribution exists and is equal to π. If g is any bounded function, then, with probability 1,*

$$\lim_{N\to\infty} \frac{1}{N} \sum_{n=1}^{N} g(X_n) \to \mathbb{E}_\pi(g) \equiv \sum_j g(j)\pi_j. \qquad (23.14)$$

Finally, there is another definition that will be useful later. We say that π satisfies **detailed balance** if

$$\pi_i p_{ij} = p_{ji}\pi_j. \qquad (23.15)$$

Detailed balance guarantees that π is a stationary distribution.

23.26 Theorem. *If π satisfies detailed balance, then π is a stationary distribution.*

PROOF. We need to show that $\pi \mathbf{P} = \pi$. The jth element of $\pi \mathbf{P}$ is $\sum_i \pi_i p_{ij} = \sum_i \pi_j p_{ji} = \pi_j \sum_i p_{ji} = \pi_j$. ∎

The importance of detailed balance will become clear when we discuss Markov chain Monte Carlo methods in Chapter 24.

Warning! Just because a chain has a stationary distribution does not mean it converges.

23.27 Example. Let

$$\mathbf{P} = \begin{bmatrix} 0 & 1 & 0 \\ 0 & 0 & 1 \\ 1 & 0 & 0 \end{bmatrix}.$$

Let $\pi = (1/3, 1/3, 1/3)$. Then $\pi P = \pi$ so π is a stationary distribution. If the chain is started with the distribution π it will stay in that distribution. Imagine simulating many chains and checking the marginal distribution at each time n. It will always be the uniform distribution π. But this chain does not have a limit. It continues to cycle around forever. ∎

EXAMPLES OF MARKOV CHAINS.

23.28 Example. Let $\mathcal{X} = \{1, 2, 3, 4, 5, 6\}$. Let

$$\mathbf{P} = \begin{bmatrix} \frac{1}{2} & \frac{1}{2} & 0 & 0 & 0 & 0 \\ \frac{1}{4} & \frac{3}{4} & 0 & 0 & 0 & 0 \\ \frac{1}{4} & \frac{1}{4} & \frac{1}{4} & \frac{1}{4} & 0 & 0 \\ \frac{1}{4} & 0 & \frac{1}{4} & \frac{1}{4} & 0 & \frac{1}{4} \\ 0 & 0 & 0 & 0 & \frac{1}{2} & \frac{1}{2} \\ 0 & 0 & 0 & 0 & \frac{1}{2} & \frac{1}{2} \end{bmatrix}$$

Then $C_1 = \{1, 2\}$ and $C_2 = \{5, 6\}$ are irreducible closed sets. States 3 and 4 are transient because of the path $3 \to 4 \to 6$ and once you hit state 6 you cannot return to 3 or 4. Since $p_{ii}(1) > 0$, all the states are aperiodic. In summary, 3 and 4 are transient while 1, 2, 5, and 6 are ergodic. ∎

23.29 Example (Hardy-Weinberg). Here is a famous example from genetics. Suppose a gene can be type A or type a. There are three types of people (called genotypes): AA, Aa, and aa. Let (p, q, r) denote the fraction of people of each genotype. We assume that everyone contributes one of their two copies of the gene at random to their children. We also assume that mates are selected at random. The latter is not realistic however, it is often reasonable to assume that you do not choose your mate based on whether they are AA, Aa, or aa. (This would be false if the gene was for eye color and if people chose mates based on eye color.) Imagine if we pooled everyone's genes together. The proportion of A genes is $P = p + (q/2)$ and the proportion of a genes is

$Q = r + (q/2)$. A child is AA with probability P^2, aA with probability $2PQ$, and aa with probability Q^2. Thus, the fraction of A genes in this generation is

$$P^2 + PQ = \left(p + \frac{q}{2}\right)^2 + \left(p + \frac{q}{2}\right)\left(r + \frac{q}{2}\right).$$

However, $r = 1 - p - q$. Substitute this in the above equation and you get $P^2 + PQ = P$. A similar calculation shows that the fraction of "a" genes is Q. We have shown that the proportion of type A and type a is P and Q and this remains stable after the first generation. The proportion of people of type AA, Aa, aa is thus $(P^2, 2PQ, Q^2)$ from the second generation and on. This is called the Hardy-Weinberg law.

Assume everyone has exactly one child. Now consider a fixed person and let X_n be the genotype of their n^{th} descendant. This is a Markov chain with state space $\mathcal{X} = \{AA, Aa, aa\}$. Some basic calculations will show you that the transition matrix is

$$\begin{bmatrix} P & Q & 0 \\ \frac{P}{2} & \frac{P+Q}{2} & \frac{Q}{2} \\ 0 & P & Q \end{bmatrix}.$$

The stationary distribution is $\pi = (P^2, 2PQ, Q^2)$. ∎

23.30 Example (Markov chain Monte Carlo). In Chapter 24 we will present a simulation method called Markov chain Monte Carlo (MCMC). Here is a brief description of the idea. Let $f(x)$ be a probability density on the real line and suppose that $f(x) = cg(x)$ where $g(x)$ is a known function and $c > 0$ is unknown. In principle, we can compute c since $\int f(x)dx = 1$ implies that $c = 1/\int g(x)dx$. However, it may not be feasible to perform this integral, nor is it necessary to know c in the following algorithm. Let X_0 be an arbitrary starting value. Given $X_0, \ldots, X_i$, draw X_{i+1} as follows. First, draw $W \sim N(X_i, b^2)$ where $b > 0$ is some fixed constant. Let

$$r = \min\left\{\frac{g(W)}{g(X_i)}, 1\right\}.$$

Draw $U \sim \text{Uniform}(0, 1)$ and set

$$X_{i+1} = \begin{cases} W & \text{if } U < r \\ X_i & \text{if } U \geq r. \end{cases}$$

We will see in Chapter 24 that, under weak conditions, $X_0, X_1, \ldots$, is an ergodic Markov chain with stationary distribution f. Hence, we can regard the draws as a sample from f. ∎

INFERENCE FOR MARKOV CHAINS. Consider a chain with finite state space $\mathcal{X} = \{1, 2, \ldots, N\}$. Suppose we observe n observations $X_1, \ldots, X_n$ from this chain. The unknown parameters of a Markov chain are the initial probabilities $\mu_0 = (\mu_0(1), \mu_0(2), \ldots,)$ and the elements of the transition matrix $\mathbf{P}$. Each row of $\mathbf{P}$ is a multinomial distribution. So we are essentially estimating N distributions (plus the initial probabilities). Let n_{ij} be the observed number of transitions from state i to state j. The likelihood function is

$$\mathcal{L}(\mu_0, \mathbf{P}) = \mu_0(x_0) \prod_{r=1}^{n} p_{X_{r-1}, X_r} = \mu_0(x_0) \prod_{i=1}^{N} \prod_{j=1}^{N} p_{ij}^{n_{ij}}.$$

There is only one observation on μ_0 so we can't estimate that. Rather, we focus on estimating $\mathbf{P}$. The MLE is obtained by maximizing $\mathcal{L}(\mu_0, \mathbf{P})$ subject to the constraint that the elements are non-negative and the rows sum to 1. The solution is

$$\widehat{p}_{ij} = \frac{n_{ij}}{n_i}$$

where $n_i = \sum_{j=1}^{N} n_{ij}$. Here we are assuming that $n_i > 0$. If not, then we set $\widehat{p}_{ij} = 0$ by convention.

23.31 Theorem (Consistency and Asymptotic Normality of the MLE). *Assume that the chain is ergodic. Let $\widehat{p}_{ij}(n)$ denote the MLE after n observations. Then $\widehat{p}_{ij}(n) \xrightarrow{\text{P}} p_{ij}$. Also,*

$$\left[\sqrt{N_i(n)}(\widehat{p}_{ij} - p_{ij}) \right] \rightsquigarrow N(0, \Sigma)$$

where the left-hand side is a matrix, $N_i(n) = \sum_{r=1}^{n} I(X_r = i)$ and

$$\Sigma_{ij,k\ell} = \begin{cases} p_{ij}(1 - p_{ij}) & (i, j) = (k, \ell) \\ -p_{ij}p_{i\ell} & i = k, j \neq \ell \\ 0 & \text{otherwise.} \end{cases}$$

23.3   Poisson Processes

The Poisson process arises when we count occurrences of events over time, for example, traffic accidents, radioactive decay, arrival of email messages, etc. As the name suggests, the Poisson process is intimately related to the Poisson distribution. Let's first review the Poisson distribution.

Recall that X has a Poisson distribution with parameter λ — written $X \sim$ Poisson(λ) — if

$$\mathbb{P}(X = x) \equiv p(x; \lambda) = \frac{e^{-\lambda}\lambda^x}{x!}, \quad x = 0, 1, 2, \ldots$$

Also recall that $\mathbb{E}(X) = \lambda$ and $\mathbb{V}(X) = \lambda$. If $X \sim \text{Poisson}(\lambda)$, $Y \sim \text{Poisson}(\nu)$ and $X \amalg Y$, then $X + Y \sim \text{Poisson}(\lambda + \nu)$. Finally, if $N \sim \text{Poisson}(\lambda)$ and $Y|N = n \sim \text{Binomial}(n, p)$, then the marginal distribution of Y is $Y \sim \text{Poisson}(\lambda p)$.

Now we describe the Poisson process. Imagine that you are at your computer. Each time a new email message arrives you record the time. Let X_t be the number of messages you have received up to and including time t. Then, $\{X_t : t \in [0, \infty)\}$ is a stochastic process with state space $\mathcal{X} = \{0, 1, 2, \ldots\}$. A process of this form is called a **counting process**. A Poisson process is a counting process that satisfies certain conditions. In what follows, we will sometimes write $X(t)$ instead of X_t. Also, we need the following notation. Write $f(h) = o(h)$ if $f(h)/h \to 0$ as $h \to 0$. This means that $f(h)$ is smaller than h when h is close to 0. For example, $h^2 = o(h)$.

23.32 Definition. *A* **Poisson process** *is a stochastic process* $\{X_t : t \in [0, \infty)\}$ *with state space* $\mathcal{X} = \{0, 1, 2, \ldots\}$ *such that*

1. $X(0) = 0$.

2. For any $0 = t_0 < t_1 < t_2 < \cdots < t_n$, *the increments*

$$X(t_1) - X(t_0), \; X(t_2) - X(t_1), \; \cdots, X(t_n) - X(t_{n-1})$$

are independent.

3. There is a function $\lambda(t)$ *such that*

$$\mathbb{P}(X(t + h) - X(t) = 1) = \lambda(t)h + o(h) \qquad (23.16)$$
$$\mathbb{P}(X(t + h) - X(t) \geq 2) = o(h). \qquad (23.17)$$

We call $\lambda(t)$ *the* **intensity function**.

The last condition means that the probability of an event in $[t, t + h]$ is approximately $h\lambda(t)$ while the probability of more than one event is small.

23.33 Theorem. *If* X_t *is a Poisson process with intensity function* $\lambda(t)$, *then*

$$X(s + t) - X(s) \sim \text{Poisson}(m(s + t) - m(s))$$

where

$$m(t) = \int_0^t \lambda(s) \, ds.$$

In particular, $X(t) \sim \text{Poisson}(m(t))$. *Hence,* $\mathbb{E}(X(t)) = m(t)$ *and* $\mathbb{V}(X(t)) = m(t)$.

23.34 Definition. *A Poisson process with intensity function* $\lambda(t) \equiv \lambda$ *for some* $\lambda > 0$ *is called a* **homogeneous Poisson process** *with rate* λ. *In this case,*

$$X(t) \sim \text{Poisson}(\lambda t).$$

Let $X(t)$ be a homogeneous Poisson process with rate λ. Let W_n be the time at which the nth event occurs and set $W_0 = 0$. The random variables $W_0, W_1, \ldots$, are called **waiting times**. Let $S_n = W_{n+1} - W_n$. Then $S_0, S_1, \ldots$, are called **sojourn times** or **interarrival times**.

23.35 Theorem. *The sojourn times* $S_0, S_1, \ldots$ *are* IID *random variables. Their distribution is exponential with mean* $1/\lambda$, *that is, they have density*

$$f(s) = \lambda e^{-\lambda s}, \quad s \geq 0.$$

The waiting time $W_n \sim \text{Gamma}(n, 1/\lambda)$ *i.e., it has density*

$$f(w) = \frac{1}{\Gamma(n)} \lambda^n w^{n-1} e^{-\lambda t}.$$

Hence, $\mathbb{E}(W_n) = n/\lambda$ *and* $\mathbb{V}(W_n) = n/\lambda^2$.

PROOF. First, we have

$$\mathbb{P}(S_1 > t) = \mathbb{P}(X(t) = 0) = e^{-\lambda t}$$

with shows that the CDF for S_1 is $1 - e^{-\lambda t}$. This shows the result for S_1. Now,

$$
\begin{aligned}
\mathbb{P}(S_2 > t | S_1 = s) &= \mathbb{P}(\text{no events in } (s, s+t] | S_1 = s) \\
&= \mathbb{P}(\text{no events in } (s, s+t]) \quad \text{(increments are independent)} \\
&= e^{-\lambda t}.
\end{aligned}
$$

Hence, S_2 has an exponential distribution and is independent of S_1. The result follows by repeating the argument. The result for W_n follows since a sum of exponentials has a Gamma distribution. ∎

23.36 Example. Figure 23.3 shows requests to a WWW server in Calgary.[1] Assuming that this is a homogeneous Poisson process, $N \equiv X(T) \sim \text{Poisson}(\lambda T)$. The likelihood is

$$\mathcal{L}(\lambda) \propto e^{-\lambda T} (\lambda T)^N$$

---

[1] See http://ita.ee.lbl.gov/html/contrib/Calgary-HTTP.html for more information.

FIGURE 23.3. Hits on a web server. Each vertical line represents one event.

which is maximized at

$$\widehat{\lambda} = \frac{N}{T} = 48.0077$$

in units per minute. Let's now test the assumption that the data follow a homogeneous Poisson process using a goodness-of-fit test. We divide the interval $[0, T]$ into 4 equal length intervals I_1, I_2, I_3, I_4. If the process is a homogeneous Poisson process then, given the total number of events, the probability that an event falls into any of these intervals must be equal. Let p_i be the probability of a point being in I_i. The null hypothesis is that $p_1 = p_2 = p_3 = p_4 = 1/4$. We can test this hypothesis using either a likelihood ratio test or a χ^2 test. The latter is

$$\sum_{i=1}^{4} \frac{(O_i - E_i)^2}{E_i}$$

where O_i is the number of observations in I_i and $E_i = n/4$ is the expected number under the null. This yields $\chi^2 = 252$ with a p-value near 0. This is strong evidence against the null so we reject the hypothesis that the data are from a homogeneous Poisson process. This is hardly surprising since we would expect the intensity to vary as a function of time. ∎

23.4   Bibliographic Remarks

This is standard material and there are many good references including Grimmett and Stirzaker (1982), Taylor and Karlin (1994), Guttorp (1995), and Ross (2002). The following exercises are from those texts.

23.5   Exercises

1. Let $X_0, X_1, \ldots$ be a Markov chain with states $\{0, 1, 2\}$ and transition matrix

$$\mathbf{P} = \begin{bmatrix} 0.1 & 0.2 & 0.7 \\ 0.9 & 0.1 & 0.0 \\ 0.1 & 0.8 & 0.1 \end{bmatrix}$$

Assume that $\mu_0 = (0.3, 0.4, 0.3)$. Find $\mathbb{P}(X_0 = 0, X_1 = 1, X_2 = 2)$ and $\mathbb{P}(X_0 = 0, X_1 = 1, X_2 = 1)$.

2. Let $Y_1, Y_2, \ldots$ be a sequence of iid observations such that $\mathbb{P}(Y = 0) = 0.1$, $\mathbb{P}(Y = 1) = 0.3$, $\mathbb{P}(Y = 2) = 0.2$, $\mathbb{P}(Y = 3) = 0.4$. Let $X_0 = 0$ and let

$$X_n = \max\{Y_1, \ldots, Y_n\}.$$

Show that $X_0, X_1, \ldots$ is a Markov chain and find the transition matrix.

3. Consider a two-state Markov chain with states $\mathcal{X} = \{1, 2\}$ and transition matrix

$$\mathbf{P} = \begin{bmatrix} 1 - a & a \\ b & 1 - b \end{bmatrix}$$

where $0 < a < 1$ and $0 < b < 1$. Prove that

$$\lim_{n \to \infty} \mathbf{P}^n = \begin{bmatrix} \frac{b}{a+b} & \frac{a}{a+b} \\ \frac{b}{a+b} & \frac{a}{a+b} \end{bmatrix}.$$

4. Consider the chain from question 3 and set $a = .1$ and $b = .3$. Simulate the chain. Let

$$\widehat{p}_n(1) = \frac{1}{n} \sum_{i=1}^{n} I(X_i = 1)$$

$$\widehat{p}_n(2) = \frac{1}{n} \sum_{i=1}^{n} I(X_i = 2)$$

be the proportion of times the chain is in state 1 and state 2. Plot $\widehat{p}_n(1)$ and $\widehat{p}_n(2)$ versus n and verify that they converge to the values predicted from the answer in the previous question.

5. An important Markov chain is the **branching process** which is used in biology, genetics, nuclear physics, and many other fields. Suppose that an animal has Y children. Let $p_k = \mathbb{P}(Y = k)$. Hence, $p_k \geq 0$ for all k and $\sum_{k=0}^{\infty} p_k = 1$. Assume each animal has the same lifespan and

that they produce offspring according to the distribution p_k. Let X_n be the number of animals in the nth generation. Let $Y_1^{(n)}, \ldots, Y_{X_n}^{(n)}$ be the offspring produced in the nth generation. Note that

$$X_{n+1} = Y_1^{(n)} + \cdots + Y_{X_n}^{(n)}.$$

Let $\mu = \mathbb{E}(Y)$ and $\sigma^2 = \mathbb{V}(Y)$. Assume throughout this question that $X_0 = 1$. Let $M(n) = \mathbb{E}(X_n)$ and $V(n) = \mathbb{V}(X_n)$.

(a) Show that $M(n+1) = \mu M(n)$ and $V(n+1) = \sigma^2 M(n) + \mu^2 V(n)$.

(b) Show that $M(n) = \mu^n$ and that $V(n) = \sigma^2 \mu^{n-1}(1 + \mu + \cdots + \mu^{n-1})$.

(c) What happens to the variance if $\mu > 1$? What happens to the variance if $\mu = 1$? What happens to the variance if $\mu < 1$?

(d) The population goes extinct if $X_n = 0$ for some n. Let us thus define the extinction time N by

$$N = \min\{n : X_n = 0\}.$$

Let $F(n) = \mathbb{P}(N \leq n)$ be the CDF of the random variable N. Show that

$$F(n) = \sum_{k=0}^{\infty} p_k (F(n-1))^k, \quad n = 1, 2, \ldots$$

Hint: Note that the event $\{N \leq n\}$ is the same as event $\{X_n = 0\}$. Thus, $\mathbb{P}(\{N \leq n\}) = \mathbb{P}(\{X_n = 0\})$. Let k be the number of offspring of the original parent. The population becomes extinct at time n if and only if each of the k sub-populations generated from the k offspring goes extinct in $n - 1$ generations.

(e) Suppose that $p_0 = 1/4$, $p_1 = 1/2$, $p_2 = 1/4$. Use the formula from (5d) to compute the CDF $F(n)$.

6. Let

$$\mathbf{P} = \begin{bmatrix} 0.40 & 0.50 & 0.10 \\ 0.05 & 0.70 & 0.25 \\ 0.05 & 0.50 & 0.45 \end{bmatrix}$$

Find the stationary distribution π.

7. Show that if i is a recurrent state and $i \leftrightarrow j$, then j is a recurrent state.

8. Let

$$P = \begin{bmatrix} \frac{1}{3} & 0 & \frac{1}{3} & 0 & 0 & \frac{1}{3} \\ \frac{1}{2} & \frac{1}{4} & \frac{1}{4} & 0 & 0 & 0 \\ 0 & 0 & 0 & 0 & 1 & 0 \\ \frac{1}{4} & \frac{1}{4} & \frac{1}{4} & 0 & 0 & \frac{1}{4} \\ 0 & 0 & 1 & 0 & 0 & 0 \\ 0 & 0 & 0 & 0 & 0 & 1 \end{bmatrix}$$

Which states are transient? Which states are recurrent?

9. Let

$$P = \begin{bmatrix} 0 & 1 \\ 1 & 0 \end{bmatrix}$$

Show that $\pi = (1/2, 1/2)$ is a stationary distribution. Does this chain converge? Why/why not?

10. Let $0 < p < 1$ and $q = 1 - p$. Let

$$P = \begin{bmatrix} q & p & 0 & 0 & 0 \\ q & 0 & p & 0 & 0 \\ q & 0 & 0 & p & 0 \\ q & 0 & 0 & 0 & p \\ 1 & 0 & 0 & 0 & 0 \end{bmatrix}$$

Find the limiting distribution of the chain.

11. Let $X(t)$ be an inhomogeneous Poisson process with intensity function $\lambda(t) > 0$. Let $\Lambda(t) = \int_0^t \lambda(u)du$. Define $Y(s) = X(t)$ where $s = \Lambda(t)$. Show that $Y(s)$ is a homogeneous Poisson process with intensity $\lambda = 1$.

12. Let $X(t)$ be a Poisson process with intensity λ. Find the conditional distribution of $X(t)$ given that $X(t + s) = n$.

13. Let $X(t)$ be a Poisson process with intensity λ. Find the probability that $X(t)$ is odd, i.e. $\mathbb{P}(X(t) = 1, 3, 5, \ldots)$.

14. Suppose that people logging in to the University computer system is described by a Poisson process $X(t)$ with intensity λ. Assume that a person stays logged in for some random time with CDF G. Assume these times are all independent. Let $Y(t)$ be the number of people on the system at time t. Find the distribution of $Y(t)$.

15. Let $X(t)$ be a Poisson process with intensity λ. Let $W_1, W_2, \ldots$, be the waiting times. Let f be an arbitrary function. Show that

$$\mathbb{E}\left(\sum_{i=1}^{X(t)} f(W_i) \right) = \lambda \int_0^t f(w)dw.$$

16. A two-dimensional Poisson point process is a process of random points on the plane such that (i) for any set A, the number of points falling in A is Poisson with mean $\lambda\mu(A)$ where $\mu(A)$ is the area of A, (ii) the number of events in non-overlapping regions is independent. Consider an arbitrary point x_0 in the plane. Let X denote the distance from x_0 to the nearest random point. Show that

$$\mathbb{P}(X > t) = e^{-\lambda \pi t^2}$$

and

$$\mathbb{E}(X) = \frac{1}{2\sqrt{\lambda}}.$$

24
Simulation Methods

In this chapter we will show how simulation can be used to approximate integrals. Our leading example is the problem of computing integrals in Bayesian inference but the techniques are widely applicable. We will look at three integration methods: (i) basic Monte Carlo integration, (ii) importance sampling, and (iii) Markov chain Monte Carlo (MCMC).

24.1 Bayesian Inference Revisited

Simulation methods are especially useful in Bayesian inference so let us briefly review the main ideas in Bayesian inference. See Chapter 11 for more details.

Given a prior $f(\theta)$ and data $X^n = (X_1, \ldots, X_n)$ the posterior density is

$$f(\theta|X^n) = \frac{\mathcal{L}(\theta)f(\theta)}{c}$$

where $\mathcal{L}(\theta)$ is the likelihood function and

$$c = \int \mathcal{L}(\theta)f(\theta)\, d\theta$$

is the **normalizing constant**. The posterior mean is

$$\overline{\theta} = \int \theta f(\theta|X^n)d\theta = \frac{\int \theta \mathcal{L}(\theta)f(\theta)d\theta}{c}.$$

If $\theta = (\theta_1, \ldots, \theta_k)$ is multidimensional, then we might be interested in the posterior for one of the components, θ_1, say. This marginal posterior density is

$$f(\theta_1 | X^n) = \int \int \cdots \int f(\theta_1, \ldots, \theta_k | X^n) d\theta_2 \cdots d\theta_k$$

which involves high-dimensional integration.

When θ is high-dimensional, it may not be feasible to calculate these integrals analytically. Simulation methods will often be helpful.

24.2   Basic Monte Carlo Integration

Suppose we want to evaluate the integral

$$I = \int_a^b h(x) \, dx$$

for some function h. If h is an "easy" function like a polynomial or trigonometric function, then we can do the integral in closed form. If h is complicated there may be no known closed form expression for I. There are many numerical techniques for evaluating I such as Simpson's rule, the trapezoidal rule and Gaussian quadrature. Monte Carlo integration is another approach for approximating I which is notable for its simplicity, generality and scalability.

Let us begin by writing

$$I = \int_a^b h(x) dx = \int_a^b w(x) f(x) dx \tag{24.1}$$

where $w(x) = h(x)(b-a)$ and $f(x) = 1/(b-a)$. Notice that f is the probability density for a uniform random variable over (a, b). Hence,

$$I = \mathbb{E}_f(w(X))$$

where $X \sim \text{Unif}(a, b)$. If we generate $X_1, \ldots, X_N \sim \text{Unif}(a, b)$, then by the law of large numbers

$$\widehat{I} \equiv \frac{1}{N} \sum_{i=1}^N w(X_i) \xrightarrow{\text{P}} \mathbb{E}(w(X)) = I. \tag{24.2}$$

This is the basic **Monte Carlo integration method**. We can also compute the standard error of the estimate

$$\widehat{\text{se}} = \frac{s}{\sqrt{N}}$$

where

$$s^2 = \frac{\sum_{i=1}^{N}(Y_i - \widehat{I})^2}{N-1}$$

where $Y_i = w(X_i)$. A $1 - \alpha$ confidence interval for I is $\widehat{I} \pm z_{\alpha/2}\widehat{se}$. We can take N as large as we want and hence make the length of the confidence interval very small.

24.1 Example. Let $h(x) = x^3$. Then, $I = \int_0^1 x^3 dx = 1/4$. Based on $N = 10,000$ observations from a Uniform$(0,1)$ we get $\widehat{I} = .248$ with a standard error of .0028. ∎

A generalization of the basic method is to consider integrals of the form

$$I = \int h(x)f(x)dx \qquad (24.3)$$

where $f(x)$ is a probability density function. Taking f to be a Uniform (a,b) gives us the special case above. Now we draw $X_1, \ldots, X_N \sim f$ and take

$$\widehat{I} \equiv \frac{1}{N}\sum_{i=1}^{N}h(X_i)$$

as before.

24.2 Example. Let

$$f(x) = \frac{1}{\sqrt{2\pi}}e^{-x^2/2}$$

be the standard Normal PDF. Suppose we want to compute the CDF at some point x:

$$I = \int_{-\infty}^{x} f(s)ds = \Phi(x).$$

Write

$$I = \int h(s)f(s)ds$$

where

$$h(s) = \begin{cases} 1 & s < x \\ 0 & s \geq x. \end{cases}$$

Now we generate $X_1, \ldots, X_N \sim N(0,1)$ and set

$$\widehat{I} = \frac{1}{N}\sum_i h(X_i) = \frac{\text{number of observations} \leq x}{N}.$$

For example, with $x = 2$, the true answer is $\Phi(2) = .9772$ and the Monte Carlo estimate with $N = 10,000$ yields .9751. Using $N = 100,000$ we get .9771. ∎

24.3 Example (Bayesian Inference for Two Binomials). Let $X \sim \text{Binomial}(n, p_1)$ and $Y \sim \text{Binomial}(m, p_2)$. We would like to estimate $\delta = p_2 - p_1$. The MLE is $\widehat{\delta} = \widehat{p}_2 - \widehat{p}_1 = (Y/m) - (X/n)$. We can get the standard error $\widehat{\text{se}}$ using the delta method which yields

$$\widehat{\text{se}} = \sqrt{\frac{\widehat{p}_1(1 - \widehat{p}_1)}{n} + \frac{\widehat{p}_2(1 - \widehat{p}_2)}{m}}$$

and then construct a 95 percent confidence interval $\widehat{\delta} \pm 2\,\widehat{\text{se}}$. Now consider a Bayesian analysis. Suppose we use the prior $f(p_1, p_2) = f(p_1)f(p_2) = 1$, that is, a flat prior on (p_1, p_2). The posterior is

$$f(p_1, p_2 | X, Y) \propto p_1^X (1 - p_1)^{n-X} p_2^Y (1 - p_2)^{m-Y}.$$

The posterior mean of δ is

$$\overline{\delta} = \int_0^1 \int_0^1 \delta(p_1, p_2) f(p_1, p_2 | X, Y) = \int_0^1 \int_0^1 (p_2 - p_1) f(p_1, p_2 | X, Y).$$

If we want the posterior density of δ we can first get the posterior CDF

$$F(c | X, Y) = P(\delta \leq c | X, Y) = \int_A f(p_1, p_2 | X, Y)$$

where $A = \{(p_1, p_2) : p_2 - p_1 \leq c\}$. The density can then be obtained by differentiating F.

To avoid all these integrals, let's use simulation. Note that $f(p_1, p_2 | X, Y) = f(p_1 | X)f(p_2 | Y)$ which implies that p_1 and p_2 are independent under the posterior distribution. Also, we see that $p_1 | X \sim \text{Beta}(X+1, n-X+1)$ and $p_2 | Y \sim \text{Beta}(Y+1, m-Y+1)$. Hence, we can simulate $(P_1^{(1)}, P_2^{(1)}), \ldots, (P_1^{(N)}, P_2^{(N)})$ from the posterior by drawing

$$P_1^{(i)} \sim \text{Beta}(X + 1, n - X + 1)$$
$$P_2^{(i)} \sim \text{Beta}(Y + 1, m - Y + 1)$$

for $i = 1, \ldots, N$. Now let $\delta^{(i)} = P_2^{(i)} - P_1^{(i)}$. Then,

$$\overline{\delta} \approx \frac{1}{N} \sum_i \delta^{(i)}.$$

We can also get a 95 percent posterior interval for δ by sorting the simulated values, and finding the .025 and .975 quantile. The posterior density $f(\delta | X, Y)$ can be obtained by applying density estimation techniques to $\delta^{(1)}, \ldots, \delta^{(N)}$ or, simply by plotting a histogram. For example, suppose that $n = m = 10$,

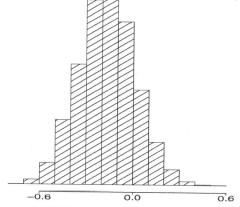

FIGURE 24.1. Posterior of δ from simulation.

$X = 8$ and $Y = 6$. From a posterior sample of size 1000 we get a 95 percent posterior interval of (-0.52,0.20). The posterior density can be estimated from a histogram of the simulated values as shown in Figure 24.1. ∎

24.4 Example (Bayesian Inference for Dose Response). Suppose we conduct an experiment by giving rats one of ten possible doses of a drug, denoted by $x_1 < x_2 < \ldots < x_{10}$. For each dose level x_i we use n rats and we observe Y_i, the number that survive. Thus we have ten independent binomials $Y_i \sim$ Binomial(n, p_i). Suppose we know from biological considerations that higher doses should have higher probability of death. Thus, $p_1 \leq p_2 \leq \cdots \leq p_{10}$. We want to estimate the dose at which the animals have a 50 percent chance of dying. This is called the LD50. Formally, $\delta = x_j$ where

$$j = \min\{i : p_i \geq .50\}.$$

Notice that δ is implicitly a (complicated) function of $p_1, \ldots, p_{10}$ so we can write $\delta = g(p_1, \ldots, p_{10})$ for some g. This just means that if we know $(p_1, \ldots, p_{10})$ then we can find δ. The posterior mean of δ is

$$\int \int \cdots \int_A g(p_1, \ldots, p_{10}) f(p_1, \ldots, p_{10} | Y_1, \ldots, Y_{10}) dp_1 dp_2 \ldots dp_{10}.$$

The integral is over the region

$$A = \{(p_1, \ldots, p_{10}) : p_1 \leq \cdots \leq p_{10}\}.$$

The posterior CDF of δ is

$$
\begin{aligned}
F(c | Y_1, \ldots, Y_{10}) &= \mathbb{P}(\delta \leq c | Y_1, \ldots, Y_{10}) \\
&= \int \int \cdots \int_B f(p_1, \ldots, p_{10} | Y_1, \ldots, Y_{10}) dp_1 dp_2 \ldots dp_{10}
\end{aligned}
$$

where

$$B = A \bigcap \left\{ (p_1, \ldots, p_{10}) : \ g(p_1, \ldots, p_{10}) \leq c \right\}.$$

We need to do a 10-dimensional integral over a restricted region A. Instead, we will use simulation. Let us take a flat prior truncated over A. Except for the truncation, each P_i has once again a Beta distribution. To draw from the posterior we do the following steps:

(1) Draw $P_i \sim \text{Beta}(Y_i + 1, n - Y_i + 1), i = 1, \ldots, 10$.

(2) If $P_1 \leq P_2 \leq \cdots \leq P_{10}$ keep this draw. Otherwise, throw it away and draw again until you get one you can keep.

(3) Let $\delta = x_j$ where

$$j = \min\{i : \ P_i > .50\}.$$

We repeat this N times to get $\delta^{(1)}, \ldots, \delta^{(N)}$ and take

$$\mathbb{E}(\delta|Y_1, \ldots, Y_{10}) \approx \frac{1}{N} \sum_i \delta^{(i)}.$$

δ is a discrete variable. We can estimate its probability mass function by

$$\mathbb{P}(\delta = x_j|Y_1, \ldots, Y_{10}) \approx \frac{1}{N} \sum_{i=1}^{N} I(\delta^{(i)} = j).$$

For example, consider the following data:

Dose	1	2	3	4	5	6	7	8	9	10
Number of animals n_i	15	15	15	15	15	15	15	15	15	15
Number of survivors Y_i	0	0	2	2	8	10	12	14	15	14

The posterior draws for $p_1, \ldots, p_{10}$ are shown in the second panel in the figure. We find that that $\overline{\delta} = 4.04$ with a 95 percent interval of (3,5). ■

24.3 Importance Sampling

Consider again the integral $I = \int h(x)f(x)dx$ where f is a probability density. The basic Monte Carlo method involves sampling from f. However, there are cases where we may not know how to sample from f. For example, in Bayesian inference, the posterior density density is is obtained by multiplying the likelihood $\mathcal{L}(\theta)$ times the prior $f(\theta)$. There is no guarantee that $f(\theta|x)$ will be a known distribution like a Normal or Gamma or whatever.

Importance sampling is a generalization of basic Monte Carlo which over-comes this problem. Let g be a probability density that we know how to simulate from. Then

$$I = \int h(x)f(x)dx = \int \frac{h(x)f(x)}{g(x)}g(x)dx = \mathbb{E}_g(Y) \qquad (24.4)$$

where $Y = h(X)f(X)/g(X)$ and the expectation $\mathbb{E}_g(Y)$ is with respect to g. We can simulate $X_1, \ldots, X_N \sim g$ and estimate I by

$$\widehat{I} = \frac{1}{N}\sum_i Y_i = \frac{1}{N}\sum_i \frac{h(X_i)f(X_i)}{g(X_i)}. \qquad (24.5)$$

This is called **importance sampling**. By the law of large numbers, $\widehat{I} \xrightarrow{P} I$. However, there is a catch. It's possible that $\widehat{I}$ might have an infinite standard error. To see why, recall that I is the mean of $w(x) = h(x)f(x)/g(x)$. The second moment of this quantity is

$$\mathbb{E}_g(w^2(X)) = \int \left(\frac{h(x)f(x)}{g(x)}\right)^2 g(x)dx = \int \frac{h^2(x)f^2(x)}{g(x)}dx. \qquad (24.6)$$

If g has thinner tails than f, then this integral might be infinite. To avoid this, a basic rule in importance sampling is to sample from a density g with thicker tails than f. Also, suppose that $g(x)$ is small over some set A where $f(x)$ is large. Again, the ratio of f/g could be large leading to a large variance. This implies that we should choose g to be similar in shape to f. In summary, a good choice for an importance sampling density g should be similar to f but with thicker tails. In fact, we can say what the optimal choice of g is.

24.5 Theorem. *The choice of g that minimizes the variance of $\widehat{I}$ is*

$$g^*(x) = \frac{|h(x)|f(x)}{\int |h(s)|f(s)ds}.$$

PROOF. The variance of $w = fh/g$ is

$$\begin{aligned}
\mathbb{E}_g(w^2) - (\mathbb{E}(w^2))^2 &= \int w^2(x)g(x)dx - \left(\int w(x)g(x)dx\right)^2 \\
&= \int \frac{h^2(x)f^2(x)}{g^2(x)}g(x)dx - \left(\int \frac{h(x)f(x)}{g(x)}g(x)dx\right)^2 \\
&= \int \frac{h^2(x)f^2(x)}{g^2(x)}g(x)dx - \left(\int h(x)f(x)dx\right)^2.
\end{aligned}$$

The second integral does not depend on g, so we only need to minimize the first integral. From Jensen's inequality (Theorem 4.9) we have

$$\mathbb{E}_g(W^2) \geq (\mathbb{E}_g(|W|))^2 = \left(\int |h(x)| f(x) dx \right)^2.$$

This establishes a lower bound on $\mathbb{E}_g(W^2)$. However, $\mathbb{E}_{g^*}(W^2)$ equals this lower bound which proves the claim. ∎

This theorem is interesting but it is only of theoretical interest. If we did not know how to sample from f then it is unlikely that we could sample from $|h(x)| f(x) / \int |h(s)| f(s) ds$. In practice, we simply try to find a thick-tailed distribution g which is similar to $f|h|$.

24.6 Example (Tail Probability). Let's estimate $I = \mathbb{P}(Z > 3) = .0013$ where $Z \sim N(0,1)$. Write $I = \int h(x) f(x) dx$ where $f(x)$ is the standard Normal density and $h(x) = 1$ if $x > 3$, and 0 otherwise. The basic Monte Carlo estimator is $\widehat{I} = N^{-1} \sum_i h(X_i)$ where $X_1, \ldots, X_N \sim N(0,1)$. Using $N = 100$ we find (from simulating many times) that $\mathbb{E}(\widehat{I}) = .0015$ and $\mathbb{V}(\widehat{I}) = .0039$. Notice that most observations are wasted in the sense that most are not near the right tail. Now we will estimate this with importance sampling taking g to be a Normal(4,1) density. We draw values from g and the estimate is now $\widehat{I} = N^{-1} \sum_i f(X_i) h(X_i) / g(X_i)$. In this case we find that $\mathbb{E}(\widehat{I}) = .0011$ and $\mathbb{V}(\widehat{I}) = .0002$. We have reduced the standard deviation by a factor of 20. ∎

24.7 Example (Measurement Model With Outliers). Suppose we have measurements $X_1, \ldots, X_n$ of some physical quantity θ. A reasonable model is

$$X_i = \theta + \epsilon_i.$$

If we assume that $\epsilon_i \sim N(0,1)$ then $X_i \sim N(\theta_i, 1)$. However, when taking measurements, it is often the case that we get the occasional wild observation, or outlier. This suggests that a Normal might be a poor model since Normals have thin tails which implies that extreme observations are rare. One way to improve the model is to use a density for ϵ_i with a thicker tail, for example, a t-distribution with ν degrees of freedom which has the form

$$t(x) = \frac{\Gamma\left(\frac{\nu+1}{2}\right)}{\Gamma\left(\frac{\nu}{2}\right)} \frac{1}{\nu\pi} \left(1 + \frac{x^2}{\nu} \right)^{-(\nu+1)/2}.$$

Smaller values of ν correspond to thicker tails. For the sake of illustration we will take $\nu = 3$. Suppose we observe n $X_i = \theta + \epsilon_i$, $i = 1, \ldots, n$ where ϵ_i has a t distribution with $\nu = 3$. We will take a flat prior on θ. The likelihood is $\mathcal{L}(\theta) = \prod_{i=1}^{n} t(X_i - \theta)$ and the posterior mean of θ is

$$\overline{\theta} = \frac{\int \theta \mathcal{L}(\theta) d\theta}{\int \mathcal{L}(\theta) d\theta}.$$

We can estimate the top and bottom integral using importance sampling. We draw $\theta_1, \ldots, \theta_N \sim g$ and then

$$\overline{\theta} \approx \frac{\frac{1}{N} \sum_{j=1}^{N} \frac{\theta_j \mathcal{L}(\theta_j)}{g(\theta_j)}}{\frac{1}{N} \sum_{j=1}^{N} \frac{\mathcal{L}(\theta_j)}{g(\theta_j)}}.$$

To illustrate the idea, we drew $n = 2$ observations. The posterior mean (computed numerically) is -0.54. Using a Normal importance sampler g yields an estimate of -0.74. Using a Cauchy (t-distribution with 1 degree of freedom) importance sampler yields an estimate of -0.53. ∎

24.4 MCMC Part I: The Metropolis–Hastings Algorithm

Consider once more the problem of estimating the integral $I = \int h(x) f(x) dx$. Now we introduce Markov chain Monte Carlo (MCMC) methods. The idea is to construct a Markov chain $X_1, X_2, \ldots$, whose stationary distribution is f. Under certain conditions it will then follow that

$$\frac{1}{N} \sum_{i=1}^{N} h(X_i) \xrightarrow{\text{P}} \mathbb{E}_f(h(X)) = I.$$

This works because there is a law of large numbers for Markov chains; see Theorem 23.25.

The **Metropolis–Hastings** algorithm is a specific MCMC method that works as follows. Let $q(y|x)$ be an arbitrary, friendly distribution (i.e., we know how to sample from $q(y|x)$). The conditional density $q(y|x)$ is called the **proposal distribution**. The Metropolis–Hastings algorithm creates a sequence of observations $X_0, X_1, \ldots$, as follows.

Metropolis–Hastings Algorithm

Choose X_0 arbitrarily. Suppose we have generated $X_0, X_1, \ldots, X_i$. To generate X_{i+1} do the following:

(1) Generate a **proposal** or **candidate** value $Y \sim q(y|X_i)$.

(2) Evaluate $r \equiv r(X_i, Y)$ where

$$r(x, y) = \min\left\{ \frac{f(y)}{f(x)} \frac{q(x|y)}{q(y|x)}, 1 \right\}.$$

(3) Set

$$X_{i+1} = \begin{cases} Y & \text{with probability } r \\ X_i & \text{with probability } 1 - r. \end{cases}$$

24.8 Remark. A simple way to execute step (3) is to generate $U \sim (0, 1)$. If $U < r$ set $X_{i+1} = Y$ otherwise set $X_{i+1} = X_i$.

24.9 Remark. A common choice for $q(y|x)$ is $N(x, b^2)$ for some $b > 0$. This means that the proposal is draw from a Normal, centered at the current value. In this case, the proposal density q is symmetric, $q(y|x) = q(x|y)$, and r simplifies to

$$r = \min\left\{ \frac{f(Y)}{f(X_i)}, 1 \right\}.$$

By construction, $X_0, X_1, \ldots$ is a Markov chain. But why does this Markov chain have f as its stationary distribution? Before we explain why, let us first do an example.

24.10 Example. The Cauchy distribution has density

$$f(x) = \frac{1}{\pi} \frac{1}{1 + x^2}.$$

Our goal is to simulate a Markov chain whose stationary distribution is f. As suggested in the remark above, we take $q(y|x)$ to be a $N(x, b^2)$. So in this case,

$$r(x, y) = \min\left\{ \frac{f(y)}{f(x)}, 1 \right\} = \min\left\{ \frac{1 + x^2}{1 + y^2}, 1 \right\}.$$

So the algorithm is to draw $Y \sim N(X_i, b^2)$ and set

$$X_{i+1} = \begin{cases} Y & \text{with probability } r(X_i, Y) \\ X_i & \text{with probability } 1 - r(X_i, Y). \end{cases}$$

The simulator requires a choice of b. Figure 24.2 shows three chains of length $N = 1,000$ using $b = .1$, $b = 1$ and $b = 10$. Setting $b = .1$ forces the chain to take small steps. As a result, the chain doesn't "explore" much of the sample space. The histogram from the sample does not approximate the true density very well. Setting $b = 10$ causes the proposals to often be far in the

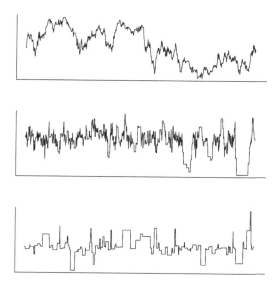

FIGURE 24.2. Three Metropolis chains corresponding to $b = .1$, $b = 1$, $b = 10$.

tails, making r small and hence we reject the proposal and keep the chain at its current position. The result is that the chain "gets stuck" at the same place quite often. Again, this means that the histogram from the sample does not approximate the true density very well. The middle choice avoids these extremes and results in a Markov chain sample that better represents the density sooner. In summary, there are tuning parameters and the efficiency of the chain depends on these parameters. We'll discuss this in more detail later. ■

If the sample from the Markov chain starts to "look like" the target distribution f quickly, then we say that the chain is "mixing well." Constructing a chain that mixes well is somewhat of an art.

WHY IT WORKS. Recall from Chapter 23 that a distribution π satisfies **detailed balance** for a Markov chain if

$$p_{ij}\pi_i = p_{ji}\pi_j.$$

We showed that if π satisfies detailed balance, then it is a stationary distribution for the chain.

Because we are now dealing with continuous state Markov chains, we will change notation a little and write $p(x, y)$ for the probability of making a transition from x to y. Also, let's use $f(x)$ instead of π for a distribution. In this new notation, f is a stationary distribution if $f(x) = \int f(y)p(y,x)dy$ and detailed balance holds for f if

$$f(x)p(x,y) = f(y)p(y,x). \tag{24.7}$$

Detailed balance implies that f is a stationary distribution since, if detailed balance holds, then

$$\int f(y)p(y,x)dy = \int f(x)p(x,y)dy = f(x)\int p(x,y)dy = f(x)$$

which shows that $f(x) = \int f(y)p(y,x)dy$ as required. Our goal is to show that f satisfies detailed balance which will imply that f is a stationary distribution for the chain.

Consider two points x and y. Either

$$f(x)q(y|x) < f(y)q(x|y) \quad \text{or} \quad f(x)q(y|x) > f(y)q(x|y).$$

We will ignore ties (which occur with probability zero for continuous distributions). Without loss of generality, assume that $f(x)q(y|x) > f(y)q(x|y)$. This implies that

$$r(x,y) = \frac{f(y)}{f(x)}\frac{q(x|y)}{q(y|x)}$$

and that $r(y,x) = 1$. Now $p(x,y)$ is the probability of jumping from x to y. This requires two things: (i) the proposal distribution must generate y, and (ii) you must accept y. Thus,

$$p(x,y) = q(y|x)r(x,y) = q(y|x)\frac{f(y)}{f(x)}\frac{q(x|y)}{q(y|x)} = \frac{f(y)}{f(x)}q(x|y).$$

Therefore,

$$f(x)p(x,y) = f(y)q(x|y). \tag{24.8}$$

On the other hand, $p(y,x)$ is the probability of jumping from y to x. This requires two things: (i) the proposal distribution must generate x, and (ii) you must accept x. This occurs with probability $p(y,x) = q(x|y)r(y,x) = q(x|y)$. Hence,

$$f(y)p(y,x) = f(y)q(x|y). \tag{24.9}$$

Comparing (24.8) and (24.9), we see that we have shown that detailed balance holds.

24.5   MCMC Part II: Different Flavors

There are different types of MCMC algorithm. Here we will consider a few of the most popular versions.

RANDOM-WALK-METROPOLIS–HASTINGS. In the previous section we considered drawing a proposal Y of the form

$$Y = X_i + \epsilon_i$$

where ϵ_i comes from some distribution with density g. In other words, $q(y|x) = g(y - x)$. We saw that in this case,

$$r(x, y) = \min\left\{1, \frac{f(y)}{f(x)}\right\}.$$

This is called a **random-walk-Metropolis–Hastings** method. The reason for the name is that, if we did not do the accept–reject step, we would be simulating a random walk. The most common choice for g is a $N(0, b^2)$. The hard part is choosing b so that the chain mixes well. A good rule of thumb is: choose b so that you accept the proposals about 50 percent of the time.

Warning! This method doesn't make sense unless X takes values on the whole real line. If X is restricted to some interval then it is best to transform X. For example, if $X \in (0, \infty)$ then you might take $Y = \log X$ and then simulate the distribution for Y instead of X.

INDEPENDENCE-METROPOLIS–HASTINGS. This is an importance-sampling version of MCMC. We draw the proposal from a fixed distribution g. Generally, g is chosen to be an approximation to f. The acceptance probability becomes

$$r(x, y) = \min\left\{1, \frac{f(y)}{f(x)}\frac{g(x)}{g(y)}\right\}.$$

GIBBS SAMPLING. The two previous methods can be easily adapted, in principle, to work in higher dimensions. In practice, tuning the chains to make them mix well is hard. Gibbs sampling is a way to turn a high-dimensional problem into several one-dimensional problems.

Here's how it works for a bivariate problem. Suppose that (X, Y) has density $f_{X,Y}(x, y)$. First, suppose that it is possible to simulate from the conditional distributions $f_{X|Y}(x|y)$ and $f_{Y|X}(y|x)$. Let (X_0, Y_0) be starting values. Assume we have drawn $(X_0, Y_0), \ldots, (X_n, Y_n)$. Then the Gibbs sampling algorithm for getting (X_{n+1}, Y_{n+1}) is:

Gibbs Sampling

$$X_{n+1} \sim f_{X|Y}(x|Y_n)$$

$$Y_{n+1} \sim f_{Y|X}(y|X_{n+1})$$

repeat

This generalizes in the obvious way to higher dimensions.

24.11 Example (Normal Hierarchical Model). Gibbs sampling is very useful for a class of models called **hierarchical models**. Here is a simple case. Suppose we draw a sample of k cities. From each city we draw n_i people and observe how many people Y_i have a disease. Thus, $Y_i \sim \text{Binomial}(n_i, p_i)$. We are allowing for different disease rates in different cities. We can also think of the $p_i's$ as random draws from some distribution F. We can write this model in the following way:

$$P_i \sim F$$

$$Y_i|P_i = p_i \sim \text{Binomial}(n_i, p_i).$$

We are interested in estimating the $p_i's$ and the overall disease rate $\int p \, dF(p)$.

To proceed, it will simplify matters if we make some transformations that allow us to use some Normal approximations. Let $\widehat{p}_i = Y_i/n_i$. Recall that $\widehat{p}_i \approx N(p_i, s_i)$ where $s_i = \sqrt{\widehat{p}_i(1 - \widehat{p}_i)/n_i}$. Let $\psi_i = \log(p_i/(1 - p_i))$ and define $Z_i \equiv \widehat{\psi}_i = \log(\widehat{p}_i/(1 - \widehat{p}_i))$. By the delta method,

$$\widehat{\psi}_i \approx N(\psi_i, \sigma_i^2)$$

where $\sigma_i^2 = 1/(n\widehat{p}_i(1 - \widehat{p}_i))$. Experience shows that the Normal approximation for ψ is more accurate than the Normal approximation for p so we shall work with ψ. We shall treat σ_i as known. Furthermore, we shall take the distribution of the $\psi_i's$ to be Normal. The hierarchical model is now

$$\psi_i \sim N(\mu, \tau^2)$$

$$Z_i|\psi_i \sim N(\psi_i, \sigma_i^2).$$

As yet another simplification we take $\tau = 1$. The unknown parameter are $\theta = (\mu, \psi_1, \ldots, \psi_k)$. The likelihood function is

$$\mathcal{L}(\theta) \propto \prod_i f(\psi_i|\mu) \prod_i f(Z_i|\psi)$$

$$\propto \prod_i \exp\left\{-\frac{1}{2}(\psi_i - \mu)^2\right\} \exp\left\{-\frac{1}{2\sigma_i^2}(Z_i - \psi_i)^2\right\}.$$

If we use the prior $f(\mu) \propto 1$ then the posterior is proportional to the likelihood. To use Gibbs sampling, we need to find the conditional distribution of each parameter conditional on all the others. Let us begin by finding $f(\mu|\text{rest})$ where "rest" refers to all the other variables. We can throw away any terms that don't involve μ. Thus,

$$f(\mu|\text{rest}) \quad \propto \quad \prod_i \exp\left\{-\frac{1}{2}(\psi_i - \mu)^2\right\}$$

$$\propto \quad \exp\left\{-\frac{k}{2}(\mu - b)^2\right\}$$

where

$$b = \frac{1}{k}\sum_i \psi_i.$$

Hence we see that $\mu|\text{rest} \sim N(b, 1/k)$. Next we will find $f(\psi|\text{rest})$. Again, we can throw away any terms not involving ψ_i leaving us with

$$f(\psi_i|\text{rest}) \quad \propto \quad \exp\left\{-\frac{1}{2}(\psi_i - \mu)^2\right\}\exp\left\{-\frac{1}{2\sigma_i^2}(Z_i - \psi_i)^2\right\}$$

$$\propto \quad \exp\left\{-\frac{1}{2d_i^2}(\psi_i - e_i)^2\right\}$$

where

$$e_i = \frac{\frac{Z_i}{\sigma_i^2} + \mu}{1 + \frac{1}{\sigma_i^2}} \quad \text{and} \quad d_i^2 = \frac{1}{1 + \frac{1}{\sigma_i^2}}$$

and so $\psi_i|\text{rest} \sim N(e_i, d_i^2)$. The Gibbs sampling algorithm then involves iterating the following steps N times:

$$\begin{aligned}
\text{draw } \mu \quad &\sim \quad N(b, v^2)\\
\text{draw } \psi_1 \quad &\sim \quad N(e_1, d_1^2)\\
&\vdots\\
\text{draw } \psi_k \quad &\sim \quad N(e_k, d_k^2).
\end{aligned}$$

It is understood that at each step, the most recently drawn version of each variable is used.

We generated a numerical example with $k = 20$ cities and $n = 20$ people from each city. After running the chain, we can convert each ψ_i back into p_i by way of $p_i = e^{\psi_i}/(1 + e^{\psi_i})$. The raw proportions are shown in Figure 24.4. Figure 24.3 shows "trace plots" of the Markov chain for p_1 and μ. Figure 24.4 shows the posterior for μ based on the simulated values. The second panel of Figure 24.4 shows the raw proportions and the Bayes estimates. Note that the Bayes estimates are "shrunk" together. The parameter τ controls the amount of shrinkage. We set $\tau = 1$ but, in practice, we should treat τ as another unknown parameter and let the data determine how much shrinkage is needed. ∎

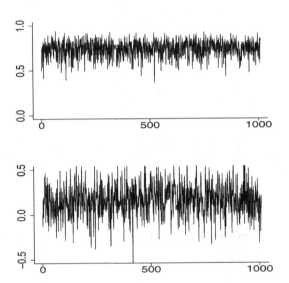

FIGURE 24.3. Posterior simulation for Example 24.11. The top panel shows simulated values of p_1. The top panel shows simulated values of μ.

So far we assumed that we know how to draw samples from the conditionals $f_{X|Y}(x|y)$ and $f_{Y|X}(y|x)$. If we don't know how, we can still use the Gibbs sampling algorithm by drawing each observation using a Metropolis–Hastings step. Let q be a proposal distribution for x and let $\widetilde{q}$ be a proposal distribution for y. When we do a Metropolis step for X, we treat Y as fixed. Similarly, when we do a Metropolis step for Y, we treat X as fixed. Here are the steps:

Metropolis within Gibbs

(1a) Draw a proposal $Z \sim q(z|X_n)$.

(1b) Evaluate
$$r = \min\left\{\frac{f(Z, Y_n)}{f(X_n, Y_n)}\frac{q(X_n|Z)}{q(Z|X_n)},\ 1\right\}.$$

(1c) Set
$$X_{n+1} = \begin{cases} Z & \text{with probability } r \\ X_n & \text{with probability } 1-r. \end{cases}$$

(2a) Draw a proposal $Z \sim \widetilde{q}(z|Y_n)$.

(2b) Evaluate
$$r = \min\left\{\frac{f(X_{n+1}, Z)}{f(X_{n+1}, Y_n)}\frac{\widetilde{q}(Y_n|Z)}{\widetilde{q}(Z|Y_n)},\ 1\right\}.$$

(2c) Set
$$Y_{n+1} = \begin{cases} Z & \text{with probability } r \\ Y_n & \text{with probability } 1-r. \end{cases}$$

Again, this generalizes to more than two dimensions.

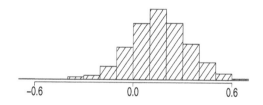

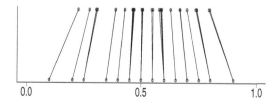

FIGURE 24.4. Example 24.11. Top panel: posterior histogram of μ. Lower panel: raw proportions and the Bayes posterior estimates. The Bayes estimates have been shrunk closer together than the raw proportions.

24.6   Bibliographic Remarks

MCMC methods go back to the effort to build the atomic bomb in World War II. They were used in various places after that, especially in spatial statistics. There was a new surge of interest in the 1990s that still continues. My main reference for this chapter was Robert and Casella (1999). See also Gelman et al. (1995) and Gilks et al. (1998).

24.7   Exercises

1. Let
$$I = \int_1^2 \frac{e^{-x^2/2}}{\sqrt{2\pi}} \, dx.$$

(a) Estimate I using the basic Monte Carlo method. Use $N = 100,000$. Also, find the estimated standard error.

(b) Find an (analytical) expression for the standard error of your estimate in (a). Compare to the estimated standard error.

(c) Estimate I using importance sampling. Take g to be $N(1.5, v^2)$ with $v = .1$, $v = 1$ and $v = 10$. Compute the (true) standard errors in each case. Also, plot a histogram of the values you are averaging to see if there are any extreme values.

(d) Find the optimal importance sampling function g^*. What is the standard error using g^*?

2. Here is a way to use importance sampling to estimate a marginal density. Let $f_{X,Y}(x, y)$ be a bivariate density and let $(X_1, X_2), \ldots, (X_N, Y_N) \sim f_{X,Y}$.

(a) Let $w(x)$ be an arbitrary probability density function. Let
$$\widehat{f}_X(x) = \frac{1}{N} \sum_{i=1}^N \frac{f_{X,Y}(x, Y_i) w(X_i)}{f_{X,Y}(X_i, Y_i)}.$$

Show that, for each x,
$$\widehat{f}_X(x) \xrightarrow{P} f_X(x).$$

Find an expression for the variance of this estimator.

(b) Let $Y \sim N(0, 1)$ and $X|Y = y \sim N(y, 1 + y^2)$. Use the method in (a) to estimate $f_X(x)$.

3. Here is a method called **accept–reject sampling** for drawing observations from a distribution.

(a) Suppose that f is some probability density function. Let g be any other density and suppose that $f(x) \leq Mg(x)$ for all x, where M is a known constant. Consider the following algorithm:

(step 1): Draw $X \sim g$ and $U \sim \text{Unif}(0, 1)$;

(step 2): If $U \leq f(X)/(Mg(X))$ set $Y = X$, otherwise go back to step 1. (Keep repeating until you finally get an observation.)

Show that the distribution of Y is f.

(b) Let f be a standard Normal density and let $g(x) = 1/(1 + x^2)$ be the Cauchy density. Apply the method in (a) to draw 1,000 observations from the Normal distribution. Draw a histogram of the sample to verify that the sample appears to be Normal.

4. A random variable Z has a **inverse Gaussian distribution** if it has density

$$f(z) \propto z^{-3/2} \exp \left\{ -\theta_1 z - \frac{\theta_2}{z} + 2\sqrt{\theta_1 \theta_2} + \log \left(\sqrt{2\theta_2} \right) \right\}, \quad z > 0$$

where $\theta_1 > 0$ and $\theta_2 > 0$ are parameters. It can be shown that

$$\mathbb{E}(Z) = \sqrt{\frac{\theta_2}{\theta_1}} \quad \text{and} \quad \mathbb{E}\left(\frac{1}{Z}\right) = \sqrt{\frac{\theta_1}{\theta_2}} + \frac{1}{2\theta_2}.$$

(a) Let $\theta_1 = 1.5$ and $\theta_2 = 2$. Draw a sample of size 1,000 using the independence-Metropolis–Hastings method. Use a Gamma distribution as the proposal density. To assess the accuracy, compare the mean of Z and $1/Z$ from the sample to the theoretical means Try different Gamma distributions to see if you can get an accurate sample.

(b) Draw a sample of size 1,000 using the random-walk-Metropolis–Hastings method. Since $z > 0$ we cannot just use a Normal density. One strategy is this. Let $W = \log Z$. Find the density of W. Use the random-walk-Metropolis–Hastings method to get a sample $W_1, \ldots, W_N$ and let $Z_i = e^{W_i}$. Assess the accuracy of the simulation as in part (a).

5. Get the heart disease data from the book web site. Consider a Bayesian analysis of the logistic regression model

$$\mathbb{P}(Y = 1|X = x) = \frac{e^{\beta_0 + \sum_{j=1}^{k} \beta_j x_j}}{1 + e^{\beta_0 + \sum_{j=1}^{k} \beta_j x_j}}.$$

Use the flat prior $f(\beta_0, \ldots, \beta_k) \propto 1$. Use the Gibbs–Metropolis algorithm to draw a sample of size 10,000 from the posterior $f(\beta_0, \beta_1|\text{data})$. Plot histograms of the posteriors for the β_j's. Get the posterior mean and a 95 percent posterior interval for each β_j.

(b) Compare your analysis to a frequentist approach using maximum likelihood.

Bibliography

AGRESTI, A. (1990). *Categorical Data Analysis.* Wiley.

AKAIKE, H. (1973). Information theory and an extension of the maximum likelihood principle. *Second International Symposium on Information Theory* 267–281.

ANDERSON, T. W. (1984). *An Introduction to Multivariate Statistical Analysis (Second Edition).* Wiley.

BARRON, A., SCHERVISH, M. J. and WASSERMAN, L. (1999). The consistency of posterior distributions in nonparametric problems. *The Annals of Statistics* **27** 536–561.

BEECHER, H. (1959). *Measurement of Subjective Responses.* Oxford University Press.

BENJAMINI, Y. and HOCHBERG, Y. (1995). Controlling the false discovery rate: A practical and powerful approach to multiple testing. *Journal of the Royal Statistical Society, Series B, Methodological* **57** 289–300.

BERAN, R. (2000). REACT scatterplot smoothers: Superefficiency through basis economy. *Journal of the American Statistical Association* **95** 155–171.

BERAN, R. and DÜMBGEN, L. (1998). Modulation of estimators and confidence sets. *The Annals of Statistics* **26** 1826–1856.

BERGER, J. and WOLPERT, R. (1984). *The Likelihood Principle.* Institute of Mathematical Statistics.

BERGER, J. O. (1985). *Statistical Decision Theory and Bayesian Analysis (Second Edition).* Springer-Verlag.

BERGER, J. O. and DELAMPADY, M. (1987). Testing precise hypotheses (c/r: P335-352). *Statistical Science* **2** 317–335.

BERLINER, L. M. (1983). Improving on inadmissible estimators in the control problem. *The Annals of Statistics* **11** 814–826.

BICKEL, P. J. and DOKSUM, K. A. (2000). *Mathematical Statistics: Basic Ideas and Selected Topics, Vol. I (Second Edition).* Prentice Hall.

BILLINGSLEY, P. (1979). *Probability and Measure.* Wiley.

BISHOP, Y. M. M., FIENBERG, S. E. and HOLLAND, P. W. (1975). *Discrete Multivariate Analyses: Theory and Practice.* MIT Press.

BREIMAN, L. (1992). *Probability.* Society for Industrial and Applied Mathematics.

BRINEGAR, C. S. (1963). Mark Twain and the Quintus Curtius Snodgrass letters: A statistical test of authorship. *Journal of the American Statistical Association* **58** 85–96.

CARLIN, B. P. and LOUIS, T. A. (1996). *Bayes and Empirical Bayes Methods for Data Analysis.* Chapman & Hall.

CASELLA, G. and BERGER, R. L. (2002). *Statistical Inference.* Duxbury Press.

CHAUDHURI, P. and MARRON, J. S. (1999). Sizer for exploration of structures in curves. *Journal of the American Statistical Association* **94** 807–823.

COX, D. and LEWIS, P. (1966). *The Statistical Analysis of Series of Events.* Chapman & Hall.

COX, D. D. (1993). An analysis of Bayesian inference for nonparametric regression. *The Annals of Statistics* **21** 903–923.

COX, D. R. and HINKLEY, D. V. (2000). *Theoretical statistics.* Chapman & Hall.

DAVISON, A. C. and HINKLEY, D. V. (1997). *Bootstrap Methods and Their Application*. Cambridge University Press.

DeGROOT, M. and SCHERVISH, M. (2002). *Probability and Statistics (Third Edition)*. Addison-Wesley.

DEVROYE, L., GYÖRFI, L. and LUGOSI, G. (1996). *A Probabilistic Theory of Pattern Recognition*. Springer-Verlag.

DIACONIS, P. and FREEDMAN, D. (1986). On inconsistent Bayes estimates of location. *The Annals of Statistics* **14** 68–87.

DOBSON, A. J. (2001). *An introduction to generalized linear models*. Chapman & Hall.

DONOHO, D. L. and JOHNSTONE, I. M. (1994). Ideal spatial adaptation by wavelet shrinkage. *Biometrika* **81** 425–455.

DONOHO, D. L. and JOHNSTONE, I. M. (1995). Adapting to unknown smoothness via wavelet shrinkage. *Journal of the American Statistical Association* **90** 1200–1224.

DONOHO, D. L. and JOHNSTONE, I. M. (1998). Minimax estimation via wavelet shrinkage. *The Annals of Statistics* **26** 879–921.

DONOHO, D. L., JOHNSTONE, I. M., KERKYACHARIAN, G. and PICARD, D. (1995). Wavelet shrinkage: Asymptopia? (Disc: p 337–369). *Journal of the Royal Statistical Society, Series B, Methodological* **57** 301–337.

DUNSMORE, I., DALY, F. ET AL. (1987). *M345 Statistical Methods, Unit 9: Categorical Data*. The Open University.

EDWARDS, D. (1995). *Introduction to graphical modelling*. Springer-Verlag.

EFROMOVICH, S. (1999). *Nonparametric Curve Estimation: Methods, Theory and Applications*. Springer-Verlag.

EFRON, B. (1979). Bootstrap methods: Another look at the jackknife. *The Annals of Statistics* **7** 1–26.

EFRON, B., TIBSHIRANI, R., STOREY, J. D. and TUSHER, V. (2001). Empirical Bayes analysis of a microarray experiment. *Journal of the American Statistical Association* **96** 1151–1160.

EFRON, B. and TIBSHIRANI, R. J. (1993). *An Introduction to the Bootstrap.* Chapman & Hall.

FERGUSON, T. (1967). *Mathematical Statistics : a Decision Theoretic Approach.* Academic Press.

FISHER, R. (1921). On the probable error of a coefficient of correlation deduced from a small sample. *Metron* **1** 1–32.

FREEDMAN, D. (1999). Wald lecture: On the Bernstein-von Mises theorem with infinite-dimensional parameters. *The Annals of Statistics* **27** 1119–1141.

FRIEDMAN, J. H. (1997). On bias, variance, 0/1-loss, and the curse-of-dimensionality. *Data Mining and Knowledge Discovery* **1** 55–77.

GELMAN, A., CARLIN, J. B., STERN, H. S. and RUBIN, D. B. (1995). *Bayesian Data Analysis.* Chapman & Hall.

GHOSAL, S., GHOSH, J. K. and VAN DER VAART, A. W. (2000). Convergence rates of posterior distributions. *The Annals of Statistics* **28** 500–531.

GILKS, W. R., RICHARDSON, S. and SPIEGELHALTER, D. J. (1998). *Markov Chain Monte Carlo in Practice.* Chapman & Hall.

GRIMMETT, G. and STIRZAKER, D. (1982). *Probability and Random Processes.* Oxford University Press.

GUTTORP, P. (1995). *Stochastic Modeling of Scientific Data.* Chapman & Hall.

HALL, P. (1992). *The Bootstrap and Edgeworth Expansion.* Springer-Verlag.

HALVERSON, N., LEITCH, E., PRYKE, C., KOVAC, J., CARLSTROM, J., HOLZAPFEL, W., DRAGOVAN, M., CARTWRIGHT, J., MASON, B., PADIN, S., PEARSON, T., SHEPHERD, M. and READHEAD, A. (2002). DASI first results: A measurement of the cosmic microwave background angular power spectrum. *Astrophysics Journal* **568** 38–45.

HARDLE, W. (1990). *Applied nonparametric regression.* Cambridge University Press.

HÄRDLE, W., KERKYACHARIAN, G., PICARD, D. and TSYBAKOV, A. (1998). *Wavelets, Approximation, and Statistical Applications.* Springer-Verlag.

HASTIE, T., TIBSHIRANI, R. and FRIEDMAN, J. H. (2001). *The Elements of Statistical Learning: Data Mining, Inference, and Prediction.* Springer-Verlag.

HERBICH, R. (2002). *Learning Kernel Classifiers: Theory and Algorithms.* MIT Press.

JOHNSON, R. A. and WICHERN, D. W. (1982). *Applied Multivariate Statistical Analysis.* Prentice-Hall.

JOHNSON, S. and JOHNSON, R. (1972). *New England Journal of Medicine* **287** 1122–1125.

JORDAN, M. (2004). *Graphical models.* In Preparation.

KARR, A. (1993). *Probability.* Springer-Verlag.

KASS, R. E. and RAFTERY, A. E. (1995). Bayes factors. *Journal of the American Statistical Association* **90** 773–795.

KASS, R. E. and WASSERMAN, L. (1996). The selection of prior distributions by formal rules (corr: 1998 v93 p 412). *Journal of the American Statistical Association* **91** 1343–1370.

LARSEN, R. J. and MARX, M. L. (1986). *An Introduction to Mathematical Statistics and Its Applications (Second Edition).* Prentice Hall.

LAURITZEN, S. L. (1996). *Graphical Models.* Oxford University Press.

LEE, A. T. ET AL. (2001). A high spatial resolution analysis of the maxima-1 cosmic microwave background anisotropy data. *Astrophys. J.* **561** L1–L6.

LEE, P. M. (1997). *Bayesian Statistics: An Introduction.* Edward Arnold.

LEHMANN, E. L. (1986). *Testing Statistical Hypotheses (Second Edition).* Wiley.

LEHMANN, E. L. and CASELLA, G. (1998). *Theory of Point Estimation.* Springer-Verlag.

LOADER, C. (1999). *Local regression and likelihood.* Springer-Verlag.

MARRON, J. S. and WAND, M. P. (1992). Exact mean integrated squared error. *The Annals of Statistics* **20** 712–736.

MORRISON, A., BLACK, M., LOWE, C., MACMAHON, B. and YUSA, S. (1973). Some international differences in histology and survival in breast cancer. *International Journal of Cancer* **11** 261–267.

NETTERFIELD, C. B. ET AL. (2002). A measurement by boomerang of multiple peaks in the angular power spectrum of the cosmic microwave background. *Astrophys. J.* **571** 604–614.

OGDEN, R. T. (1997). *Essential Wavelets for Statistical Applications and Data Analysis.* Birkhäuser.

PEARL, J. (2000). *Casuality: models, reasoning, and inference.* Cambridge University Press.

PHILLIPS, D. and KING, E. (1988). Death takes a holiday: Mortality surrounding major social occasions. *Lancet* **2** 728–732.

PHILLIPS, D. and SMITH, D. (1990). Postponement of death until symbolically meaningful occasions. *Journal of the American Medical Association* **263** 1947–1961.

QUENOUILLE, M. (1949). Approximate tests of correlation in time series. *Journal of the Royal Statistical Society B* **11** 18–84.

RICE, J. A. (1995). *Mathematical Statistics and Data Analysis (Second Edition).* Duxbury Press.

ROBERT, C. P. (1994). *The Bayesian Choice: A Decision-theoretic Motivation.* Springer-Verlag.

ROBERT, C. P. and CASELLA, G. (1999). *Monte Carlo Statistical Methods.* Springer-Verlag.

ROBINS, J., SCHEINES, R., SPIRTES, P. and WASSERMAN, L. (2003). Uniform convergence in causal inference. *Biometrika* (to appear).

ROBINS, J. M. and RITOV, Y. (1997). Toward a curse of dimensionality appropriate (CODA) asymptotic theory for semi-parametric models. *Statistics in Medicine* **16** 285–319.

ROSENBAUM, P. (2002). *Observational Studies.* Springer-Verlag.

ROSS, S. (2002). *Probability Models for Computer Science.* Academic Press.

ROUSSEAUW, J., DU PLESSIS, J., BENADE, A., JORDAAN, P., KOTZE, J., JOOSTE, P. and FERREIRA, J. (1983). Coronary risk factor screening in three rural communities. *South African Medical Journal* **64** 430–436.

SCHERVISH, M. J. (1995). *Theory of Statistics.* Springer-Verlag.

SCHOLKOPF, B. and SMOLA, A. (2002). *Learning with Kernels: Support Vector Machines, Regularization, Optimization, and Beyond.* MIT Press.

SCHWARZ, G. (1978). Estimating the dimension of a model. *The Annals of Statistics* **6** 461–464.

SCOTT, D., GOTTO, A., COLE, J. and GORRY, G. (1978). Plasma lipids as collateral risk factors in coronary artery disease: a study of 371 males with chest pain. *Journal of Chronic Diseases* **31** 337–345.

SCOTT, D. W. (1992). *Multivariate Density Estimation: Theory, Practice, and Visualization.* Wiley.

SHAO, J. and TU, D. (1995). *The Jackknife and Bootstrap (German).* Springer-Verlag.

SHEN, X. and WASSERMAN, L. (2001). Rates of convergence of posterior distributions. *The Annals of Statistics* **29** 687–714.

SHORACK, G. R. and WELLNER, J. A. (1986). *Empirical Processes With Applications to Statistics.* Wiley.

SILVERMAN, B. W. (1986). *Density Estimation for Statistics and Data Analysis.* Chapman & Hall.

SPIRTES, P., GLYMOUR, C. N. and SCHEINES, R. (2000). *Causation, prediction, and search.* MIT Press.

TAYLOR, H. M. and KARLIN, S. (1994). *An Introduction to Stochastic Modeling.* Academic Press.

VAN DER LAAN, M. and ROBINS, J. (2003). *Unified Methods for Censored Longitudinal Data and Causality.* Springer Verlag.

VAN DER VAART, A. W. (1998). *Asymptotic Statistics.* Cambridge University Press.

VAN DER VAART, A. W. and WELLNER, J. A. (1996). *Weak Convergence and Empirical Processes: With Applications to Statistics.* Springer-Verlag.

VAPNIK, V. N. (1998). *Statistical Learning Theory.* Wiley.

WEISBERG, S. (1985). *Applied Linear Regression.* Wiley.

WHITTAKER, J. (1990). *Graphical Models in Applied Multivariate Statistics.* Wiley.

WRIGHT, S. (1934). The method of path coefficients. *The Annals of Mathematical Statistics* **5** 161–215.

ZHAO, L. H. (2000). Bayesian aspects of some nonparametric problems. *The Annals of Statistics* **28** 532–552.

ZHENG, X. and LOH, W.-Y. (1995). Consistent variable selection in linear models. *Journal of the American Statistical Association* **90** 151–156.

List of Symbols

Convergence Symbols

$\xrightarrow{\text{P}}$	convergence in probability
$\rightsquigarrow$	convergence in distribution
$\xrightarrow{\text{qm}}$	convergence in quadratic mean
$X_n \approx N(\mu, \sigma_n^2)$	$(X_n - \mu)/\sigma_n \rightsquigarrow N(0,1)$
$x_n = o(a_n)$	$x_n/a_n \to 0$
$x_n = O(a_n)$	$|x_n/a_n|$ is bounded for large n
$X_n = o_P(a_n)$	$X_n/a_n \xrightarrow{\text{P}} 0$
$X_n = O_P(a_n)$	$|X_n/a_n|$ is bounded in probability for large n

Statistical Models

$\mathfrak{F}$	statistical model; a set of distribution functions, density functions or regression functions
θ	parameter
$\widehat{\theta}$	estimate of parameter
$T(F)$	statistical functional (the mean, for example)
$\mathcal{L}_n(\theta)$	likelihood function

Useful Math Facts

$e^x = \sum_{k=0}^{\infty} \frac{x^k}{k!} = 1 + x + \frac{x^2}{2!} + \cdots$

$\sum_{j=k}^{\infty} r^j = \frac{r^k}{1-r}$ for $0 < r < 1$

$\lim_{n \to \infty} \left(1 + \frac{a}{n}\right)^n = e^a$

Stirling's approximation: $n! \approx n^n e^{-n} \sqrt{2\pi n}$

THE GAMMA FUNCTION. The Gamma function is defined by

$$\Gamma(\alpha) = \int_0^{\infty} y^{\alpha-1} e^{-y} dy$$

for $\alpha \geq 0$. If $\alpha > 1$ then $\Gamma(\alpha) = (\alpha - 1)\Gamma(\alpha - 1)$. If n is a positive integer then $\Gamma(n) = (n-1)!$. Some special values are: $\Gamma(1) = 1$ and $\Gamma(1/2) = \sqrt{\pi}$.

Table of Distributions

Distribution	PDF or probability function	mean	variance	MGF
Point mass at a	$I(x=a)$	a	0	e^{at}
Bernoulli(p)	$p^x(1-p)^{1-x}$	p	$p(1-p)$	$pe^t+(1-p)$
Binomial(n,p)	$\binom{n}{x}p^x(1-p)^{n-x}$	np	$np(1-p)$	$(pe^t+(1-p))^n$
Geometric(p)	$p(1-p)^{x-1}I(x\geq1)$	$1/p$	$\frac{1-p}{p^2}$	$\frac{pe^t}{1-(1-p)e^t}$ $(t<-\log(1-p))$
Poisson(λ)	$\frac{\lambda^x e^{-\lambda}}{x!}$	λ	λ	$e^{\lambda(e^t-1)}$
Uniform(a,b)	$I(a<x<b)/(b-a)$	$\frac{a+b}{2}$	$\frac{(b-a)^2}{12}$	$\frac{e^{bt}-e^{at}}{(b-a)t}$
Normal(μ,σ^2)	$\frac{1}{\sigma\sqrt{2\pi}}e^{-(x-\mu)^2/(2\sigma^2)}$	μ	σ^2	$\exp\left\{\mu t+\frac{\sigma^2t^2}{2}\right\}$
Exponential(β)	$\frac{e^{-x/\beta}}{\beta}$	β	β^2	$\frac{1}{1-\beta t}$ $(t<1/\beta)$
Gamma(α,β)	$\frac{x^{\alpha-1}e^{-x/\beta}}{\Gamma(\alpha)\beta^\alpha}$	$\alpha\beta$	$\alpha\beta^2$	$\left(\frac{1}{1-\beta t}\right)^\alpha$ $(t<1/\beta)$
Beta(α,β)	$\frac{\Gamma(\alpha+\beta)}{\Gamma(\alpha)\Gamma(\beta)}x^{\alpha-1}(1-x)^{\beta-1}$	$\frac{\alpha}{\alpha+\beta}$	$\frac{\alpha\beta}{(\alpha+\beta)^2(\alpha+\beta+1)}$	$1+\sum_{k=1}^\infty\left(\prod_{r=0}^{k-1}\frac{\alpha+r}{\alpha+\beta+r}\right)\frac{t^k}{k!}$
t_ν	$\frac{\Gamma\left(\frac{\nu+1}{2}\right)}{\Gamma\left(\frac{\nu}{2}\right)}\frac{1}{\left(1+\frac{x^2}{\nu}\right)^{(\nu+1)/2}}$	0 (if $\nu>1$)	$\frac{\nu}{\nu-2}$ (if $\nu>2$)	does not exist
χ^2_p	$\frac{1}{\Gamma(p/2)2^{p/2}}x^{(p/2)-1}e^{-x/2}$	p	$2p$	$\left(\frac{1}{1-2t}\right)^{p/2}$ $(t<1/2)$

Index

Springer Texts in Statistics *(continued from page ii)*

Lehmann: Testing Statistical Hypotheses, Second Edition

Lehmann and Casella: Theory of Point Estimation, Second Edition

Lindman: Analysis of Variance in Experimental Design

Lindsey: Applying Generalized Linear Models

Madansky: Prescriptions for Working Statisticians

McPherson: Applying and Interpreting Statistics: A Comprehensive Guide, Second Edition

Mueller: Basic Principles of Structural Equation Modeling: An Introduction to LISREL and EQS

Nguyen and Rogers: Fundamentals of Mathematical Statistics: Volume I: Probability for Statistics

Nguyen and Rogers: Fundamentals of Mathematical Statistics: Volume II: Statistical Inference

Noether: Introduction to Statistics: The Nonparametric Way

Nolan and Speed: Stat Labs: Mathematical Statistics Through Applications

Peters: Counting for Something: Statistical Principles and Personalities

Pfeiffer: Probability for Applications

Pitman: Probability

Rawlings, Pantula and Dickey: Applied Regression Analysis

Robert: The Bayesian Choice: From Decision-Theoretic Foundations to Computational Implementation, Second Edition

Robert and Casella: Monte Carlo Statistical Methods, Second Edition

Rose and Smith: Mathematical Statistics with *Mathematica*

Ruppert: Statistics and Finance: An Introduction

Santner and Duffy: The Statistical Analysis of Discrete Data

Saville and Wood: Statistical Methods: The Geometric Approach

Sen and Srivastava: Regression Analysis: Theory, Methods, and Applications

Shao: Mathematical Statistics, Second Edition

Shorack: Probability for Statisticians

Shumway and Stoffer: Time Series Analysis and Its Applications

Simonoff: Analyzing Categorical Data

Terrell: Mathematical Statistics: A Unified Introduction

Timm: Applied Multivariate Analysis

Toutenburg: Statistical Analysis of Designed Experiments, Second Edition

Wasserman: All of Statistics: A Concise Course in Statistical Inference

Whittle: Probability via Expectation, Fourth Edition

Zacks: Introduction to Reliability Analysis: Probability Models and Statistical Methods